de Gruyter Expositions in Mathematics 52

Editors

V. P. Maslov, Academy of Sciences, Moscow
W. D. Neumann, Columbia University, New York
R. O. Wells, Jr., International University, Bremen

de Gruyter Expositions in Mathematics

Stability Analysis of Impulsive Functional Differential Equations

by

Ivanka Stamova

Walter de Gruyter · Berlin · New York

Author

Ivanka Stamova
Department of Mathematics
Bourgas Free University
62 San Stefano Str.
8001 Bourgas, Bulgaria
E-mail: stamova@bfu.bg

Mathematics Subject Classification 2000: 37-02, 34K20, 34K45, 34K60, 93D30

Keywords: Impulsive functional differential equations, Lyapunov method, stability, boundedness.

♾ Printed on acid-free paper which falls within the guidelines
of the ANSI to ensure permanence and durability.

ISSN 0938-6572
ISBN 978-3-11-022181-7

Bibliographic information published by the Deutsche Nationalbibliothek

The Deutsche Nationalbibliothek lists this publication in the Deutsche Nationalbibliografie;
detailed bibliographic data are available in the Internet at http://dnb.d-nb.de.

Typeset using the author's LaTeX files: Catherine Rollet, Berlin.
Printing and binding: Hubert & Co. GmbH & Co. KG, Göttingen.
Cover design: Thomas Bonnie, Hamburg.

*To my husband, Gani,
and our sons, Trayan and Alex,
for their support and encouragement*

Preface

The mathematical investigations of the impulsive ordinary differential equations mark their beginning with the work of Mil'man and Myshkis [159], 1960. In it some general concepts are given about the systems with impulse effect and the first results on stability of such systems solutions are obtained. In recent years the fundamental and qualitative theory of such equations has been extensively studied. A number of results on existence, uniqueness, continuability, stability, boundedness, oscillations, asymptotic properties, etc. were published [17, 18, 28–30, 32, 33, 83, 89, 90, 97, 103, 126, 129, 131, 132, 133, 164, 167, 172–174, 178–181, 189, 195, 231].

Scientists have been aware of the fact that many applicable problems are pointless unless the dependence on previous states is being taken into account. But until Volterra's work [212], a bigger part of the obtained results refers to several specific properties of a narrow type of equations. This work marks the beginning of the development of the functional differential equations theory [4, 58, 60, 63, 64, 82, 84, 91, 94, 95, 98, 99, 108, 114, 118–124, 134, 163, 168, 169, 225].

Impulsive functional differential equations are a natural generalization of impulsive ordinary differential equations (without delay) and of functional differential equations (without impulses). At the present time the qualitative theory of such equations undergoes rapid development. Many results on the stability and boundedness of their solutions are obtained. It is natural to ask whether we can find a systematic account of recent developments in the stability and boundedness theory for impulsive functional differential equations. This is precisely what is planned in this book. Its aim is to present the main results on stability theory for impulsive functional differential equations by means of the second method of Lyapunov and provide a unified general structure applicable to study the dynamics of mathematical models based on such equations.

Some important features of the monograph are as follows:

(i) It is the first book that is dedicated to a systematic development of *stability theory* for *impulsive functional differential equations.*

(ii) It fills a void by making available a source book which describes existing literature on the *extensions* of stability theory for impulsive functional differential equations.

(iii) It shows the manifestations of *Lyapunov–Razumikhin method* by demonstrating how this effective technique can be applied to investigate stability and boundedness of the solutions of impulsive functional differential equations and provides interesting *applications* of many practical problems of diverse interest.

The book consists of four chapters.

Chapter 1 has an introductory character. In it a description of the systems of impulsive functional differential equations and the main results on the fundamental theory are given: conditions for absence of the phenomenon beating, theorems for existence, uniqueness, continuability of the solutions. The class of piecewise continuous Lyapunov functions, which are a basic apparatus in the stability and boundedness theory, is introduced. Some comparison lemmas and auxiliary assertions, which are used in the remaining three chapters, are exposed.

In Chapter 2 the main definitions on the Lyapunov stability and boundedness of the solutions of the impulsive functional differential equations are given. Using the Lyapunov–Razumikhin technique and comparison technique theorems on Lyapunov stability, boundedness and global stability are proved. Many examples are considered to illustrate the feasibility of the results.

Chapter 3 is dedicated to some extensions of Lyapunov stability and boundedness. Theorems on stability and boundedness of sets, conditional stability, parametric stability, eventual stability and boundedness, practical stability, Lipschitz stability, stability and boundedness in terms of two measures are presented. Many interesting results are considered in which assumptions allowing the derivatives of Lyapunov function to be positive are used to impulsively stabilize functional differential equations.

Finally, in Chapter 4, the applications of stability and boundedness theory to Lotka–Volterra models, neural networks and economic models are presented. The impulses are considered either as means of perturbations or as control.

Each chapter is supplied with notes and comments.

The book is addressed to a wide audience of professionals such as mathematicians, applied researches and practitioners.

The author has the pleasure to express her sincere gratitude to Prof. Drumi Bainov and Prof. Angel Dishliev for their help while she was making her first steps in this field, and also to Prof. Dr. Mihail Konstantinov, Prof. Angel Dishliev and Prof. Gani Stamov for their valuable comments and suggestions during the preparation of the manuscript. She is also thankful to all her co-authors, the work with whom expanded her experience and gave her opportunities for perfection. In addition, the author is indebted to Dr. Robert Plato and Ms. Catherine Rollet from Walter de Gruyter for all their very professional work.

Bourgas, August 2009   *I. M. Stamova*

Contents

Chapter 1

Introduction

The problems of existence, uniqueness, and continuability of the solutions will be discussed. The piecewise continuous Lyapunov functions will be introduced and some main comparison results will be given.

1.1 Preliminary notes

The necessity to study impulsive functional differential equations is due to the fact that these equations are an useful mathematical machinery in modelling many real processes and phenomena studied in optimal control, biology, mechanics, medicine, bio-technologies, electronics, economics, etc.

For instance, impulsive interruptions are observed in mechanics [11, 57, 61], in radio engineering [11], in communication security [115, 116], in Lotka–Volterra models [7, 8, 26, 109, 111, 135, 141, 144, 145, 218, 220, 222, 226], in control theory [110, 136, 142, 145, 158, 185], in neural networks [10, 15, 69, 101, 192, 221], in economics [81, 83, 177, 206]. Indeed, the states of many evolutionary processes are often subject to instantaneous perturbations and experience abrupt changes at certain moments of time. The duration of the changes is very short and negligible in comparison with the duration of the process considered, and can be thought of as "momentary" changes or as impulses. Systems with short-term perturbations are often naturally described by impulsive differential equations [30, 32, 33, 129].

On the other hand many models of dynamical systems with delays have been investigated intensively in population dynamics [4, 8, 85, 91, 105, 113, 127, 140, 157, 165, 166, 176, 211, 214, 230], in medicine [102, 137, 161], in neural networks, signal theory and control theory [16, 66, 67, 68, 73, 76, 79, 92, 106, 107, 117, 154, 160, 170, 192, 227, 232, 233], in economics and social sciences [62, 86, 96, 105, 151, 162, 175, 176]. Mathematical models with delay take into account the memory (aftereffect, hereditary effects) of the dynamic system when a sequence of its past states impacts its future evolution.

Impulsive functional differential equations may be used for the mathematical simulation of processes which are characterized by the fact that their state changes by jumps and by the dependence of the process on its history at each moment of time.

The next examples give a more concrete notion of processes that can be described by impulsive functional differential equations.

Example 1.1. One of the first mathematical models which incorporate interaction between two species (predator-prey, or herbivore-plant, or parasitoid-host) was proposed by Alfred Lotka [147] and Vito Volterra [212]. The classical "predator-prey" model is based on the following system of two differential equations:

$$
\begin{cases}
\dot{H}(t) = H(t)[r_1 - bP(t)] \\
\dot{P}(t) = P(t)[-r_2 + cH(t)],
\end{cases}
\tag{1.1}
$$

where $H(t)$ and $P(t)$ represent the population densities of prey and predator at time t, respectively; $t \geq 0$; $r_1 > 0$ is the intrinsic growth rate of the prey; $r_2 > 0$ is the death rate of the predator or consumer; b and c are the interaction constants. More concrete, the constant b is the per-capita rate of the predator predation and the constant c is the product of the predation per-capita rate and the rate of converting the prey into the predator.

The product $p = p(H) = bH$ of b and H is the predator's functional response (response function) of type I, or rate of prey capture as a function of prey abundance.

The model (1.1) is derived by making the following assumptions: 1) the prey population will grow exponentially when the predator is absent; 2) the predator population will starve in the absence of the prey population (as opposed to switching to another type of prey); 3) predators can consume infinite quantities of prey; and 4) there is no environmental complexity (in other words, both populations are moving randomly through a homogeneous environment).

It is generally recognized that some kinds of time delays are inevitable in population interactions. Time delay, due to gestation, is a common example, because generally the consumption of prey by the predator throughout its past history governs the present birth rate of the predator. If we take into account the effect of time delays of population interactions, we will have more realistic Lotka–Volterra models. The model (1.1) can be improved by the following predator-prey system with distributed delays:

$$
\begin{cases}
\dot{H}(t) = H(t)\left[r_1 - a \int_{-\tau_1}^{0} H(t+s)d\mu_1(s) - bP(t)\right] \\
\dot{P}(t) = P(t)\left[-r_2 + cH(t) - d \int_{-\tau_2}^{0} P(t+s)d\mu_2(s)\right],
\end{cases}
\tag{1.2}
$$

where $\tau_i \geq 0$; $\mu_i : [-\tau_i, 0] \to \mathbb{R}$ is non-decreasing on $[-\tau_i, 0]$, $i = 1, 2$; a, d are the intra-species competition coefficients.

There have been many studies in literatures that investigate the population dynamics of the type (1.2) models [91, 105, 127, 157, 211]. However, in the study of the dynamic relationship between species, the effect of some impulsive factors has been ignored, which exists widely in the real world. For example, the birth of many species is an annual birth pulse or harvesting. Moreover, the human beings have been harvesting or stocking species at some time, then the species is affected by another impulsive type. Also, impulsive reduction of the population density of a given species is possible after its partial destruction by catching or poisoning with chemicals used at some transitory slots in fishing or agriculture. Such factors have a great impact on the population growth. If we incorporate these impulsive factors into the model of population interaction, the model must be governed by impulsive functional differential system.

For example, if at the moment $t = t_k$ the population density of the predator is changed, then we can assume that

$$\Delta P(t_k) = P(t_k + 0) - P(t_k - 0) = g_k P(t_k), \tag{1.3}$$

where $P(t_k - 0) = P(t_k)$ and $P(t_k + 0)$ are the population densities of the predator before and after impulsive perturbation, respectively, and $g_k \in \mathbb{R}$ are constants which characterize the magnitude of the impulsive effect at the moment t_k. If $g_k > 0$, then the population density increases and if $g_k < 0$, then the population density decreases at the moment t_k.

Relations (1.2) and (1.3) determine the following impulsive functional differential system:

$$\begin{cases} \dot{H}(t) = H(t)\Big[r_1 - a \int_{-\tau_1}^{0} H(t+s)d\mu_1(s) - bP(t)\Big], & t \neq t_k \\[2mm] \dot{P}(t) = P(t)\Big[-r_2 + cH(t) - d \int_{-\tau_2}^{0} P(t+s)d\mu_2(s)\Big], & t \neq t_k \\[2mm] H(t_k + 0) = H(t_k), \quad P(t_k + 0) = P(t_k) + g_k P(t_k), \end{cases} \tag{1.4}$$

where t_k are fixed moments of time, $0 < t_1 < t_2 < \cdots$, $\lim_{k \to \infty} t_k = \infty$.

In mathematical ecology the system (1.4) denotes a model of the dynamics of a predator-prey system, which is subject to impulsive effects at certain moments of time. By means of such models, it is possible to take into account the possible environmental changes or other exterior effects due to which the population density of the predator is changed momentary.

Example 1.2. The most important and useful functional response is the Holling type II function of the form

$$p(H) = \frac{CH}{m + H},$$

where $C > 0$ is the maximal growth rate of the predator, and $m > 0$ is the half-saturation constant. Since the function $p(H)$ depends solely on prey density, it is usually called a *prey-dependent* response function. Predator-prey systems with prey-dependent response have been studied extensively and the dynamics of such systems are now very well understood [112, 127, 165, 176, 211, 219, 222].

Recently, the traditional prey-dependent predator-prey models have been challenged, based on the fact that functional and numerical responses over typical ecological timescales ought to depend on the densities of both prey and predators, especially when predators have to search for food (and therefore have to share or compete for food). Such a functional response is called a *ratio-dependent* response function. Based on the Holling type II function, several biologists (see [219] and the references cited therein) proposed a ratio-dependent function of the form

$$p\left(\frac{H}{P}\right) = \frac{C\frac{H}{P}}{m + \frac{H}{P}} = \frac{CH}{mP + H}$$

and the following ratio-dependent Lotka–Volterra model

$$\begin{cases} \dot{H}(t) = H(t)\left[r_1 - aH(t) - \dfrac{CP(t)}{mP(t) + H(t)}\right] \\ \dot{P}(t) = P(t)\left[-r_2 + \dfrac{KH(t)}{mP(t) + H(t)}\right], \end{cases} \tag{1.5}$$

where K is the conversion rate.

If we introduce time delays in model (1.5), we will obtain a more realistic approach to the understanding of predator-prey dynamics. Time delay plays an important role in many biological dynamical systems, being particularly relevant in ecology, where time delays have been recognized to contribute critically to the stable or unstable outcome of prey densities due to predation. Also, the population of given species depends on their maturity and on the natural growth rate of the proceeding generations. Therefore, it is interesting and important to study the following delayed modified ratio-dependent Lotka–Volterra system:

$$\begin{cases} \dot{H}(t) = H(t)\left[r_1 - a\displaystyle\int_{-\infty}^{t} k(t-u)H(u)du - \dfrac{CP(t - \tau(t))}{mP(t) + H(t)}\right] \\ \dot{P}(t) = P(t)\left[-r_2 + \dfrac{KH(t - \tau(t))}{mP(t) + H(t - \tau(t))}\right], \end{cases} \tag{1.6}$$

where $k : \mathbb{R}_+ \to \mathbb{R}_+$ is a measurable function, corresponding to a delay kernel or a weighting factor, which says how much emphasis should be given to the size of the prey population at earlier times to determine the present effect on resource availability; $\tau \in C[\mathbb{R}, \mathbb{R}_+]$.

However, the ecological system is often affected by environmental changes and other human activities. In many practical situations, it is often the case that predator or parasites are released at some transitory time slots and harvest or stock of the species is seasonal or occurs in regular pulses. By means of exterior effects we can control population densities of the prey and predator.

If at certain moments of time biotic and anthropogeneous factors act on the two populations "momentary", then the population numbers vary by jumps. In this case we will study Lotka–Volterra models with impulsive perturbations of the type

$$
\begin{cases}
\dot{H}(t) = H(t)\left[r_1 - a\int_{-\infty}^{t} k(t-u)H(u)du - \dfrac{CP(t-\tau(t))}{mP(t)+H(t)}\right], & t \neq t_k \\[3mm]
\dot{P}(t) = P(t)\left[-r_2 + \dfrac{KH(t-\tau(t))}{mP(t)+H(t-\tau(t))}\right], & t \neq t_k \\[3mm]
H(t_k+0) = (1+h_k)H(t_k), & k = 1,2,\ldots \\[1mm]
P(t_k+0) = (1+g_k)P(t_k), & k = 1,2,\ldots,
\end{cases}
$$

(1.7)

where $h_k, g_k \in \mathbb{R}$ and $t_k, k = 1, 2, \ldots$ are fixed moments of impulse effects, $0 < t_1 < t_2 < \cdots$, $\lim_{k\to\infty} t_k = \infty$.

By means of the type (1.7) models it is possible to investigate one of the most important problems of the mathematical ecology – the problem for stability of the ecosystems and respectively the problem of the optimal control of such systems.

Example 1.3. Mathematical modelling of plankton population is an important alternative method of improving our knowledge of the physical and biological processes relating to plankton ecology. One of the first mathematical representations of allelopathic interactions was proposed by Maynard–Smith [157]. The author considered a two species Lotka–Volterra competition model and introduced a term to take into account the effect of a toxic substance, which is released at a constant rate by one species when the other is present.

Motivated by his work, Xia [218] proposes the following neutral Lotka–Volterra competition system:

$$
\begin{cases}
\dot{N}_i(t) = N_i(t)\left[r_i(t) - \sum_{j=1}^{n} a_{ij}(t)N_j(t) - \sum_{j=1}^{n} b_{ij}(t)N_j(t-\tau_{ij}(t))\right. \\[3mm]
\left. \qquad\quad - \sum_{j=1}^{n} c_{ij}(t)\dot{N}_j(t-\gamma_{ij}(t)) - \sum_{j=1}^{n} d_{ij}(t)N_i(t)N_j(t)\right],
\end{cases}
$$

(1.8)

where $i = 1,\ldots,n, n \geq 2; t \geq 0; N_i(t)$ is the population number of the ith species at the moment t; $r_i(t)$ are intrinsic growth rates at the moment t; $a_{ij}(t), b_{ij}(t), c_{ij}(t)$, $d_{ij}, \tau_{ij}(t)$ and $\gamma_{ij}(t)$ are non-negative continuous functions.

If at certain moments of time the population densities $N_i(t), i = 1,\ldots,n$ are subject to pulse perturbations, it is not reasonable to expect a regular solution. Instead, the solution must have some jumps at these moments and the jumps follow a specific pattern. An adequate mathematical model of the population dynamics in this situation is the following impulsive neutral type functional differential Lotka–Volterra system:

$$
\left\{
\begin{aligned}
& \dot{N}_i(t) = N_i(t)\left[r_i(t) - \sum_{j=1}^{n} a_{ij}(t)N_j(t) - \sum_{j=1}^{n} b_{ij}(t)N_j(t - \tau_{ij}(t)) \right. \\
& \qquad\qquad \left. - \sum_{j=1}^{n} c_{ij}(t)\dot{N}_j(t - \gamma_{ij}(t)) - \sum_{j=1}^{n} d_{ij}(t)N_i(t)N_j(t) \right], \quad t \neq t_k \\
& N_i(t_k + 0) = N_i(t_k) + h_{ik}N_i(t_k), \quad k = 1, 2, \ldots,
\end{aligned}
\right.
$$

$$(1.9)$$

where t_k $(0 < t_1 < t_2 < \cdots < t_k < \cdots)$ are fixed impulsive points, $\lim_{k\to\infty} t_k = \infty$; $N_i(t_k - 0) = N_i(t_k)$ and $N_i(t_k + 0)$ are the population densities of ith species before and after impulsive perturbation at the moment t_k, respectively; h_{ik} are constants which characterize the magnitude of the impulsive effect of ith species at the moments t_k.

Example 1.4. One contemporary application of the functional differential equations is their use for the study of processes involved in HIV infection. The mathematical model for HIV/AIDS, with explicit incubation period, is presented in [161] as a system of discrete time delay differential equations. Time delay is due to the long incubation period (period from the point of infection to the appearance of disease symptoms). HIV/AIDS models have been studied by several authors [102, 137, 156, 161].

The model in [161] classifies the sexually active population into three classes that are: susceptibles, infectives and AIDS cases, with population numbers in each class denoted as functions of time by $S(t)$, $I(t)$ and $A(t)$ respectively. Sexually mature susceptibles $S(t)$ contain sexually mature people in the population, who have had no contact with the virus. This compartment increases through maturation of individuals into a sexually mature age group and decreases by contagion going to the next compartment, emigration to other countries and natural death. Sexually mature infectives $I(t)$ contain sexually mature individuals, who are infected with the virus but have not developed AIDS symptoms yet. The number of the people in this group would decrease through natural death, emigration to other countries and development to AIDS after a certain stay in this class (develop symptomatic AIDS). AIDS cases $A(t)$ are those individuals, who have developed fully symptomatic AIDS and exhibit specific clinical features and this class would decrease by natural death and AIDS-related death. The population is assumed to be uniform and homogeneously mixing. The total adult population and sexually interacting adult population are denoted by $N_T(t) = S(t) + I(t) + A(t)$ and $N(t) = S(t) + I(t)$, respectively.

A recruitment-death demographic structure is assumed. Individuals enter the susceptibles class at a constant rate $b > 0$. The natural death rate is assumed to be proportional to the population number in each class, with rate constant $\mu > 0$. The model assumes a constant emigration rate $m > 0$ of individuals to other countries. This assumption makes the model more appropriate for the developing countries, where a significant part of the population emigrates to other countries for better educational facilities and in search of employment. In addition, there is an AIDS-related death in the AIDS class which is assumed to be proportional to the population number in that class, with rate constant $v > 0$.

The model also assumes: 1) a standard incidence of the form $\beta c S I/N$ where $\beta > 0$ is probability of being infected by a new sexual partner, $c > 0$ is the rate at which an individual acquires a new sexual partner; and 2) a constant incubation period $\tau > 0$. The probability that an individual remains in the incubation period at least t time units before developing AIDS is given by a step function with value 1 for $0 \le t \le \tau$ and value 0 for $t > \tau$. The probability that an individual in the incubation period time t units has survived to develop AIDS and did not emigrate is $e^{-(\mu+m)\tau}$, $\tau > 0$. A latent period for HIV is not assumed since the latent period is negligible, compared with the period of infectivity. AIDS cases are taken to be sexually inactive so that there are no new infections due to the AIDS class. HIV/AIDS is assumed to have been in the population for at least time $\tau > 0$, such that initial perturbations have died out.

The assumptions result in the following HIV/AIDS model for $t > \tau$,

$$
\begin{cases}
\dot{S}(t) = b - \beta c \dfrac{S(t)I(t)}{N(t)} - (\mu + m)S(t) \\[2mm]
I(t) = \displaystyle\int_{t-\tau}^{t} \beta c \dfrac{S(u)I(u)}{N(u)} e^{-(\mu+m)(t-u)} du \\[2mm]
\dot{A}(t) = \beta c \dfrac{S(t-\tau)I(t-\tau)}{N(t-\tau)} e^{-(\mu+m)\tau} - (\mu + v)A(t),
\end{cases}
\tag{1.10}
$$

where the integral in the second equation represents the summation over the interval $[t - \tau, t]$ of those individuals who become infectives at time $u \ge 0$ and have neither developed AIDS nor died. One equivalent form of the model (1.10) is the following model:

$$
\begin{cases}
\dot{S}(t) = b - \beta c \dfrac{S(t)I(t)}{N(t)} - (\mu + m)S(t) \\[2mm]
\dot{I}(t) = \beta c \dfrac{S(t)I(t)}{N(t)} - \beta c q \dfrac{S(t-\tau)I(t-\tau)}{N(t-\tau)} - (\mu + m)I(t) \\[2mm]
\dot{A}(t) = \beta c q \dfrac{S(t-\tau)I(t-\tau)}{N(t-\tau)} - (\mu + v)A(t),
\end{cases}
\tag{1.11}
$$

where $q = e^{-(\mu+m)\tau}$.

The model (1.11) does not take into account possible exterior or interior impulsive effects on the population numbers of the individuals from the three groups. For example, we notice that it is reasonable to regard the birth of individuals an impulse to the population number. Also, there are some other perturbations in the real world such as fires and floods that are not suitable to be considered continually. These perturbations

bring sudden changes to the system. A more realistic HIV/AIDS model should take
into account the impulsive effects.

If at the moment t_k the population of the individuals from the first class is increased
with magnitude $h_k > 0$, then the adequate model will be the following impulsive
system:

$$
\begin{cases}
\dot{S}(t) = b - \beta c \dfrac{S(t)I(t)}{N(t)} - (\mu + m)S(t), \quad t \neq t_k \\[2mm]
\dot{I}(t) = \beta c \dfrac{S(t)I(t)}{N(t)} - \beta cq \dfrac{S(t-\tau)I(t-\tau)}{N(t-\tau)} - (\mu + m)I(t), \quad t \neq t_k \\[2mm]
\dot{A}(t) = \beta cq \dfrac{S(t-\tau)I(t-\tau)}{N(t-\tau)} - (\mu + v)A(t), \quad t \neq t_k \\[2mm]
S(t_k + 0) = (1 + h_k)S(t_k) \\[1mm]
I(t_k + 0) = I(t_k), \quad A(t_k + 0) = A(t_k), \quad k = 1, 2, \dots .
\end{cases}
\tag{1.12}
$$

Example 1.5. Chua and Yang [74, 75] proposed a novel class of information-
processing system called Cellular Neural Networks (CNN) in 1988. Like neural networks, it is a large-scale nonlinear analog circuit which processes signals in real time.
Like cellular automata [216] it is made of a massive aggregate of regularly spaced circuit clones, called cells, which communicate with each other directly only through its
nearest neighbours.

The key features of neural networks are asynchronous parallel processing and global
interaction of network elements. For the circuit diagram and connection pattern, implementation the CNN can be referred to [74]. Impressive applications of neural networks
have been proposed for various fields such as optimization, linear and nonlinear pro-
gramming, associative memory, pattern recognition and computer vision [72–76, 104,
216]. However, it is necessary to solve some dynamic image processing and pattern
recognition problems by using Delayed Cellular Neural Networks (DCNN) [66, 67,
68, 76, 79, 92, 106, 107, 117, 154, 160, 170, 192, 227, 232, 233].

Zhou and Cao [233] considered the following DCNN:

$$
\dot{x}_i(t) = -c_i x_i(t) + \sum_{j=1}^{n} a_{ij} f_j\left(x_j(t)\right) + \sum_{j=1}^{n} b_{ij} f_j\left(x_j(t - \tau_j(t))\right) + I_i, \tag{1.13}
$$

where $i = 1, 2, \dots, n$, n corresponds to the numbers of units in the neural network;
$x_i(t)$ corresponds to the state of the ith unit at time t; $f_j(x_j(t))$ denotes the output of
the jth unit at time t; a_{ij} denotes the strength of the jth unit on the ith unit at time
t; b_{ij} denotes the strength of the jth unit on the ith unit at time $t - \tau_j(t)$; I_i denotes
the external bias on the ith unit; $\tau_j(t)$ corresponds to the transmission delay along the
axon of the jth unit and satisfies $0 \leq \tau_j(t) \leq \tau$ ($\tau = $ const); c_i represents the rate

with which the ith unit will reset its potential to the resting state in isolation when disconnected from the network and external inputs.

On the other hand, the state of DCNN is often subject to instantaneous perturbations and experiences abrupt changes at certain instants which may be caused by switching phenomenon, frequency change or other sudden noise, that is, do exhibit impulsive effects. For instance, according to Arbib [15] and Haykin [101], when a stimulus from the body or the external environment is received by receptors, the electrical impulses will be conveyed to the neural net and impulsive effects arise naturally in the net.

Therefore, neural network model with delay and impulsive effects should be more accurate to describe the evolutionary process of the systems. Since delays and impulses can affect the dynamical behaviors of the system, it is necessary to investigate both delay and impulsive effects on neural networks stability.

Let at fixed moments t_k the system (1.13) be subject to shock effects due to which the state of the ith unit gets momentary changes. The adequate mathematical model in such situation is the following impulsive CNNs with time-varying delays:

$$\begin{cases} \dot{x}_i(t) = -c_i x_i(t) + \sum_{j=1}^{n} a_{ij} f_j\left(x_j(t)\right) \\ \qquad + \sum_{j=1}^{n} b_{ij} f_j\left(x_j(t - \tau_j(t))\right) + I_i, \quad t \neq t_k, \, t \geq 0 \\ \Delta x_i(t_k) = x_i(t_k + 0) - x_i(t_k) = P_{ik}(x_i(t_k)), \quad k = 1, 2, \ldots, \end{cases} \tag{1.14}$$

where t_k, $k = 1, 2, \ldots$ are the moments of impulsive perturbations and satisfy $0 < t_1 < t_2 < \cdots$, $\lim_{k \to \infty} t_k = \infty$ and $P_{ik}(x_i(t_k))$ represents the abrupt change of the state $x_i(t)$ at the impulsive moment t_k.

Such a generalization of the DCNN notion should enable us to study different types of classical problems as well as to "control" the solvability of the differential equations (without impulses).

In the examples considered the systems of impulsive functional differential equations are given by means of a system of functional differential equations and conditions of jumps.

Let $\mathbb{R}^n$ be the n-dimensional Euclidean space with norm $\|.\|$, and let $\mathbb{R}_+ = [0, \infty)$. We shall make a brief description of the systems of impulsive functional differential equations.

Let Ω be the phase space of some evolutionary process, i.e. the set of its states. Denote by P_t the point mapping the process at the moment t and assume that the state of the process is determined by n parameters. Then the mapping point P_t can be interpreted as a point (t, x) of the $(n + 1)$-dimensional space $\mathbb{R}^{n+1}$ and Ω as a set in $\mathbb{R}^n$. The set $\mathbb{R} \times \Omega$ will be called an extended phase space of the evolutionary process considered. Assume that the evolution law of the process is described by:

(a) the system of functional differential equations

$$\dot{x}(t) = f(t, x_t), \tag{1.15}$$

where $t_0 \in \mathbb{R}$; $x = \mathrm{col}(x_1, x_2, \ldots, x_n) \in \Omega$; $f : [t_0, \infty) \times D \to \mathbb{R}^n$; $r > 0$; $D = \{\varphi : [-r, 0] \to \Omega\}$; and for $t \geq t_0$, $x_t \in D$ is defined by $x_t(s) = x(t + s)$, $-r \leq s \leq 0$;

(b) sets M_t, N_t of arbitrary topological structure contained in $\mathbb{R} \times \Omega$;

(c) the operator $A_t : M_t \to N_t$.

Let $\varphi_0 \in D$, $x_{t_0} = \varphi_0$, and $x(t_0) = \varphi_0(0)$. The motion of the point P_t in the extended phase space is performed in the following way: the point P_t begins its motion from the point $(t_0, x(t_0))$ and moves along the curve $(t, x(t))$ described by the solution $x(t)$ of system (1.15) with initial conditions $x_{t_0} = \varphi_0$, $x(t_0) = \varphi_0(0)$ till the moment $t_1 > t_0$ when P_t meets the set M_t. At the moment t_1 the operator A_{t_1} "instantly" transfers the point P_t from the position $P_{t_1} = (t_1, x(t_1))$ into the position $(t_1, x_1^+) \in N_{t_1}$, $x_1^+ = A_{t_1} x(t_1)$. Then the point P_t goes on moving along the curve $(t, y(t))$ described by the solution $y(t)$ of system (1.15) with initial conditions $y(t) = x(t)$ for $t_1 - r \leq t \leq t_1$ and $y(t_1) = x_1^+$ till a new meeting with the set M_t, etc.

The union of relations (a), (b), (c) characterizing the evolutionary process will be called a *system of impulsive functional differential equations*, the curve described by the point P_t in the extended phase space – an *integral curve* and the function defining this curve – a *solution* of the system of impulsive functional differential equations. The moments $t_1, t_2, \ldots$ when the mapping point P_t meets the set M_t will be called *moments of impulse effect* and the operator $A_t : M_t \to N_t$ a *jump operator*.

We shall assume that the solution $x(t)$ of the impulsive system is a left continuous function at the moments of impulse effect, i.e. that $x(t_k - 0) = x(t_k)$, $k = 1, 2, \ldots$.

The freedom of choice of the sets M_t, N_t and the operator A_t leads to the great variety of the impulsive systems. The solution of the systems of impulsive functional differential equations may be:

– a continuous function if the integral curve does not intersect the set M_t or intersects it at the fixed points of the operator A_t;

– a piecewise continuous function with a finite number of points of discontinuity of first type if the integral curve intersects M_t at a finite number of points which are not fixed points of the operator A_t;

– a piecewise continuous function with a countable set of points of discontinuity of first type if the integral curve intersects M_t at a countable set of points which are not fixed points of the operator A_t.

In the present book systems of impulsive functional differential equations will be considered for which the moments of impulse effect come when some spatial-temporal relation $\Phi(t, x) = 0$, $(t, x) \in \mathbb{R} \times \Omega$ is satisfied, i.e. when the mapping point (t, x) meets the surface with the equation $\Phi(t, x) = 0$. Such systems can be written in the form

$$\dot{x}(t) = f(t, x_t), \quad \Phi(t, x) \neq 0$$

$$\Delta x(t) = I(t, x), \quad \Phi(t, x) = 0.$$

The sets M_t, N_t and the operator A_t are defined by the relations

$$M_t = \{(t, x) \in \mathbb{R} \times \Omega : \Phi(t, x) = 0\}, \quad N_t = \mathbb{R} \times \Omega,$$

$$A_t : M_t \to N_t, \quad (t, x) \to (t, x + I(t, x)),$$

where $I : \mathbb{R} \times \Omega \to \Omega$ and $t = t_k$ is a moment of impulse effect for the solution $x(t)$ if $\Phi(t_k, x(t_k)) = 0$. Then

$$\Delta x(t_k) = I(t_k, x(t_k)).$$

Let $\varphi_0 \in D$ be a piecewise continuous function with points of discontinuity of the first kind in the interval $(-r, 0)$ at which it is continuous from the left. In a particular case, it is possible φ_0 to be a continuous function.

We shall give a more detailed description of the following two classes of systems of impulsive functional differential equations which have particular interest.

I. Systems with fixed moments of impulse effect. For these systems, the set M_t is represented by a sequence of hyperplanes $t = t_k$ where $\{t_k\}$ is a given sequence of impulse effect moments. The operator A_t is defined only for $t = t_k$ giving the sequence of operators $A_k : \Omega \to \Omega, x \to A_k x = x + I_k(x)$.

The systems of this class are written as follows:

$$\dot{x}(t) = f(t, x_t), \quad t \neq t_k, \quad t > t_0 \tag{1.16}$$

$$\Delta x(t_k) = I_k(x(t_k)), \quad t_k > t_0, \quad k = 1, 2, \dots, \tag{1.17}$$

where $\Delta x(t_k) = x(t_k^+) - x(t_k); I_k : \Omega \to \mathbb{R}^n, k = 1, 2, \dots$.

Let $t_0 < t_1 < t_2 < \cdots$ and $\lim_{k \to \infty} t_k = \infty$. Denote by $x(t) = x(t; t_0, \varphi_0)$ the solution of system (1.16), (1.17), satisfying the initial conditions

$$\begin{cases} x(t; t_0, \varphi_0) = \varphi_0(t - t_0), \quad t_0 - r \leq t \leq t_0 \\ x(t_0 + 0; t_0, \varphi_0) = \varphi_0(0). \end{cases} \tag{1.18}$$

The solution $x(t) = x(t; t_0, \varphi_0)$ of the initial value problem (1.16), (1.17), (1.18) is characterized by the following:

(1) For $t_0 - r \leq t \leq t_0$ the solution $x(t)$ satisfies the initial conditions (1.18).

(2) For $t_0 < t \leq t_1$, $x(t)$ coincides with the solution of the problem

$$\dot{x}(t) = f(t, x_t), \quad t > t_0, \quad t \neq t_k$$

$$x_{t_0} = \varphi_0(s), \quad -r \leq s \leq 0.$$

At the moment $t = t_1$ the mapping point $(t, x(t; t_0, \varphi_0))$ of the extended phase space jumps momentarily from the position $(t_1, x(t_1; t_0, \varphi_0))$ to the position $(t_1, x(t_1; t_0, \varphi_0) + I_1(x(t_1; t_0, \varphi_0)))$.

(3) For $t_1 < t \le t_2$ the solution $x(t)$ coincides with the solution of

$$\begin{cases} \dot{y}(t) = f(t, y_t), & t > t_1 \\ y_{t_1} = \varphi_1, & \varphi_1 \in D, \end{cases}$$

where

$$\varphi_1(t - t_1) = \begin{cases} \varphi_0(t - t_1), & t \in [t_0 - r, t_0] \cap [t_1 - r, t_1] \\ x(t; t_0, \varphi_0), & t \in (t_0, t_1) \cap [t_1 - r, t_1] \\ x(t; t_0, \varphi_0) + I_1(x(t; t_0, \varphi_0)), & t = t_1. \end{cases}$$

At the moment $t = t_2$ the mapping point $(t, x(t))$ jumps momentarily, etc.

Thus the solution $x(t; t_0, \varphi_0)$ of problem (1.16), (1.17), (1.18) is a piecewise continuous function for $t > t_0$ with points of discontinuity of the first kind $t = t_k, k = 1, 2, \ldots$ at which it is continuous from the left, i.e. the following relations are satisfied:

$$x(t_k^-) = x(t_k), \ t_k > t_0, \quad k = 1, 2, \ldots$$
$$x(t_k^+) = x(t_k) + I_k(x(t_k)), \quad t_k > t_0, \quad k = 1, 2, \ldots .$$

II. Systems with variable impulsive perturbations. For these systems, the set M_t is represented by a sequence of hypersurfaces $\sigma_k : t = \tau_k(x), \ k = 1, 2, \ldots$ (or $k = 0, \pm 1, \pm 2, \ldots$). Assume that $\tau_k(x) < \tau_{k+1}(x)$ for $x \in \Omega, \ k = 1, 2, \ldots$ (or $k = 0, \pm 1, \pm 2, \ldots$) and $\lim_{k \to \infty} \tau_k(x) = \infty$ for $x \in \Omega$ ($\lim_{k \to -\infty} \tau_k(x) = -\infty$).

We shall assume that the restriction of the operator A_t to the hypersurface σ_k is given by the operator $A_k x = x + I_k(x)$ where $I_k : \Omega \to \mathbb{R}^n$. The systems of this class are written in the form

$$\dot{x}(t) = f(t, x_t), \ t \ne \tau_k(x(t)), \quad t \ge t_0 \tag{1.19}$$
$$\Delta x(t) = I_k(x(t)), \ t = \tau_k(x(t)), \quad k = 1, 2, \ldots . \tag{1.20}$$

Denote by $x(t) = x(t; t_0, \varphi_0)$ the solution of system (1.19), (1.20), satisfying the initial conditions (1.18).

The solution $x(t) = x(t; t_0, \varphi_0)$ of the initial value problem (1.19), (1.20), (1.18) is characterized by the following:

(1) For $t_0 - r \le t \le t_0$ the solution $x(t)$ satisfies the initial conditions (1.18).

(2) Denote by $\varphi(t; t_0, \varphi_0)$ the solution of the respective problem without impulses (1.19), (1.18).

Let $t_1 = \tau_{l_1}(\varphi(t_1; t_0, \varphi_0)) = \min\{t : t = \tau_k(\varphi(t; t_0, \varphi_0)), t > t_0, k = 1, 2, \ldots\}$, i.e. t_1 be the first moment when the integral curve of the problem without impulses (1.19), (1.18) meets some of the hypersurfaces σ_k. The number of this hypersurface is l_1.

Then we have $x(t; t_0, \varphi_0) = \varphi(t; t_0, \varphi_0)$ for $t_0 < t \le t_1$. At the moment $t = t_1$ the mapping point $(t, x(t; t_0, \varphi_0))$ jumps momentarily from the position $(t_1, x(t_1; t_0, \varphi_0))$ to the position $(t_1, x(t_1; t_0, \varphi_0) + I_{l_1}(x(t_1; t_0, \varphi_0))) = (t_1, x_1^+)$.

(3) Let $\varphi_1 \in D$, $\varphi_1(0) = x_1^+$ and $t_2 = \tau_{l_2}(\varphi(t_2; t_1, \varphi_1)) = \min\{t : t = \tau_k(\varphi(t; t_1, \varphi_1)), t > t_1, k = 1, 2, \ldots\}$, i.e. t_2 be the first bigger than t_1 moment when the integral curve of (1.19) with initial conditions

$$\begin{cases} \varphi(t; t_1, \varphi_1) = \varphi_1(t - t_1), & t_1 - r \le t \le t_1 \\ \varphi(t_1 + 0; t_1, \varphi_1) = \varphi_1(0) \end{cases}$$

meets some of the hypersurfaces σ_k. The number of this hypersurface is denoted by l_2. Then we have $x(t) = x(t; t_0, \varphi_0) = \varphi(t; t_1, \varphi_1)$ for $t_1 < t \le t_2$.

At the moment $t = t_2$ the mapping point $(t, x(t; t_0, \varphi_0))$ jumps momentarily, etc.

The points $t_1, t_2, \ldots$ $(t_0 < t_1 < t_2)$ are the impulsive moments. Let us note that, in general, $k \ne l_k$. In other words, it is possible that the integral curve of the problem under consideration does not meet the hypersurface σ_k at the moment t_k.

The solutions of systems with variable impulsive perturbations are piecewise continuous functions but unlike the solutions of systems with fixed moments of impulse effect, they have points of discontinuity depending on the solutions, i.e. the different solutions have different points of discontinuity. This leads to a number of difficulties in the investigation of systems with variable impulsive perturbations. One of the phenomena occurring with such systems is the so called "beating" of the solutions. This is the phenomenon when the mapping point $(t, x(t))$ meets one and the same hypersurface σ_k several or infinitely many times [21, 22]. Part of the difficulties are related to the possibilities of "merging" of different integral curves after a given moment, loss of the property of autonomy, etc.

It is clear that systems of impulsive functional differential equations with fixed moments of impulse effect can be considered as a particular case of the systems with variable impulsive perturbations. Indeed, if $t = t_k, k = 1, 2, \ldots$ are fixed moments of time and we introduce the notation $\sigma_k = \{(t, x) \in [t_0, \infty) \times \Omega : t = t_k\}$, then the systems of the first class are reduced to the systems of the second class.

Early research results on the theory of impulsive functional differential equations were published by Anokhin [12], Bainov, Covachev and Stamova [19, 20] and Gopalsamy and Zhang [93]. In many papers, interesting results on the impulsive functional differential equations with constant delays have been obtained [20, 23–27, 34–38, 41–47, 49, 51, 53, 54, 190, 191, 193, 194, 198, 205]. However, in practical evolutionary processes, absolute constant delay may be scarce and delays are frequently varied with time.

In spite of the great possibilities for application, the theory of the impulsive functional differential equations is developing rather slowly at the beginning, due to obstacles of theoretical and technical character. The presence of delay and impulses requires introduction of new and modification of the standard methods for investigations. Some of the properties, such as existence, continuability, and stability may be changed greatly by impulses [21, 22, 24, 25, 39, 125, 143, 189, 195, 215, 217].

In recent years there has been a growing interest in the area of impulsive functional differential equations by many mathematicians, specialists in the theory of optimal control, physics, chemical technologies, population dynamics, biotechnologies, industrial robotics, and economics.

An interesting and fruitful technique that has gained increasing significance and has given decisive impetus for modern development of stability theory of impulsive functional differential equations is the second method of Lyapunov. A manifest advantage of this method is that it does not require the knowledge of solutions and therefore has great power in applications.

The study of stability theory of impulsive systems with delay is usually more challenging than that of systems without delay. Significant progress on stability of impulsive functional differential equations has been made during the past decades. Results have been obtained by using the Lyapunov functions as well as by the Lyapunov–Krasovskii functionals [13, 14, 59, 69, 146, 155]. When Lyapunov functions are used, the method is coupled with the Razumikhin technique [169]. This technique is more appropriate for applications. By means of the Lyapunov–Razumikhin method, many stability results for impulsive functional differential equations were obtained [24, 26, 27, 34–38, 40–42, 44–46, 48–54, 131, 133, 148, 149, 155, 184, 194, 196, 198–203, 205–210, 213, 229].

Boundedness theory has played a significant role in the existence of periodic solutions and it has many applications in areas such as neural network, biological population management, secure communication and chaos control. The theory has been greatly developed during the past decades (see [69, 70, 87, 111, 126, 148, 193, 197, 204, 220, 228] and the references cited therein).

The aim of this book is to present a systematic account of the recent developments in the stability theory for impulsive functional differential equations. Also, we would like to show the manifestations of Lyapunov–Razumikhin method by demonstrating how this effective technique can be applied to investigate stability and boundedness of many practical problems of diverse interest.

1.2   Existence, uniqueness and continuability

Let $t_0 \in \mathbb{R}$, $r = \text{const} > 0$, $\Omega \subseteq \mathbb{R}^n$, $\Omega \neq \emptyset$. Let $J \subseteq \mathbb{R}$. Define the following class of functions:

$PC[J, \Omega] = \{\sigma : J \to \Omega : \sigma(t)$ is a piecewise continuous function with
points of discontinuity $\tilde{t} \in J$ at which $\sigma(\tilde{t} - 0)$ and $\sigma(\tilde{t} + 0)$ exist
and $\sigma(\tilde{t} - 0) = \sigma(\tilde{t})\}$.

Consider the following system of impulsive functional differential equations with variable impulsive perturbations:

$$\dot{x}(t) = f(t, x_t), \quad t \neq \tau_k(x(t)) \tag{1.21}$$

$$\Delta x(t) = I_k(x(t)), \quad t = \tau_k(x(t)), \quad k = 1, 2, \ldots, \tag{1.22}$$

where $f : [t_0, \infty) \times PC[[-r, 0], \Omega] \to \mathbb{R}^n$; $\tau_k : \Omega \to (t_0, \infty)$, $I_k : \Omega \to \mathbb{R}^n$, $k = 1, 2, \ldots$; $\Delta x(t) = x(t + 0) - x(t - 0)$; and for $t \geq t_0$, $x_t \in PC[[-r, 0], \Omega]$ is defined by $x_t(s) = x(t + s)$, $-r \leq s \leq 0$.

Let $\varphi_0 \in PC[[-r, 0], \Omega]$. Denote by $x(t) = x(t; t_0, \varphi_0)$ the solution of system (1.21), (1.22), satisfying the initial conditions

$$\begin{cases} x(t; t_0, \varphi_0) = \varphi_0(t - t_0), & t_0 - r \leq t \leq t_0 \\ x(t_0 + 0; t_0, \varphi_0) = \varphi_0(0). \end{cases} \tag{1.23}$$

Let $J_1 = [t_0, \omega)$, $J_2 = [t_0, \tilde{\omega})$, and $J_1 \subseteq J_2$.

Definition 1.6. If:

(1) $x(t) = x(t; t_0, \varphi_0)$ and $y(t) = y(t; t_0, \varphi_0)$ are two solutions of the system (1.21), (1.22) on the intervals J_1 and J_2, respectively;

(2) $x(t) = y(t)$ for $t \in J_1$;

then $y(t)$ is said to be a *continuation* of $x(t)$ on the interval J_2 (*continuation to the right*).

The solution $x(t) = x(t; t_0, \varphi_0)$ is said to be *continuable* on the interval J_2 if there exists its continuation $y(t)$ on J_2. Otherwise $x(t) = x(t; t_0, \varphi_0)$ is said to be *noncontinuable* and the interval J_1 is called a *maximal interval of existence* of $x(t)$.

Definition 1.7. The solution $x(t) = x(t; t_0, \varphi_0)$ of the system (1.21), (1.22) is said to be *unique* if given any other solution $y(t) = y(t; t_0, \varphi_0)$ of the system (1.21), (1.22), $x(t) = y(t)$ on their common interval of existence.

The next example illustrates the phenomenon "beating".

Example 1.8. The initial value problem (1.21), (1.22), (1.23) with:

(a) $n = 1$, $t_0 = 0$, $\Omega = (0, \infty)$;

(b) $f(t, x_t) = x(t)[x(t - r(t)) - 1]$, $0 < r(t) \leq r$;

(c) $x(t) = \varphi_0(t) = 1$, $t \in [-r, 0]$;

(d) $\tau_k(x) = \arctan x + k\pi$, $x \in \Omega$, $k = 1, 2, \ldots$;

(e) $I_k \in C[\Omega, \mathbb{R}]$, $k = 1, 2, \ldots$

has a solution $x(t) = x(t; 0, \varphi_0) = 1$ for $t > 0$ till the first meeting with the hypersurface $t = \tau_1(x)$. Then, for any choice of the functions $I_k(x)$, $k = 1, 2, \ldots$ so that $I_k(x) > 0$ for $x \in \Omega$, the integral curve $(t, x(t; 0, \varphi_0))$ meets one and the same hypersurface (in the present case this is the curve τ_1). Moreover, the solution is not continuable for $t \geq \frac{3\pi}{2}$.

It is clear that the presence of the phenomenon "beating" for impulsive systems considerably complicates their investigation. Efficient sufficient conditions, which guarantee the absence of this phenomenon for impulsive functional differential equations with variable impulsive perturbations, were found by Bainov and Dishliev in [21, 22].

Definition 1.9. The solution $x(t) = x(t; t_0, \varphi_0)$ of the problem (1.21), (1.22), (1.23) is said to be *quasiunique* if the solution of the corresponding problem without impulses (1.21), (1.23) is unique for $t \geq t_0$.

We specially emphasize that if the solutions of (1.21), (1.22) are quasiunique, then it is possible for two distinct integral curves to merge after some impulse. We shall illustrate merging by the following example:

Example 1.10 ([21]). Consider the initial value problem (1.21), (1.22), (1.23) with:

(a) $n = 1$, $\Omega = \mathbb{R}$;

(b) $f(t, x_t) = 0, t \in \mathbb{R}_+$;

(c) $\tau_k(x) = 2k - \dfrac{1}{1 + x^2}, x \in \mathbb{R}, k = 1, 2, \ldots$;

(d) $I_k(x) = -x, x \in \mathbb{R}, k = 1, 2, \ldots$.

From assumptions (b) and (d) it follows immediately that the impulsive system (1.21), (1.22) has zero solution. Moreover, for any initial function $\varphi_0 \in PC[[-r, 0], \mathbb{R}]$ and for any $t_0 \in \mathbb{R}_+$, the solution of problem (1.21), (1.22), (1.23) merges with the zero solution after the first impulse.

Let $\tau_0(x) \equiv t_0$ for $x \in \Omega$. Introduce the following notation:

$$\sigma_k = \{(t, x) : t = \tau_k(x), \ x \in \Omega\}, \quad k = 1, 2, \ldots, \tag{1.24}$$

i.e. $\sigma_k, k = 1, 2, \ldots$ are hypersurfaces with equations $t = \tau_k(x)$.

Introduce the following conditions:

H1.1. The function f is continuous in $[t_0, \infty) \times PC[[-r, 0], \Omega]$.

H1.2. The function f is locally Lipschitz continuous with respect to its second argument in $[t_0, \infty) \times PC[[-r, 0], \Omega]$.

H1.3. There exists a constant $P > 0$ such that

$$\|f(t, x_t)\| \leq P < \infty \text{ for } (t, x_t) \in [t_0, \infty) \times PC[[-r, 0], \Omega].$$

H1.4. The functions τ_k are Lipschitz continuous with respect to $x \in \Omega$ with Lipschitz constants $L_k, 0 \leq L_k < \frac{1}{P}, k = 1, 2, \ldots$.

H1.5. $t_0 < \tau_1(x) < \tau_2(x) < \cdots, x \in \Omega$.

H1.6. $\tau_k(x) \to \infty$ as $k \to \infty$, uniformly on $x \in \Omega$.

H1.7. $\tau_k(x + I_k(x)) \leq \tau_k(x)$ for $x \in \Omega, k = 1, 2, \ldots$.

H1.8. For each $(t_0, \varphi_0) \in \mathbb{R} \times PC[[-r, 0], \Omega]$, the solution of initial value problem without impulses (1.21), (1.23) does not leave the domain Ω for $t \geq t_0$.

H1.9. $(E + I_k) : \Omega \to \Omega, k = 1, 2, \ldots$, where E is the identity in Ω.

H1.10. The functions I_k are Lipschitz continuous with respect to $x \in \Omega$ with Lipschitz constants $\Lambda_k, 0 \leq \Lambda_k < 1 - L_k P, k = 1, 2, \ldots$.

Theorem 1.11 ([21, 22]). *Let conditions H1.1–H1.5 and H1.7 hold. Then the integral curve $(t, x(t))$ of problem* (1.21), (1.22), (1.23) *meets each one of the hypersurfaces* (1.24) *at most once.*

The absence of the phenomenon "beating" does not guarantee the continuability of the solution of the initial value problem (1.21), (1.22), (1.23) for $t \geq t_0$. In the subsequent example the following situation is considered: the solutions of the corresponding system without impulses (1.21) are continuable for all $t \geq t_0$ for any choice of the initial data $(t_0, \varphi_0) \in \mathbb{R} \times PC[[-r, 0], \Omega]$. Any solution of the system with impulses (1.21), (1.22) meets each one of the hypersurfaces (1.24) at most once. In spite of this, some solutions of system (1.21), (1.22) are noncontinuable from a certain time on.

Example 1.12 ([21]). Consider the initial value problem (1.21), (1.22), (1.23) under the following assumptions:

(a) $n = 1, \Omega = \mathbb{R}$;

(b) $\tau_k(x) = 2 - 2^{-k} - \dfrac{1}{1+x^2}$, $x \in \mathbb{R}$, $k = 1, 2, \ldots$;

(c) conditions H1.1, H1.2 and H1.3 hold with constant $P < \dfrac{8}{3\sqrt{3}}$;

(d) for any $x \in \mathbb{R}$ and any number $k = 1, 2, \ldots$ the following inequalities are valid:

$$x I_k(x) < 0, \quad |I_k(x)| < 2|x|;$$

(e) $\tau_k(x) < \tau_{k+1}(x + I_k(x))$ for $x \in \mathbb{R}$, $k = 1, 2, \ldots$.

It is easy to check that the functions τ_k are Lipschitz continuous with respect to $x \in \mathbb{R}$ with Lipschitz constants $L_k = \frac{3\sqrt{3}}{8}$, $k = 1, 2, \ldots$.

Indeed, we can set

$$L_k = \sup\{|\dot{\tau}_k(x)|, \ x \in \mathbb{R}\} = \max\left\{\frac{2|x|}{(1+x^2)^2}, \ x \in \mathbb{R}\right\}$$

$$= \frac{2|x|}{(1+x^2)^2}\Big|_{x=1/\sqrt{3}} = \frac{3\sqrt{3}}{8}.$$

Condition H1.5 holds. It follows from (c) that for this choice of the constant P condition H1.4 holds too. The two inequalities in (d) immediately imply condition H1.7.

By Theorem 1.11, the integral curve of the problem considered meets each one of the curves τ_k at most once. If we suppose that $0 < t_0 < \tau_1(\varphi_0(0))$, then by condition (e) we conclude that the integral curve $(t, x(t; t_0, \varphi_0))$ meets each one of the curves τ_k exactly once. This means that the solution of the problem considered is noncontinuable for $t \geq 2$.

Under the assumption that conditions H1.4, H1.5 and H1.6 hold, we define the following notation:

$$G_k = \{(t, x) : \tau_{k-1}(x) < t < \tau_k(x), \ x \in \Omega\}, \ k = 1, 2, \ldots.$$

Lemma 1.13 ([21]). *Let conditions* H1.1–H1.5 *and* H1.8 *hold, and* $(t_0, \varphi_0(0)) \in G_k \cup \sigma_{k-1}$. *Then for* $t > t_0$ *the integral curve of problem* (1.21), (1.22), (1.23) *meets first the hypersurface* σ_k.

Let $t_1, t_2, \ldots$ ($t_0 < t_1 < t_2 < \cdots$) be the moments in which the integral curve $(t, x(t; t_0, \varphi_0))$ of the problem (1.21), (1.22), (1.23) meets the hypersurfaces σ_k, $k = 1, 2, \ldots$, i.e. each of the points t_k is a solution of some of the equations $t = \tau_k(x(t))$, $k = 1, 2, \ldots$.

Theorem 1.14 ([21]). *Let conditions* H1.1–H1.9 *hold. Then for each* $(t_0, \varphi_0) \in \mathbb{R} \times PC[[-r, 0], \Omega]$:

(1) *The integral curve* $(t, x(t; t_0, \varphi_0))$ *meets infinitely many hypersurfaces of* (1.24).

(2) $t_k \to \infty$ *as* $k \to \infty$.

(3) *The solution of problem* (1.21), (1.22), (1.23) *is quasiunique and continuable for all* $t \geq t_0$.

Theorem 1.15 ([21]). *Assume that:*

(1) *Conditions* H1.1–H1.4 *and* H1.10 *hold.*

(2) *The integral curves* $(t, x(t; t_0, \varphi_0))$, $\varphi_0 \in PC[[-r, 0], \Omega]$ *meet successively the same hypersurfaces of* (1.24).

Then the solution of problem (1.21), (1.22), (1.23) *is unique.*

Theorem 1.16. *Let conditions* H1.1–H1.10 *and condition* (2) *of Theorem 1.15 hold. Then for each* $(t_0, \varphi_0) \in \mathbb{R} \times PC[[-r, 0], \Omega]$ *the solution of problem* (1.21), (1.22), (1.23) *is unique and continuable for all* $t \geq t_0$.

Theorem 1.16 follows from Theorem 1.14 and Theorem 1.15.

Consider the system of functional differential equations with fixed moments of impulse effect

$$\dot{x}(t) = f(t, x_t), \ t \neq t_k, \ t \geq t_0, \tag{1.25}$$

$$\Delta x(t_k) = I_k(x(t_k)), \ t_k > t_0, \ k = 1, 2, \ldots . \tag{1.26}$$

In the present case, $\tau_k(x) \equiv t_k$, $k = 1, 2, \ldots$ and σ_k are hyperplanes in $\mathbb{R}^{n+1}$.

Denote by $x(t) = x(t; t_0, \varphi_0)$ the solution of the initial value problem (1.25), (1.26), (1.23) and by $J^+(t_0, \varphi_0)$ the maximal interval of type $[t_0, \omega)$ in which the solution $x(t; t_0, \varphi_0)$ is defined.

Introduce the following conditions:

H1.11. $I_k \in C[\Omega, \Omega]$, $k = 1, 2, \ldots$.

H1.12. $t_0 < t_1 < t_2 < \cdots < t_k < t_{k+1} < \cdots$.

H1.13. $\lim\limits_{k \to \infty} t_k = \infty$.

Theorem 1.17. *Let conditions* H1.1, H1.3, H1.9, H1.11, H1.12 *and* H1.13 *hold.*
   Then for each $(t_0, \varphi_0) \in \mathbb{R} \times PC[[-r, 0], \Omega]$:

(1) *There exists a solution* $x(t) = x(t; t_0, \varphi_0)$ *of the initial value problem* (1.25),
    (1.26), (1.23) *defined on* $J^+(t_0, \varphi_0)$.

(2) $J^+(t_0, \varphi_0) = [t_0, \infty)$.

(3) *If, moreover, condition* H1.2 *is met then the solution* $x(t; t_0, \varphi_0)$ *is unique.*

Proof of Assertion 1. Assertion 1 follows from Theorem 1.14. Indeed, the validity of
H1.1, H1.3, as well as, the existence theorem applied to problem (1.25), (1.23) (cf.
[98, 99]) imply that for each $(t_0, \varphi_0) \in \mathbb{R} \times C[[-r, 0], \Omega]$ there exists a solution $\Phi_1(t)$
of problem without impulses (1.25), (1.23) for $t \geq t_0$. Moreover, $\Phi_1(t) = \varphi_0(t - t_0)$
as $t \in [t_0 - r, t_0]$, and this solution does not leave the domain Ω. Let t_1 be the first
moment of impulsive perturbation. Setting $x(t; t_0, \varphi_0) = \Phi_1(t)$ as $t \in [t_0, t_1]$, we have
$\Phi_1(t_1 + 0) = I_1(\Phi_1(t_1)) + \Phi_1(t_1) = \Phi_1^+$.
   Now the above mentioned existence theorem applied to the system (1.25) in the
interval (t_1, t_2) ensures that there exists a solution $\Phi_2(t)$ such that $\Phi_2(t) = \Phi_1(t)$ for
$t_1 - r \leq t \leq t_1$ and $\Phi_2(t_1) = \Phi_1^+$. The solution $x(t)$ of problem (1.25), (1.26), (1.23)
can be extended to the moment $t = t_2$ by setting $x(t; t_0, \varphi_0) = \Phi_2(t)$ for $t_1 < t \leq t_2$.
   In the same way, let us denote by $\Phi_k(t)$ the solutions of the system (1.25) in the
intervals $(t_{k-1}, t_k]$, $k = 3, 4, \ldots$, respectively. Then for $t = t_k$ we have

$$\Phi_k(t_k + 0) = I_k(\Phi_k(t_k)) + \Phi_k(t_k) = \Phi_k^+.$$

   It follows from the existence theorem for problem (1.25), (1.23) on the interval
$(t_k, t_{k+1}]$ that there exists a solution $\Phi_{k+1}(t)$ such that $\Phi_{k+1}(t) = \Phi_k(t)$ for $t_k - r \leq$
$t \leq t_k$ and $\Phi_{k+1}(t_k) = \Phi_k^+$. Thus, the solution $x(t; t_0, \varphi_0)$ of problem (1.25), (1.26),
(1.23) can be extended to the moment t_{k+1}, $k = 2, 3, \ldots$, by setting $x(t; t_0, \varphi_0) =$
$\Phi_{k+1}(t)$ for $t_k < t \leq t_{k+1}$.
   Finally, by means of condition H1.12, solution $x(t; t_0, \varphi_0)$ of problem (1.25), (1.26),
(1.23) is defined for $t \in J^+(t_0, \varphi_0)$.
   In the case when $(t_0, \varphi_0) \in \mathbb{R} \times PC[[-r, 0], \Omega]$ and $\theta_1, \theta_2, \ldots, \theta_s \in (t_0 - r, t_0)$ are
the points of discontinuity of first kind of the function φ_0 at which it is continuous
from the left, the proof of Assertion 1 is similar. We shall note that, in this case, it is
possible that $t_k = \theta_l + r$ for some $k = 1, 2, \ldots$ and $l = 1, 2, \ldots, s$.

Proof of Assertion 2. Since the solution $x(t) = x(t; t_0, \varphi_0)$ is defined on $[t_0, t_1] \cup$
$(t_k, t_{k+1}]$, $k = 1, 2, \ldots$, then from H1.12 and H1.13, we conclude that it can be
continued for all $t \geq t_0$, i.e. $J^+(t_0, \varphi_0) = [t_0, \infty)$.

Proof of Assertion 3. The validity of condition H1.2 ensures that the above defined
solutions $\Phi_1(t), \Phi_2(t), \ldots$ are unique and therefore the solution $x(t; t_0, \varphi_0)$ of problem
(1.25), (1.26), (1.23) is unique. $\square$

Now we consider an initial value problem for the linear system of functional differential equations with impulse effects at fixed moments:

$$
\begin{cases}
\dot{x}(t) = A(t)x(t) + B(t)x_t, \ t \geq t_0, \ t \neq t_k \\
x(t) = \varphi_0(t - t_0), \ t \in [t_0 - r, t_0] \\
x(t_0 + 0) = \varphi_0(0) \\
\Delta x(t_k) = B_k x(t_k), \ t_k > t_0, \ k = 1, 2, \ldots,
\end{cases}
\tag{1.27}
$$

where $A(t)$, $B(t)$ and B_k, $k = 1, 2, \ldots$ are $(n \times n)$ matrices.

Theorem 1.18. *Let the matrix functions $A(t)$ and $B(t)$ are continuous for $t \geq t_0$, $t \neq t_k$, $k = 1, 2, \ldots$ with points of discontinuity at $t_1, t_2, \ldots$ where they are left continuous. Then for each $(t_0, \varphi_0) \in \mathbb{R} \times PC[[-r, 0], \Omega]$ there exists a unique solution $x(t) = x(t; t_0, \varphi_0)$ of problem (1.27) that is defined for all $t \geq t_0$.*

Theorem 1.18 is a consequence of the theorem on existence and uniqueness for the solutions of a linear system of functional differential equations [98, 99].

The problem on left-continuability of solutions will be considered now for systems of type (1.25), (1.26) only.

Assume that $x(t)$ is a solution of (1.25), (1.26) defined on interval (γ, ω).

If $\gamma \neq t_k$, then the problem on continuability of $x(t)$ on the left of γ can be solved in the same way as for functional differential equations without impulses. In this case the solution $x(t)$ is continuable on the left of γ and $J^- = J^-(t_0, \varphi_0) = (\alpha, t_0)$.

Straightforward calculations show that the solution $x(t)$ of problem (1.25), (1.26) satisfies the equation

$$
x(t) =
\begin{cases}
\varphi_0(0) + \displaystyle\sum_{t_0 < t_k < t} I_k(x(t_k)) + \int_{t_0}^{t} f(s, x_s)ds, \ t \in J^+ \\
\varphi_0(0) - \displaystyle\sum_{t < t_k < t_0} I_k(x(t_k)) + \int_{t_0}^{t} f(s, x_s)ds, \ t \in J^-.
\end{cases}
$$

The solution of the linear system (1.27) can be extended to the left of t_k if the below conditions are met:

$$
\det(E + B_k) \neq 0, \ k = 1, 2, \ldots,
\tag{1.28}
$$

where E is the $(n \times n)$ identity matrix.

Let $U_k(t, s)$ $(t, s \in (t_{k-1}, t_k])$ be the Cauchy matrix [100] for the linear system

$$
\dot{x}(t) = A(t)x(t), \ t_{k-1} < t \leq t_k, \ k = 1, 2, \ldots .
$$

Then by virtue of Theorem 1.18, the solution of the initial problem (1.27) can be decomposed as

$$
x(t; t_0, \varphi_0) = x(t) = W(t, t_0 + 0)\varphi_0(0) + \int_{t_0}^{t} W(t, s)B(s)x_s ds, \ t > t_0,
\tag{1.29}
$$

where

$$W(t,s) = \begin{cases} U_k(t,s) \quad \text{as} \quad t,s \in (t_{k-1}, t_k] \\[2mm] U_{k+1}(t, t_k + 0)(E + B_k)U_k(t_k, s) \quad \text{as} \quad t_{k-1} < s \leq t_k < t \leq t_{k+1} \\[2mm] U_k(t, t_k)(E + B_k)^{-1}U_{k+1}(t_k + 0, s) \quad \text{as} \quad t_{k-1} < t \leq t_k < s \leq t_{k+1} \\[2mm] U_{k+1}(t, t_k + 0)\prod_{j=k}^{i+1}(E + B_j)U_j(t_j, t_{j-1} + 0)(E + B_i)U_i(t_i, s) \\[2mm] \quad \text{as} \quad t_{i-1} < s \leq t_i < t_k < t \leq t_{k+1} \\[2mm] U_i(t, t_i)\prod_{j=i}^{k-1}(E + B_j)^{-1}U_{j+1}(t_j + 0, t_{j+1})(E + B_k)^{-1}U_{k+1}(t_k + 0, s) \\[2mm] \quad \text{as} \quad t_{i-1} < t \leq t_i < t_k < s \leq t_{k+1}, \end{cases}$$

is the solving operator of the system

$$\begin{cases} \dot{x}(t) = A(t)x(t), \ t \neq t_k \\ \Delta x(t_k) = B_k x(t_k). \end{cases}$$

1.3   Piecewise continuous Lyapunov functions

The second method of Lyapunov is one of the universal methods for investigating the dynamical systems from a different type. The method is also known as a direct method of Lyapunov or a method of the Lyapunov functions. Put forward in the end of the XIX century by Lyapunov [150], this method hasn't lost its popularity today. It has been applied initially to ordinary differential equations, and in his first work Lyapunov standardized the definition for stability and generalized the Lagrange's work [128] on potential energy. The essence of the method is the investigation of the qualitative properties of the solutions without an explicit formula. For this purpose we need auxiliary functions – the so-called Lyapunov functions.

The first more significant development of the Lyapunov second method were made in the 1930s in the works of Barbashin [55, 56] as well as in the works of Chatayev, Malkin, and Marachkov [71, 152, 153]. The researches in that period were focused on simplifying the conditions of Lyapunov theorems and proving the inverse theorems for ordinary differential equations. Different aspects of the Lyapunov second method applications for ordinary differential equations are given in [55, 56, 61, 100, 124, 152, 153, 171, 182, 183, 223].

Gradually, there has been an expansion both in the class of the studied objects and in the mathematical problems investigated by means of the method.

One of the directions in which the method is used is the study of the qualitative properties of the solutions of functional differential equations. This method allows to make a conclusion about the stability (asymptotic stability, instability) of the investigated systems. Also, the well-constructed Lyapunov function allows the area of stability to be evaluated in the phase space as well as in the space of the parameters. The reason for the development of this direction, observed in the middle of the 20th century, is infiltration of mathematical methods in the study of technical, biological, ecological and other systems. Today there is a significant amount of articles on the stability and boundedness of differential equations with constants delays, neutral type differential equations, integro-differential equations, linear and non-linear functional differential equations [4, 58, 60, 63, 64, 66–68, 76, 79, 82, 84, 91, 92, 94, 95, 98, 99, 106–108, 114, 118, 119, 131–134, 140, 154, 163, 168–170, 214, 219, 225, 227, 232, 233].

There are two main approaches, when the second method of Lyapunov is applied, for investigating the stability and boundedness of the solutions of functional differential equations. The first one is the method of *Lyapunov functions*. The direct transfer of the Lyapunov theorems to functional differential equations leads to significant difficulties when the sign of the derivative of the Lyapunov function with respect to the system has to be determined. Therefore, in the works [98, 99, 119, 131–134, 169], it is offered that the derivative of the Lyapunov function should be estimated by the elements of the minimal subsets of the integral curves of the investigated system. This technique is known as *Razumikhin technique* [169].

When using the direct method of Lyapunov for functional differential equations, Krasovskii [124] approached from a functional analysis point of view. He replaced the Lyapunov function with a *Lyapunov functional*. The method of Lyapunov–Krasovskii functionals has been used by many researchers on the stability theory for functional differential equations [4, 13, 14, 58, 63, 64, 66, 68, 91, 98, 99, 118, 119, 160, 166].

Gurgulla and Perestyuk were the first who applied the Lyapunov direct method for impulsive systems. In the work [97] they used classical (continuous) Lyapunov functions. The application of continuous Lyapunov functions to the investigation of impulsive systems restricts the possibilities of Lyapunov second method. The fact that the solutions of impulsive systems are piecewise continuous functions requires introducing some analogous of the classical Lyapunov functions which have discontinuities of the first kind [30]. By means of such functions it becomes possible to solve basic problems related to the application of Lyapunov second method to impulsive systems.

The presence of impulses as well as the delay in the impulsive functional differential equations require a combination between the method of piecewise continuous Lyapunov function and Razumikhin technique. By means of such approach, many interesting results of the stability and boundedness theory of these equations have been obtained [24, 26, 27, 34–38, 40–46, 48–54, 87, 131, 133, 148, 149, 155, 184, 185, 193, 194, 196–210, 213, 228, 229]. The results obtained in such manner can be applied more easily in comparison with the method of Lyapunov–Krasovskii functionals.

Scalar Lyapunov functions

Consider the system (1.21), (1.22). Introduce the following notation:

$$G = \bigcup_{k=1}^{\infty} G_k.$$

Definition 1.19. A function $V : [t_0, \infty) \times \Omega \to \mathbb{R}_+$ belongs to the class V_0, if:

(1) V is continuous in G and locally Lipschitz continuous with respect to its second argument on each of the sets G_k, $k = 1, 2, \ldots$.

(2) For each $k = 1, 2, \ldots$ and $(t_0^*, x_0^*) \in \sigma_k$ there exist the finite limits

$$V(t_0^* - 0, x_0^*) = \lim_{\substack{(t,x) \to (t_0^*, x_0^*) \\ (t,x) \in G_k}} V(t, x), \quad V(t_0^* + 0, x_0^*) = \lim_{\substack{(t,x) \to (t_0^*, x_0^*) \\ (t,x) \in G_{k+1}}} V(t, x)$$

and the equality $V(t_0^* - 0, x_0^*) = V(t_0^*, x_0^*)$ holds.

Let $V \in V_0$. For $(t, x) \in G$, we define

$$\dot{V}_{(1.21),(1.22)}(t, x) = \lim_{h \to 0^+} \sup \frac{1}{h}[V(t + h, x + hf(t, x_t)) - V(t, x)].$$

Note that if $x = x(t)$ is a solution of system (1.21), (1.22), then for $t > t_0$, $t \neq \tau_k(x(t))$, $k = 1, 2, \ldots$ we have $\dot{V}_{(1.21),(1.22)}(t, x) = D^+_{(1.21),(1.22)} V(t, x(t))$, where

$$D^+_{(1.21),(1.22)} V(t, x(t)) = \lim_{h \to 0^+} \sup \frac{1}{h}[V(t + h, x(t + h)) - V(t, x(t))] \qquad (1.30)$$

is *the upper right-hand Dini derivative* of $V \in V_0$ (with respect to the system (1.21), (1.22)).

For $V \in V_0$ and for some $t \geq t_0$, define the following set:

$$\Omega_1 = \{x \in PC[[t_0, \infty) \times \Omega] : V(s, x(s)) \leq V(t, x(t)), t - r \leq s \leq t\}.$$

In the present book results on stability and boundedness for the systems of the type (1.21), (1.22) are given where the upper right-hand derivatives of the Lyapunov piecewise continuous functions are estimated by means of elements of the sets of the type Ω_1.

The class of functions V_0 is also used for investigation of stability and boundedness of the systems of impulsive functional differential equations with fixed moments of impulse effect (1.25), (1.26). In this case, $\tau_k(x) \equiv t_k$, $k = 1, 2, \ldots$, σ_k are hyperplanes in $\mathbb{R}^{n+1}$, the sets G_k are

$$G_k = \{(t, x) : t_{k-1} < t < t_k, x \in \Omega\},$$

and the condition (2) of Definition 1.19 is substituted by the condition:

(2′) For each $k = 1, 2, \ldots$ and $x \in \Omega$, there exist the finite limits

$$V(t_k - 0, x) = \lim_{\substack{t \to t_k \\ t < t_k}} V(t, x), \qquad V(t_k + 0, x) = \lim_{\substack{t \to t_k \\ t > t_k}} V(t, x),$$

and the following equalities are valid:

$$V(t_k - 0, x) = V(t_k, x).$$

Vector Lyapunov functions

It is well known that employing several Lyapunov functions, in the investigation of the qualitative properties of the differential equations solutions, is more useful than employing a single one since each function can satisfy less rigid requirements. Hence, the corresponding theory, known as the method of vector Lyapunov functions, offers a very flexible mechanism (see [132] and references therein).

Moreover, by means of the method of vector Lyapunov functions we can prove the results in some cases in which using the scalar Lyapunov functions is impossible.

In the present book we shall use vector Lyapunov functions $V : [t_0, \infty) \times \Omega \to \mathbb{R}_+^m$, $V = \mathrm{col}(V_1, V_2, \ldots, V_m)$ such that $V_j \in V_0$, $j = 1, 2, \ldots, m$.

1.4 Comparison theorems

In this section we shall present the main comparison results we use. The essence of the comparison method in the stability theory is in studying the relations between the given system and a comparison system so that the stability properties of the solutions of comparison system should imply the corresponding stability properties of the solutions of system under consideration. These relations are obtained employing differential inequalities. The comparison system is usually of lower order and its right-hand side possesses a certain type of monotonicity, which considerably simplifies the study of its solutions.

Consider the system of impulsive functional differential equations

$$\begin{cases} \dot{x}(t) = f(t, x_t), & t \neq t_k \\ \Delta x(t_k) = x(t_k + 0) - x(t_k) = I_k(x(t_k)), & t_k > t_0, \end{cases} \tag{1.31}$$

where $f : [t_0, \infty) \times PC[[-r, 0], \Omega] \to \mathbb{R}^n$; $I_k : \Omega \to \mathbb{R}^n$, $k = 1, 2, \ldots$; $t_0 < t_1 < t_2 < \cdots$, $\lim_{k \to \infty} t_k = \infty$.

Denote by $x(t) = x(t; t_0, \varphi_0)$ the solution of problem (1.31), (1.23), and by $J^+(t_0, \varphi_0)$ – the maximal interval of type $[t_0, \omega)$ in which the solution $x(t; t_0, \varphi_0)$ is defined.

Define the following classes:

$$PC^1[J, \Omega] = \{\sigma \in PC[J, \Omega] : \ \sigma(t) \text{ is continuously differentiable everywhere}$$
$$\text{except the points } t_k \text{ at which } \dot{\sigma}(t_k - 0) \text{ and } \dot{\sigma}(t_k + 0) \text{ exist and}$$
$$\dot{\sigma}(t_k - 0) = \dot{\sigma}(t_k), \ k = 1, 2, \ldots\};$$

and

$$\Omega_P = \{x \in PC[[t_0, \infty], \Omega] : \ V(s, x(s)) \le P(V(t, x(t))), \ t - r \le s \le t\},$$

where $t \ge t_0$, $V \in V_0$, $P(u)$ is continuous on $\mathbb{R}_+$, non-decreasing in u, and $P(u) > u$ for $u > 0$.

Together with system (1.31) we shall consider the comparison system

$$\begin{cases} \dot{u}(t) = F(t, u(t)), \ t \ne t_k \\ \Delta u(t_k) = u(t_k + 0) - u(t_k) = J_k(u(t_k)), \ t_k > t_0, \end{cases} \qquad (1.32)$$

where $F : [t_0, \infty) \times \mathbb{R}_+^m \to \mathbb{R}^m$; $J_k : \mathbb{R}_+^m \to \mathbb{R}^m$, $k = 1, 2, \ldots$.

Let $u_0 \in \mathbb{R}_+^m$. Denote by $u(t) = u(t; t_0, u_0)$ the solution of system (1.32) satisfying the initial condition $u(t_0 + 0) = u(t_0) = u_0$ and by $J^+(t_0, u_0)$ the maximal interval of type $[t_0, \beta)$ in which the solution $u(t; t_0, u_0)$ is defined.

We introduce into $\mathbb{R}^m$ a partial ordering in the following way: for the vectors $u, v \in \mathbb{R}^m$ we shall say that $u \ge v$ if $u_j \ge v_j$ for each $j = 1, 2, \ldots, m$ and $u > v$ if $u_j > v_j$ for each $j = 1, 2, \ldots, m$.

Definition 1.20. The solution $u^+ : \ J^+(t_0, u_0) \to \mathbb{R}_+^m$ of the system (1.32) for which $u^+(t_0; t_0, u_0) = u_0$ is said to be a *maximal solution* if any other solution $u : [t_0, \tilde{\omega}) \to \mathbb{R}_+^m$, for which $u(t_0) = u_0$ satisfies the inequality $u^+(t) \ge u(t)$ for $t \in J^+(t_0, u_0) \cap [t_0, \tilde{\omega})$.

Analogously, the *minimal solution* of system (1.32) is defined.

Definition 1.21. The function $\psi : \mathbb{R}_+^m \to \mathbb{R}^m$ is said to be:

(a) *non-decreasing* in $\mathbb{R}_+^m$ if $\psi(u) \ge \psi(v)$ for $u \ge v$, $u, v \in \mathbb{R}_+^m$.

(b) *monotone increasing* in $\mathbb{R}_+^m$ if $\psi(u) > \psi(v)$ for $u > v$ and $\psi(u) \ge \psi(v)$ for $u \ge v$, $u, v \in \mathbb{R}_+^m$.

Definition 1.22. The function $F : [t_0, \infty) \times \mathbb{R}_+^m \to \mathbb{R}^m$ is said to be *quasi-monotone increasing* in $[t_0, \infty) \times \mathbb{R}_+^m$ if for each pair of points (t, u) and (t, v) from $[t_0, \infty) \times \mathbb{R}_+^m$ and for $i \in \{1, 2, \ldots, m\}$ the inequality $F_i(t, u) \ge F_i(t, v)$ holds whenever $u_i = v_i$ and $u_j \ge v_j$ for $j = 1, 2, \ldots, m$, $i \ne j$, i.e. for any fixed $t \in [t_0, \infty)$ and any $i \in \{1, 2, \ldots, m\}$ the function $F_i(t, u)$ is non-decreasing with respect to $(u_1, u_2, \ldots, u_{i-1}, u_{i+1}, \ldots, u_m)$.

In the case when the function $F : [t_0, \infty) \times \mathbb{R}^m_+ \to \mathbb{R}^m$ is continuous and quasi-monotone increasing, all solutions of problem (1.32) starting from the point $(\overline{t_0}, u_0) \in [t_0, \infty) \times \mathbb{R}^m_+$ lie between two singular solutions – the maximal and the minimal ones.

Theorem 1.23. *Assume that:*

(1) *Conditions* H1.1, H1.3, H1.9, H1.11, H1.12 *and* H1.13 *hold.*

(2) *The function F is quasi-monotone increasing, continuous in the sets $(t_k, t_{k+1}]$ $\times \mathbb{R}^m_+$, $k \in N \cup \{0\}$ and for $k = 1, 2, \dots$ and $v \in \mathbb{R}^m_+$ there exists the finite limit*

$$\lim_{\substack{(t,u) \to (t,v) \\ t > t_k}} F(t, u).$$

(3) *The maximal solution $u^+ : J^+(t_0, u_0) \to \mathbb{R}^m_+$ of the system (1.32) is defined for $t \geq t_0$.*

(4) *The functions $\psi_k : \mathbb{R}^m_+ \to \mathbb{R}^m_+$, $\psi_k(u) = u + J_k(u)$, $k = 1, 2, \dots$ are non-decreasing in $\mathbb{R}^m_+$.*

(5) *The function $V : [t_0, \infty) \times \Omega \to \mathbb{R}^m_+$, $V = \mathrm{col}(V_1, V_2, \dots, V_m)$, $V_j \in V_0$, $j = 1, 2, \dots, m$, is such that*

$$V(t_0 + 0, \varphi_0(0)) \leq u_0,$$

$$V(t + 0, x + I_k(x)) \leq \psi_k(V(t, x)), \quad x \in \Omega, \ t = t_k, \ k = 1, 2, \dots,$$

and the inequality

$$D^+_{(1.31)} V(t, x(t)) \leq F(t, V(t, x(t))), \ t \neq t_k, \ k = 1, 2, \dots$$

is valid for $t \in [t_0, \infty)$, $x \in \Omega_P$.

Then

$$V(t, x(t; t_0, \varphi_0)) \leq u^+(t; t_0, u_0) \ \text{for } t \in [t_0, \infty). \tag{1.33}$$

Proof. From Theorem 1.17 it follows that $J^+(t_0, \varphi_0) = [t_0, \infty)$ and the solution $x = x(t; t_0, \varphi_0)$ of the problem (1.31), (1.23) is such that

$$x \in PC[(t_0 \quad r, \infty), \Omega] \cap PC^1[[t_0, \infty), \Omega].$$

The maximal solution $u^+(t; t_0, u_0)$ of the system (1.32) is defined by the equality

$$u^+(t; t_0, u_0) = \begin{cases} r_0(t; t_0, u_0^+), \ t_0 < t \leq t_1 \\ r_1(t; t_1, u_1^+), \ t_1 < t \leq t_2 \\ \quad \vdots \\ r_k(t; t_k, u_k^+), \ t_k < t \leq t_{k+1} \\ \quad \vdots \end{cases}$$

where $r_k(t; t_k, u_k^+)$ is the maximal solution of the system without impulses $\dot{u} = F(t, u)$ in the interval $(t_k, t_{k+1}]$, $k = 0, 1, 2, \ldots$, for which $u_k^+ = \psi_k(r_{k-1}(t_k; t_{k-1}, u_{k-1}^+))$, $k = 1, 2, \ldots$ and $u_0^+ = u_0$.

Let $t \in (t_0, t_1]$. Then, from the corresponding comparison theorem for the continuous case [132], it follows that

$$V(t, x(t; t_0, \varphi_0)) \leq u^+(t; t_0, u_0),$$

i.e. the inequality (1.33) is valid for $t \in (t_0, t_1]$.

Suppose that (1.33) is satisfied for $t \in (t_{k-1}, t_k]$, $k > 1$. Then, using condition (5) of Theorem 1.23 and the fact that the function ψ_k is non-decreasing, we obtain

$$V(t_k + 0, x(t_k + 0; t_0, \varphi_0)) \leq \psi_k(V(t_k, x(t_k; t_0, \varphi_0)))$$
$$\leq \psi_k(u^+(t_k; t_0, \varphi_0)) = \psi_k(r_{k-1}(t_k; t_{k-1}, u_{k-1}^+)) = u_k^+.$$

We apply again the comparison theorem for the continuous case in the interval $(t_k, t_{k+1}]$ and obtain

$$V(t, x(t; t_0, \varphi_0)) \leq r_k(t; t_k, u_k^+) = u^+(t; t_0, u_0),$$

i.e. the inequality (1.33) is valid for $t \in (t_k, t_{k+1}]$.

The proof is completed by induction.   □

The next theorem follows immediately from Theorem 1.23.

Theorem 1.24. *Assume that:*

(1) *Conditions H1.1, H1.3, H1.9, H1.11, H1.12 and H1.13 hold.*

(2) *The function $g : [t_0, \infty) \times \mathbb{R}_+ \to \mathbb{R}$ is continuous in each of the sets $(t_{k-1}, t_k] \times \mathbb{R}_+$, $k = 1, 2, \ldots$.*

(3) *$B_k \in C[\mathbb{R}_+, \mathbb{R}_+]$ and $\psi_k(u) = u + B_k(u) \geq 0, k = 1, 2, \ldots$ are non-decreasing with respect to u.*

(4) *The maximal solution $u^+(t; t_0, u_0)$ of the scalar problem*

$$\begin{cases} \dot{u}(t) = g(t, u(t)), \ t \neq t_k \\ u(t_0) = u_0 \geq 0 \\ \Delta u(t_k) = B_k(u(t_k)), \ t_k > t_0, \ k = 1, 2, \ldots \end{cases}$$

is defined in the interval $[t_0, \infty)$.

(5) *The function $V \in V_0$ is such that $V(t_0 + 0, \varphi_0(0)) \leq u_0$,*

$$V(t + 0, x + I_k(x)) \leq \psi_k(V(t, x)), \quad x \in \Omega, \ t = t_k, \ k = 1, 2, \ldots,$$

and the inequality

$$D^+_{(1.31)} V(t, x(t)) \leq g(t, V(t, x(t))), \ t \neq t_k, \ k = 1, 2, \ldots$$

is valid for $t \in [t_0, \infty)$, $x \in \Omega_P$.

Then

$$V(t, x(t; t_0, \varphi_0)) \leq u^+(t; t_0, u_0), \ t \in [t_0, \infty).$$

In the case when $g(t, u) = 0$ for $(t, u) \in [t_0, \infty) \times \mathbb{R}_+$ and $\psi_k(u) = u$ for $u \in \mathbb{R}_+$, $k = 1, 2, \ldots$, we deduce the following corollary from Theorem 1.24.

Corollary 1.25. *Assume that:*

(1) *Conditions H1.1, H1.3, H1.9, H1.11, H1.12 and H1.13 hold.*

(2) *The function $V \in V_0$ is such that*

$$V(t + 0, x + I_k(x)) \leq V(t, x), \quad x \in \Omega, \ t = t_k, \ k = 1, 2, \ldots,$$

and the inequality

$$D^+_{(1.31)} V(t, x(t)) \leq 0, \ t \neq t_k, \ k = 1, 2, \ldots$$

is valid for $t \in [t_0, \infty)$, $x \in \Omega_P$.

Then

$$V(t, x(t; t_0, \varphi_0)) \leq V(t_0 + 0, \varphi_0(0)), \ t \in [t_0, \infty).$$

Remark 1.26. All theorems in Section 1.4 are true if we substitute the set Ω_P with the set Ω_1.

Notes and comments

Conditions for absence of the phenomenon beating were first obtained by Samoilenko and Perestyuk in [179]. Theorem 1.11 is due to Bainov and Dishliev [21]. A number of sufficient conditions for the absence of the phenomenon beating for impulsive functional differential equations were obtained by Bainov and Dishliev [21, 22].

The results on the existence, uniqueness and continuability of the solutions were taken from Bainov and Dishliev [21]. Analogous results were obtained in the works [25, 47, 143, 203, 215]. In the particular case, Lakshmikantham and Rao [134] considered the impulsive integro-differential systems of the type

$$\dot{x}(t) = f(t, x(t)) + \int_{t_0}^{t} g(t, s, x(s))ds, \ t \neq t_k$$

$$\Delta x(t_k) = I_k(x(t_k)), \quad t_k > t_0, \ k = 1, 2, \ldots .$$

The systems of the type (1.25), (1.26) also include the following systems of impulsive integro-differential equations with infinite delays:

$$\dot{x}(t) = -a(t)x(t) + f(t, x(t)) + \int_{-\infty}^{t} c(t-s)x(s)ds, \quad t \neq t_k$$

$$\Delta x(t_k) = I_k(x(t_k)), \quad t_k > t_0, \ k = 1, 2, \ldots,$$

studied in [148, 149].

The method of piecewise continuous Lyapunov functions for impulsive systems was introduced by Bainov and Simeonov [30].

The corresponding comparison theorem of the Theorem 1.23 for the continuous case was proved in [132]. The applied technique is used by many authors [7, 8, 19, 31, 130–134].

Chapter 2

Lyapunov stability and boundedness

The present chapter will deal with basic stability theory for impulsive functional differential equations by Lyapunov's direct method. Applications to real world problems will also be discussed.

Section 2.1 will offer *Lyapunov stability* results. The obtained theorems are parallel to the classical theorems of Lyapunov for ordinary differential equations and show the role of delay and impulses.

Section 2.2 will deal with *boundedness properties* for impulsive functional differential equations. By means of piecewise continuous Lyapunov functions coupled with the Razumikhin technique, sufficient conditions for equi-boundedness, uniform boundedness and uniform-ultimate boundedness of the solutions of such equations will be given.

Finally, in Section 2.3, we shall continue to use Lyapunov's direct method and we shall investigate *global stability* of the solutions.

2.1 Lyapunov stability of the solutions

Let $t_0 \in \mathbb{R}$, $r = \text{const} > 0$, Ω be a domain in $\mathbb{R}^n$ containing the origin and $\|x\| = (\sum_{k=1}^{n} x_k^2)^{\frac{1}{2}}$ be the norm of the element $x \in \mathbb{R}^n$. Consider the following system of impulsive functional differential equations with variable impulsive perturbations:

$$
\begin{cases}
\dot{x}(t) = f(t, x_t), \ t \neq \tau_k(x(t)) \\
\Delta x(t) = I_k(x(t)), \ t = \tau_k(x(t)), \ k = 1, 2, \ldots,
\end{cases}
\tag{2.1}
$$

where $f : [t_0, \infty) \times PC[[-r, 0], \Omega] \to \mathbb{R}^n$; $\tau_k : \Omega \to (t_0, \infty)$, $I_k : \Omega \to \mathbb{R}^n$, $k = 1, 2, \ldots$; $\Delta x(t) = x(t + 0) - x(t - 0)$; and for $t \geq t_0$, $x_t \in PC[[-r, 0], \Omega]$ is defined by $x_t(s) = x(t + s)$, $-r \leq s \leq 0$.

Let $\varphi_0 \in PC[[-r, 0], \Omega]$. Denote by $x(t) = x(t; t_0, \varphi_0)$ the solution of system (2.1), satisfying the initial conditions

$$
\begin{cases}
x(t) = \varphi_0(t - t_0), \ t_0 - r \leq t \leq t_0 \\
x(t_0 + 0) = \varphi_0(0),
\end{cases}
\tag{2.2}
$$

and by $J^+(t_0, \varphi_0)$ the maximal interval of type $[t_0, \beta)$ in which the solution $x(t; t_0, \varphi_0)$ is defined.

Let $\tau_0(x) \equiv t_0$ for $x \in \Omega$. Introduce the following condition:

H2.1. $\tau_k \in C[\Omega, (t_0, \infty)]$, $k = 1, 2, \ldots$.

Assuming that conditions H2.1, H1.5, and H1.6 are fulfilled, we consider the hypersurfaces

$$\sigma_k = \{(t, x) : t = \tau_k(x), \; x \in \Omega\}, \quad k = 1, 2, \ldots.$$

Introduce the following notations:

$$\|\varphi\|_r = \sup_{t \in [t_0-r, t_0]} \|\varphi(t - t_0)\| \text{ is the norm of the function } \varphi \in PC[[-r, 0], \Omega];$$

$$K = \{a \in C[\mathbb{R}_+, \mathbb{R}_+] : a(r) \text{ is strictly increasing and } a(0) = 0\}.$$

In the case $r = \infty$ we have $\|\varphi\|_r = \|\varphi\|_\infty = \sup_{t \in (-\infty, t_0]} \|\varphi(t - t_0)\|$.

Introduce the following conditions:

H2.2. $f(t, 0) = 0$, $t \geq t_0$.

H2.3. $I_k(0) = 0$, $k = 1, 2, \ldots$.

H2.4. The integral curves of the system (2.1) meet successively each one of the hypersurfaces $\sigma_1, \sigma_2, \ldots$ exactly once.

Let $t_1, t_2, \ldots$ ($t_0 < t_1 < t_2 < \cdots$) be the moments in which the integral curve $(t, x(t; t_0, \varphi_0))$ of problem (2.1), (2.2) meets the hypersurfaces σ_k, $k = 1, 2, \ldots$.

It follows from Theorem 1.16 that if the conditions H1.1, H1.2, H1.3, H1.5, H1.6, H1.9, H1.11, H2.1 and H2.4 are met, then $t_k \to \infty$ as $k \to \infty$ and $J^+(t_0, \varphi_0) = [t_0, \infty)$.

We shall use the following definitions of Lyapunov like stability of the zero solution of (2.1).

Definition 2.1. The zero solution $x(t) \equiv 0$ of system (2.1) is said to be:

(a) *stable*, if

$$(\forall t_0 \in \mathbb{R})(\forall \varepsilon > 0)(\exists \delta = \delta(t_0, \varepsilon) > 0)$$

$$(\forall \varphi_0 \in PC[[-r, 0], \Omega] : \|\varphi_0\|_r < \delta)(\forall t \geq t_0) : \|x(t; t_0, \varphi_0)\| < \varepsilon;$$

(b) *uniformly stable*, if the number δ in (a) is independent of $t_0 \in \mathbb{R}$;

(c) *attractive*, if

$$(\forall t_0 \in \mathbb{R})(\exists \lambda = \lambda(t_0) > 0)(\forall \varphi_0 \in PC[[-r, 0], \Omega] : \|\varphi_0\|_r < \lambda) :$$

$$\lim_{t \to \infty} x(t; t_0, \varphi_0) = 0;$$

(d) *equi-attractive*, if

$$(\forall t_0 \in \mathbb{R})(\exists \lambda = \lambda(t_0) > 0)(\forall \varepsilon > 0)(\exists T = T(t_0, \varepsilon) > 0)$$

$$(\forall \varphi_0 \in PC[[-r, 0], \Omega] : \|\varphi_0\|_r < \lambda)(\forall t \geq t_0 + T) : \|x(t; t_0, \varphi_0)\| < \varepsilon;$$

(e) *uniformly attractive*, if the numbers λ and T in (d) are independent of $t_0 \in \mathbb{R}$;

(f) *asymptotically stable*, if it is stable and attractive;

(g) *uniformly asymptotically stable*, if it is uniformly stable and uniformly attractive;

(h) *unstable*, if

$$(\exists t_0 \in \mathbb{R})(\exists \varepsilon > 0)(\forall \delta > 0)(\exists \varphi_0 \in PC[[-r, 0], \Omega] : \|\varphi_0\|_r < \delta)$$

$$(\exists t \geq t_0) : \|x(t; t_0, \varphi_0)\| \geq \varepsilon.$$

In the proofs of our main theorems in this section we shall use piecewise continuous Lyapunov functions $V : [t_0, \infty) \times \Omega \to \mathbb{R}_+$, $V \in V_0$ for which the following condition is true:

H2.5. $V(t, 0) = 0$, $t \geq t_0$.

Theorem 2.2. *Assume that:*

(1) *Conditions* H1.1, H1.2, H1.3, H1.5, H1.6, H1.9, H1.11, H2.1–H2.4 *hold.*

(2) *There exists a function* $V \in V_0$ *such that* H2.5 *holds,*

$$a(\|x\|) \leq V(t, x), \ a \subset K, \ (t, x) \in [t_0, \infty) \times \Omega, \tag{2.3}$$

$$V(t + 0, x + I_k(x)) \leq V(t, x), \ (t, x) \in \sigma_k, \ k = 1, 2, \dots, \tag{2.4}$$

and the inequality

$$D^+_{(2.1)}V(t, x(t)) \leq 0, \ t \neq \tau_k(x(t)), \ k = 1, 2, \dots$$

is valid for $t \in [t_0, \infty)$, $x \in \Omega_1$.

Then the zero solution of system (2.1) *is stable.*

Proof. Let $\varepsilon > 0$. It follows from the properties of the function V that there exists a constant $\delta = \delta(t_0, \varepsilon) > 0$ such that if $x \in \Omega : \|x\| < \delta$, then $\sup_{\|x\| < \delta} V(t_0 + 0, x) < a(\varepsilon)$.

Let $\varphi_0 \in PC[[-r, 0], \Omega] : \|\varphi_0\|_r < \delta$. Then $\|\varphi_0(0)\| \leq \|\varphi_0\|_r < \delta$ and therefore

$$V(t_0 + 0, \varphi_0(0)) < a(\varepsilon). \tag{2.5}$$

Let $x(t) = x(t; t_0, \varphi_0)$ be the solution of problem (2.1), (2.2). Since all the conditions of Corollary 1.25 are met, then

$$V(t, x(t; t_0, \varphi_0)) \leq V(t_0 + 0, \varphi_0(0)), \ t \in [t_0, \infty). \tag{2.6}$$

From (2.3), (2.5) and (2.6) there follow the inequalities

$$a(\|x(t;t_0,\varphi_0)\|) \le V(t, x(t;t_0,\varphi_0)) \le V(t_0+0, \varphi_0(0)) < a(\varepsilon),$$

whence we obtain that $\|x(t;t_0,\varphi_0)\| < \varepsilon$ for $t \ge t_0$. This implies that the zero solution of system (2.1) is stable.   □

Theorem 2.3. *Let the conditions of Theorem 2.2 hold, and let a function $b \in K$ exist such that*

$$V(t, x) \le b(\|x\|), \ (t, x) \in (t_0, \infty) \times \Omega. \tag{2.7}$$

Then the zero solution of system (2.1) is uniformly stable.

Proof. Let $\varepsilon > 0$ be given. Choose $\delta = \delta(\varepsilon) > 0$ so that $b(\delta) < a(\varepsilon)$.
   Let $\varphi_0 \in PC[[-r, 0], \Omega] : \|\varphi_0\|_r < \delta$ and $x(t) = x(t;t_0,\varphi_0)$ be the solution of problem (2.1), (2.2).
   As in Theorem 2.2, we prove that

$$a(\|x(t;t_0,\varphi_0)\|) \le V(t, x(t;t_0,\varphi_0)) \le V(t_0+0, \varphi_0(0)), \ t \ge t_0.$$

From the above inequalities and (2.7), we get to the inequalities

$$a(\|x(t;t_0,\varphi_0)\|) \le V(t_0+0, \varphi_0(0)) \le b(\|\varphi_0(0)\|) \le b(\|\varphi_0\|_r) < b(\delta) < a(\varepsilon),$$

from which it follows that $\|x(t;t_0,\varphi_0)\| < \varepsilon$ for $t \ge t_0$. This proves the uniform stability of the zero solution of system (2.1).   □

Theorem 2.4. *Assume that:*

(1) *Condition (1) of Theorem 2.2 holds.*

(2) *There exists a function $V \in V_0$ such that H2.5 and (2.4) hold,*

$$a(\|x\|) \le V(t, x) \le b(\|x\|), \ a, b \in K, \ (t, x) \in [t_0, \infty) \times \Omega, \tag{2.8}$$

   and the inequality

$$D^+_{(2.1)}V(t, x(t)) \le -c(\|x(t)\|), \ t \ne \tau_k(x(t)), \ k = 1, 2, \ldots \tag{2.9}$$

   is valid for $c \in K$, $t \in [t_0, \infty)$ and $x \in \Omega_1$.

Then the zero solution of system (2.1) is uniformly asymptotically stable.

Proof. 1. Let $\alpha = \text{const} > 0 : \{x \in \mathbb{R}^n : \|x\| \le \alpha\} \subset \Omega$.
   For any $t \in [t_0, \infty)$ denote

$$V^{-1}_{t,\alpha} = \{x \in \Omega : V(t+0, x) \le a(\alpha)\}.$$

From (2.8), we deduce

$$V_{t,\alpha}^{-1} \subset \{x \in \mathbb{R}^n : \|x\| \le \alpha\} \subset \Omega.$$

From condition (2) of Theorem 2.4, it follows that for any $t_0 \in \mathbb{R}$ and any function $\varphi_0 \in PC[[-r, 0], \Omega] : \varphi_0(0) \in V_{t_0,\alpha}^{-1}$ we have $x(t; t_0, \varphi_0) \in V_{t,\alpha}^{-1}$, $t \ge t_0$.

Let $\varepsilon > 0$ be chosen. Choose $\eta = \eta(\varepsilon)$ so that $b(\eta) < a(\varepsilon)$, and let $T > \frac{b(\alpha)}{c(\eta)}$.

If we assume that for each $t \in [t_0, t_0 + T]$ the inequality $\|x(t; t_0, \varphi_0)\| \ge \eta$ is valid, then from (2.4) and (2.9) we get

$$V(t, x(t; t_0, \varphi_0)) \le V(t_0 + 0, \varphi_0(0))$$

$$- \int_{t_0}^t c(\|x(s; t_0, \varphi_0)\|)\, ds \le b(\alpha) - c(\eta)T < 0,$$

which contradicts (2.8). The contradiction obtained shows that there exists $t^* \in [t_0, t_0 + T]$ such that $\|x(t^*; t_0, \varphi_0)\| < \eta$.

Then from (2.4), (2.8) and (2.9) it follows that for $t \ge t^*$ (hence for any $t \ge t_0 + T$) the following inequalities hold:

$$a(\|x(t; t_0, \varphi_0)\|) \le V(t, x(t; t_0, \varphi_0)) \le V(t^*, x(t^*; t_0, \varphi_0))$$

$$\le b(\|x(t^*; t_0, \varphi_0)\|) < b(\eta) < a(\varepsilon).$$

Therefore, $\|x(t; t_0, \varphi_0)\| < \varepsilon$ for $t \ge t_0 + T$.

2. Let $\lambda = \text{const} > 0$ be such that $b(\lambda) < a(\alpha)$. Then, if $\varphi_0 \in PC[[-r, 0], \Omega]$: $\|\varphi_0\|_r < \lambda$, (2.8) implies

$$V(t_0 + 0, \varphi_0(0)) \le b(\|\varphi_0(0)\|) \le b(\|\varphi_0\|_r) < b(\lambda) < a(\alpha),$$

which shows that $\varphi_0 \in PC[[-r, 0], \Omega] : \varphi_0(0) \in V_{t_0,\alpha}^{-1}$. From what we proved in item 1, it follows that the zero solution of system (2.1) is uniformly attractive and since Theorem 2.3 implies that it is uniformly stable, then the solution $x \equiv 0$ is uniformly asymptotically stable.   □

Corollary 2.5. *If in Theorem 2.4 condition (2.9) is replaced by the condition*

$$D_{(2.1)}^+ V(t, x(t)) \le -c V(t, x(t)), \quad t \ne \tau_k(x(t)), \quad k = 1, 2, \dots, \tag{2.10}$$

where $t \in [t_0, \infty)$, $x \in \Omega_1$, $c = \text{const} > 0$, then the zero solution of system (2.1) is uniformly asymptotically stable.

Proof. The proof of Corollary 2.5 is analogous to the proof of Theorem 2.4. It uses the fact that

$$V(t, x(t; t_0, \varphi_0)) \le V(t_0 + 0, \varphi_0(0)) \exp[-c(t - t_0)]$$

for $t \ge t_0$, which is obtained from (2.10) and (2.4).

In fact, let $\alpha = \text{const} > 0 : \{x \in \mathbb{R}^n : \|x\| \leq \alpha\} \subset \Omega$. Choose $\lambda > 0$ so that $b(\lambda) < a(\alpha)$. Let $\varepsilon > 0$ and $T \geq \frac{1}{c} \ln \frac{a(\alpha)}{a(\varepsilon)}$. Then for $\varphi_0 \in PC[[-r, 0], \Omega] : \|\varphi_0\|_r < \lambda$ and $t \geq t_0 + T$ the following inequalities hold:

$$V(t, x(t; t_0, \varphi_0)) \leq V(t_0 + 0, \varphi_0(0)) \exp[-c(t - t_0)] < a(\varepsilon),$$

whence, in view of (2.8), we deduce that the solution $x \equiv 0$ of system (2.1) is uniformly attractive. □

Consider the system of impulsive functional differential equations with fixed moments of impulsive perturbations

$$\begin{cases} \dot{x}(t) = f(t, x_t), \ t \neq t_k \\ \Delta x(t) = I_k(x(t)), \ t = t_k, \ k = 1, 2, \ldots, \end{cases} \tag{2.11}$$

where $f : [t_0, \infty) \times PC[[-r, 0], \Omega] \rightarrow \mathbb{R}^n$; $I_k : \Omega \rightarrow \mathbb{R}^n$, $k = 1, 2, \ldots$; $t_0 < t_1 < t_2 < \cdots < t_k < t_{k+1} < \cdots$ and $\lim_{k \to \infty} t_k = \infty$.

Let $\varphi_1 \in PC[[-r, 0], \Omega]$. Denote by $x_1(t) = x_1(t; t_0, \varphi_1)$ the solution of system (2.11), satisfying the initial conditions

$$\begin{cases} x_1(t; t_0, \varphi_1) = \varphi_1(t - t_0), \ t_0 - r \leq t \leq t_0 \\ x_1(t_0 + 0; t_0, \varphi_1) = \varphi_1(0). \end{cases}$$

Definition 2.6. The solution $x_1(t)$ of system (2.11) is said to be:

(a) *stable*, if

$$(\forall t_0 \in \mathbb{R})(\forall \varepsilon > 0)(\exists \delta = \delta(t_0, \varepsilon) > 0)$$

$$(\forall \varphi_0 \in PC[[-r, 0], \Omega] : \|\varphi_0 - \varphi_1\|_r < \delta)$$

$$(\forall t \geq t_0) : \|x(t; t_0, \varphi_0) - x_1(t; t_0, \varphi_1)\| < \varepsilon;$$

(b) *uniformly stable*, if the number δ in (a) is independent of $t_0 \in \mathbb{R}$;

(c) *attractive*, if

$$(\forall t_0 \in \mathbb{R})(\exists \lambda = \lambda(t_0) > 0)$$

$$(\forall \varphi_0 \in PC[[-r, 0], \Omega] : \|\varphi_0 - \varphi_1\|_r < \lambda) :$$

$$\lim_{t \to \infty} x(t; t_0, \varphi_0) = x_1(t; t_0, \varphi_1);$$

(d) *equi-attractive*, if

$$(\forall t_0 \in \mathbb{R})(\exists \lambda = \lambda(t_0) > 0)(\forall \varepsilon > 0)(\exists T = T(t_0, \varepsilon) > 0)$$

$$(\forall \varphi_0 \in PC[[-r, 0], \Omega] : \|\varphi_0 - \varphi_1\|_r < \lambda)$$

$$(\forall t \geq t_0 + T) : \|x(t; t_0, \varphi_0) - x_1(t; t_0, \varphi_1)\| < \varepsilon;$$

(e) *uniformly attractive*, if the numbers λ and T in (d) are independent of $t_0 \in \mathbb{R}$;

(f) *asymptotically stable*, if it is stable and attractive;

(g) *uniformly asymptotically stable*, if it is uniformly stable and uniformly attractive;

(h) *unstable*, if (a) does not hold.

The following theorems follow directly from Theorems 2.2, 2.3 and 2.4.

Theorem 2.7. *Assume that:*

(1) *Conditions* H1.1, H1.2, H1.3, H1.9, H1.11, H1.12 *and* H1.13 *hold.*

(2) *There exists a function* $V \in V_0$ *such that*

$$V(t, x_1(t)) = 0, \quad t \in [t_0, \infty), \tag{2.12}$$

$$a(\|x - x_1(t)\|) \le V(t, x), \; a \in K, \; (t, x) \in [t_0, \infty) \times \Omega,$$

$$V(t + 0, x + I_k(x)) \le V(t, x), \quad x \in \Omega, \; t = t_k, \; k = 1, 2, \ldots, \tag{2.13}$$

and the inequality

$$D^+_{(2.11)} V(t, x(t)) \le 0, \; t \ne t_k, \; k = 1, 2, \ldots$$

is valid for $t \in [t_0, \infty)$, $x \in \Omega_1$.

Then the solution $x_1(t)$ *of system* (2.11) *is stable.*

Theorem 2.8. *Let the conditions of Theorem* 2.7 *hold, and let a function* $b \in K$ *exist such that*

$$V(t, x) \le b(\|x - x_1(t)\|), \; (t, x) \in [t_0, \infty) \times \Omega.$$

Then the solution $x_1(t)$ *of system* (2.11) *is uniformly stable.*

Theorem 2.9. *Assume that:*

(1) *Condition* (1) *of Theorem* 2.7 *holds.*

(2) *There exists a function* $V \in V_0$ *such that* (2.12), (2.13) *hold,*

$$a(\|x - x_1(t)\|) \le V(t, x) \le b(\|x - x_1(t)\|), \; a, b \in K, \; (t, x) \in [t_0, \infty) \times \Omega,$$

and the inequality

$$D^+_{(2.11)} V(t, x(t)) \le -c(\|x(t) - x_1(t)\|), \; t \ne t_k, \; k = 1, 2, \ldots$$

is valid for $c \in K$, $t \in [t_0, \infty)$ *and* $x \in \Omega_1$.

Then the solution $x_1(t)$ *of system* (2.11) *is uniformly asymptotically stable.*

We shall apply the obtained results in investigating the stability of the mathematical models from the population dynamics.

Example 2.10. Gopalsamy [91] studied the asymptotic behavior of the solutions of the linear system
$$\dot{x}(t) = Ax(t) + Bx(t - r), \quad t \geq 0,$$
where $x \in \mathbb{R}_+^n$, $r > 0$, A and B are diagonal constant $(n \times n)$ matrices.

If at certain moments of time the above system is subject to impulsive perturbations, then the adequate mathematical model is the following impulsive system:

$$\begin{cases} \dot{x}(t) = Ax(t) + Bx(t - r), \ t \neq t_k, \ t \geq 0 \\ \Delta x(t_k) = C_k x(t_k), \ k = 1, 2, \ldots, \end{cases} \tag{2.14}$$

where $C_k = \text{diag}(c_{1k}, c_{2k}, \ldots, c_{nk})$, $-1 < c_{ik} \leq 0$, $i = 1, 2, \ldots, n$, $k = 1, 2, \ldots$; $0 < t_1 < t_2 < \cdots < t_k < t_{k+1} < \cdots$ and $\lim_{k \to \infty} t_k = \infty$.

Let $V(t, x) = \|x\|^2 = \langle x, x \rangle$, where $\langle x, y \rangle = x_1 y_1 + x_2 y_2 + \cdots + x_n y_n$ is the dot product of $x, y \in \mathbb{R}^n$. Then the set

$$\Omega_1 = \{x \in PC[\mathbb{R}_+, \mathbb{R}_+^n] : \|x(s)\|^2 \leq \|x(t)\|^2, \ t - r \leq s \leq t\}.$$

Let $\varphi_0 \in PC[[-r, 0], \mathbb{R}_+^n]$. Denote by $x(t) = x(t; 0, \varphi_0)$ the solution of system (2.14) satisfying the initial conditions

$$x(s) = \varphi_0(s) \geq 0, \ s \in [-r, 0); \quad x(0) > 0.$$

For $t \neq t_k$, $k = 1, 2, \ldots$, we have

$$D_{(2.14)}^+ V(t, x(t)) = 2\langle x(t), \dot{x}(t) \rangle = 2\langle x(t), Ax(t) + Bx(t - r) \rangle.$$

Also, for $k = 1, 2, \ldots$

$$V(t_k + 0, x(t_k) + C_k x(t_k)) = \sum_{i=1}^{n} (1 + c_{ik})^2 x_i^2(t_k) \leq V(t_k, x(t_k)).$$

If $A = \text{diag}(a_1, a_2, \ldots, a_n)$, $B = \text{diag}(b_1, b_2, \ldots, b_n)$, $a_i \leq 0$, $b_i \geq 0$, $b = \max_i b_i$ and $a_i \leq -(b + c)$ for $i = 1, 2, \ldots, n$, then for $x \in \Omega_1$,

$$D_{(2.14)}^+ V(t, x(t)) \leq -2c\langle x(t), x(t) \rangle = -2c V(t, x(t)), \ t \neq t_k$$

and according to Theorem 2.9 the trivial solution of (2.14) is uniformly asymptotically stable.

Example 2.11. Consider the equation

$$\begin{cases} \dot{N}(t) = mN(t)[K - aN(t) + bN(t - \sigma(t))], \ t \neq t_k, \ t \geq 0 \\ \Delta N(t_k) = \alpha_k N(t_k) - \alpha_k \dfrac{K}{a - b}, \ t_k > 0, \ k = 1, 2, \ldots, \end{cases} \tag{2.15}$$

where $m > 0$; $K > 0$; $0 \leq \sigma(t) \leq r$; $a \geq 0$; $b > 0$; $a \neq b$; $-1 < \alpha_k \leq 0$, $k = 1, 2, \ldots$; $0 < t_1 < t_2 < \cdots$ and $\lim_{k \to \infty} t_k = \infty$.

Let $\varphi \in PC[[-r, 0], \mathbb{R}_+]$ and $N(t) = N(t; 0, \varphi)$ be the solution of equation (2.15) satisfying the initial conditions

$$N(s) = \varphi(s) \geq 0, \ s \in [-r, 0); \quad N(0) > 0.$$

One can show that if $a > b$ the point $N^* = \frac{K}{a-b}$ is a positive equilibrium of (2.15). Define the function $V(t, N) = \frac{1}{2}(N - N^*)^2$. Then the set

$$\Omega_1 = \{N \in PC[\mathbb{R}_+, (0, \infty)] : \ (N(s) - N^*)^2 \leq (N(t) - N^*)^2, \ t - r \leq s \leq t\}.$$

For $t \geq 0$, $t \neq t_k$, we have

$$D^+_{(2.15)} V(t, N(t)) = m N(t)(N(t) - N^*)[K - a N(t) + b N(t - \sigma(t))].$$

Since N^* is an equilibrium of (2.15), then

$$D^+_{(2.15)} V(t, N(t)) = m N(t)(N(t) - N^*)[-a(N(t) - N^*) + b(N(t - \sigma(t)) - N^*)].$$

For $t \neq t_k$, $k = 1, 2, \ldots$ and $N \in \Omega_1$, we obtain the estimate

$$D^+_{(2.15)} V(t, N(t)) \leq m N(t)[-a + b](N(t) - N^*)^2 \leq 0.$$

Also, for $t > 0$, $t = t_k$, we have

$$V(t_k + 0, N(t_k) + \wedge N(t_k)) - \frac{1}{2}[(1 + \alpha_k)N(t_k) - \alpha_k N^* - N^*]^2$$

$$= \frac{1}{2}(1 + \alpha_k)^2 [N(t_k) - N^*]^2 \leq V(t_k, N(t_k)).$$

Then all conditions of Theorem 2.8 are satisfied. Hence the equilibrium N^* of (2.15) is uniformly stable.

2.2   Theorems on boundedness

In this section, we shall apply Lyapunov's second method for investigating the boundedness of the solutions of system (2.1) for $\Omega = \mathbb{R}^n$, i.e. we shall consider the following system:

$$\begin{cases} \dot{x}(t) = f(t, x_t), \ t \neq \tau_k(x(t)) \\ \Delta x(t) = I_k(x(t)), \ t = \tau_k(x(t)), \ k = 1, 2, \ldots, \end{cases} \tag{2.16}$$

where $f : [t_0, \infty) \times PC[[-r, 0], \mathbb{R}^n] \to \mathbb{R}^n$; $\tau_k : \mathbb{R}^n \to (t_0, \infty)$, $I_k : \mathbb{R}^n \to \mathbb{R}^n$, $k = 1, 2, \ldots$; $\Delta x(t) = x(t + 0) - x(t - 0)$; and for $t \geq t_0$, $x_t \in PC[[-r, 0], \mathbb{R}^n]$ is defined by $x_t(s) = x(t + s)$, $-r \leq s \leq 0$.

Let $\varphi_0 \in PC[[-r, 0], \mathbb{R}^n]$. Denote by $x(t) = x(t; t_0, \varphi_0)$ the solution of system (2.16), satisfying the initial conditions

$$
\begin{cases}
x(t; t_0, \varphi_0) = \varphi_0(t - t_0), & t_0 - r \le t \le t_0 \\
x(t_0 + 0; t_0, \varphi_0) = \varphi_0(0),
\end{cases}
\tag{2.17}
$$

and by $J^+(t_0, \varphi_0)$ the maximal interval of type $[t_0, \beta)$, in which the solution $x(t; t_0, \varphi_0)$ is defined.

Let $\tau_0(x) \equiv t_0$ for $x \in \mathbb{R}^n$. Introduce the following conditions:

H2.6.   The function f is continuous on $[t_0, \infty) \times PC[[-r, 0], \mathbb{R}^n]$.

H2.7.   The function f is locally Lipschitz continuous with respect to its second argument on $[t_0, \infty) \times PC[[-r, 0], \mathbb{R}^n]$.

H2.8.   There exists a constant $P > 0$ such that

$$
\|f(t, x_t)\| \le P < \infty \text{ for } (t, x_t) \in [t_0, \infty) \times PC[[-r, 0], \mathbb{R}^n].
$$

H2.9. $\tau_k \in C[\mathbb{R}^n, (t_0, \infty)]$, $k = 1, 2, \ldots$.

H2.10. $t_0 < \tau_1(x) < \tau_2(x) < \cdots$, $x \in \mathbb{R}^n$.

H2.11. $\tau_k(x) \to \infty$ as $k \to \infty$, uniformly on $x \in \mathbb{R}^n$.

H2.12. The functions I_k, $k = 1, 2, \ldots$ are Lipschitz continuous with respect to $x \in \mathbb{R}^n$.

Assuming that conditions H2.9, H2.10 and H2.11 are fulfilled, we define the following notations:

$$
G_k = \left\{ (t, x) : \tau_{k-1}(x) < t < \tau_k(x), \ x \in \mathbb{R}^n \right\}, \quad k = 1, 2, \ldots
$$

$$
\sigma_k = \left\{ (t, x) : t = \tau_k(x), \ x \in \mathbb{R}^n \right\}, \quad k = 1, 2, \ldots.
$$

Introduce the following condition:

H2.13. The integral curves of the system (2.16) meet successively each one of the hypersurfaces $\sigma_1, \sigma_2, \ldots$ exactly once.

Let $t_1, t_2, \ldots$ $(t_0 < t_1 < t_2 < \cdots)$ be the moments in which the integral curve $(t, x(t; t_0, \varphi_0))$ of the problems (2.16), (2.17) meets the hypersurfaces $\{\sigma_k\}_{k=1}^{\infty}$.

We shall use also the following notations:

$$
S_\alpha = \{x \in \mathbb{R}^n : \|x\| < \alpha\}, \ \alpha > 0, \quad S_\alpha^c = \{x \in \mathbb{R}^n : \|x\| \ge \alpha\}.
$$

Definition 2.12. We say that the solutions of system (2.16) are:

(a) *equi-bounded*, if

$$(\forall t_0 \in \mathbb{R})(\forall \alpha > 0)(\exists \beta = \beta(t_0, \alpha) > 0)$$

$$(\forall \varphi_0 \in PC[[-r, 0], \mathbb{R}^n] : \|\varphi_0\|_r < \alpha)(\forall t \geq t_0) : \|x(t; t_0, \varphi_0)\| < \beta;$$

(b) *uniformly bounded*, if the number β in (a) is independent of $t_0 \in \mathbb{R}$;

(c) *quasi-uniformly ultimately bounded*, if

$$(\exists B > 0)(\forall \alpha > 0)(\exists T = T(\alpha) > 0)(\forall t_0 \in \mathbb{R})$$

$$(\forall \varphi_0 \in PC[[-r, 0], \mathbb{R}^n] : \|\varphi_0\|_r < \alpha)(\forall t \geq t_0 + T) : \|x(t; t_0, \varphi_0)\| < B;$$

(d) *uniformly ultimately bounded*, if (b) and (c) hold together.

In the further considerations, we shall use the class V_0 of piecewise continuous auxiliary functions $V \in V_0$ for $\Omega = \mathbb{R}^n$.

Theorem 2.13. *Assume that:*

(1) *Conditions H2.6–H2.13 hold.*

(2) *There exists a function $V \in V_0$ such that H2.5 holds,*

$$a(\|x\|) \leq V(t, x), \ (t, x) \in [t_0, \infty) \times \mathbb{R}^n, \tag{2.18}$$

where $a \subset K$ and $a(u) \to \infty$ as $u \to \infty$,

$$V(t + 0, x + I_k(x)) \leq V(t, x), \ (t, x) \in \sigma_k, \ k = 1, 2, \ldots, \tag{2.19}$$

and the inequality

$$D^+_{(2.16)} V(t, x(t)) \leq 0, \ t \neq \tau_k(x(t)), \ k = 1, 2, \ldots$$

is valid for $t \in [t_0, \infty)$, $x \in \Omega_P$.

Then the solutions of system (2.16) are equi-bounded.

Proof. Let $\alpha > 0$, $t_0 \in \mathbb{R}$, $\varphi_0 \in PC[[-r, 0], \mathbb{R}^n]$. Consider the solution $x(t) = x(t; t_0, \varphi_0)$ of (2.16) for which $\|\varphi_0\|_r < \alpha$.

By Theorem 1.16, we have $J^+(t_0, \varphi_0) = [t_0, \infty)$. From the properties of the function V, it follows that there exists a constant $\Gamma = \Gamma(t_0, \alpha) > 0$ such that if $x \in \mathbb{R}^n : \|x\| < \alpha$, then $\sup_{\|x\| < \alpha} V(t_0 + 0, x) < \Gamma(t_0, \alpha)$.

Since for the function $a \in K$ we have $a(u) \to \infty$ as $u \to \infty$, then we can choose $\beta = \beta(t_0, \alpha) > 0$ so that $\beta > \alpha$ and $a(\beta) > \Gamma(t_0, \alpha)$.

We shall prove that $\|x(t; t_0, \varphi_0)\| < \beta$ for $t \geq t_0$.

Suppose that this is not true. Then, there exists $t^* > t_0$ such that $t_k < t^* \leq t_{k+1}$ for some fixed k and

$$\|x(t^*)\| \geq \beta, \quad \text{and} \quad \|x(t; t_0, \varphi_0)\| < \beta, \; t \in [t_0, t^*). \tag{2.20}$$

Since the conditions of Corollary 1.25 are met, then

$$V(t, x(t; t_0, \varphi_0)) \leq V(t_0 + 0, \varphi_0(0)), \; t \in [t_0, \infty).$$

From the above inequality, (2.18) and (2.20) we have

$$a(\beta) \leq a(\|x(t^*; t_0, \varphi_0)\|) \leq V(t^*, x(t^*; t_0, \varphi_0)) \leq V(t_0 + 0, \varphi_0(0)) < \Gamma(t_0, \alpha),$$

which contradicts the choice of β.

Therefore, $\|x(t; t_0, \varphi_0)\| < \beta$ for $t \geq t_0$. This implies that the solutions of (2.16) are equi-bounded.   □

Theorem 2.14. *Assume that:*

(1) *Condition (1) of Theorem 2.13 holds.*

(2) *For $\rho > 0$, there exists $V \in V_0$ such that (2.19) holds,*

$$a(\|x\|) \leq V(t, x) \leq b(\|x\|), \quad (t, x) \in [t_0, \infty) \times S_\rho^c, \tag{2.21}$$

where $a, b \in K$ and $a(u) \to \infty$ as $u \to \infty$, and the inequality

$$D_{(2.16)}^+ V(t, x(t)) \leq 0, \; t \neq \tau_k(x(t)), \; k = 1, 2, \ldots \tag{2.22}$$

is valid for $t \in [t_0, \infty)$, $x \in S_\rho^c \cap \Omega_P$.

Then the solutions of system (2.16) are uniformly bounded.

Proof. Let $\alpha > 0$ and assume, without loss of generality, that $\alpha \geq \rho$. Choose $\beta = \beta(\alpha) > 0$ so that

$$\beta > \max\{\alpha, \; a^{-1}(b(\alpha))\}.$$

Let $t_0 \in \mathbb{R}$ and $\varphi_0 \in PC[[-r, 0], \mathbb{R}^n]$. Consider the solution $x(t) = x(t; t_0, \varphi_0)$ of (2.16) with $\|\varphi_0\|_r < \alpha$. Obviously,

$$\|x(t_0 + 0; t_0, \varphi_0)\| = \|\varphi_0(0)\| \leq \|\varphi_0\|_r < \alpha < \beta.$$

We claim that

$$\|x(t)\| < \beta, \quad t > t_0.$$

If it is not true, then there exists some solution $x(t) = x(t; t_0, \varphi_0)$ of (2.16) with $\|\varphi_0\|_r < \alpha$ and a $t^* > t_0$ such that $\|x(t^*; t_0, \varphi_0)\| \geq \beta$. Thus, there exist $s_1, s_2, t_0 \leq s_1 < s_2 \leq t^*$ such that

$$\|x(s_1 + 0)\| \geq \alpha, \; \|x(s_1)\| \leq \alpha, \; \|x(s_2 + 0)\| \geq \beta$$

and

$$x(t) \in S_\beta \cap \overline{S}_\alpha^c, \ t \in [s_1, s_2). \tag{2.23}$$

First, we shall show that

$$V(s_1 + 0, x(s_1 + 0)) < a(\beta).$$

If $s_1 \neq t_k$, then $\|x(s_1)\| = \alpha$, and we have by (2.21)

$$V(s_1, x(s_1)) \le b(\|x(s_1)\|) = b(\alpha) < a(\beta).$$

If $s_1 = t_k$ for some k, then $\|x(s_1)\| \le \alpha$, and

$$V(s_1, x(s_1)) \le b(\|x(s_1)\|) \le b(\alpha) < a(\beta).$$

Thus, by (2.19) we obtain

$$V(s_1 + 0, x(s_1 + 0)) < a(\beta).$$

Next, we want to show that

$$V(t + 0, x(t + 0)) < a(\beta), \ t \in [s_1, s_2]. \tag{2.24}$$

Suppose that this is not true and let

$$\mu = \inf\{s_2 \ge t > s_1 : V(t + 0, x(t + 0)) \ge a(\beta)\}.$$

We discuss two possibilities:

(A) $\mu \neq t_k, k = 1, 2, \dots$. Since $V(t, x(t))$ is continuous at μ, we have

$$V(\mu + 0, x(\mu + 0)) = V(\mu, x(\mu)) = a(\beta).$$

Thus, for $h > 0$ small enough the inequality

$$V(\mu + h, x(\mu + h)) \ge a(\beta)$$

holds which implies that

$$D^+_{(2.16)} V(\mu, x(\mu)) = \lim_{h \to 0^+} \sup h^{-1} [V(\mu + h, x(\mu + h)) - V(\mu, x(\mu))] \ge 0. \tag{2.25}$$

From the choice of μ it is clear that

$$P(V(\mu, x(\mu))) > V(\mu, x(\mu)) \ge V(s, x(s)), \quad s_1 \le s \le \mu.$$

Thus, we get using (2.22)
$$D^+_{(2.16)} V(\mu, x(\mu)) \le 0,$$

which contradicts (2.25).

(B) $\mu = t_k$ for some $k = 1, 2, \ldots$. We must have

$$V(t_k + 0, x(t_k + 0)) = a(\beta).$$

In fact, if $V(t_k + 0, x(t_k + 0)) > a(\beta)$, then by assumption (2.19) $V(t_k, x(t_k)) > a(\beta)$. Since $V(t, x(t))$ is left continuous at t_k, it follows that there exists $\tilde{\mu} < t_k$ such that $V(\tilde{\mu} + 0, x(\tilde{\mu} + 0)) \geq a(\beta)$ which contradicts the choice of μ.

Hence
$$V(t_k + 0, x(t_k + 0)) - V(t_k, x(t_k)) \geq 0. \tag{2.26}$$

Since
$$P(V(\mu, x(\mu))) > V(\mu, x(\mu)) \geq V(s, x(s)), \quad s_1 \leq s \leq \mu,$$

we obtain using (2.19)

$$V(t_k + 0, x(t_k + 0)) - V(t_k, x(t_k)) \leq 0$$

which contradicts (2.26). Therefore (2.24) holds.

On the other hand, using (2.21) we get

$$V(s_2 + 0, x(s_2 + 0)) \geq a(\|x(s_2 + 0)\|) \geq a(\beta),$$

which contradicts (2.24). Thus

$$\|x(t)\| < \beta, \quad t \geq t_0$$

for any solution $x(t) = x(t; t_0, \varphi_0)$ of (2.16) with $\|\varphi_0\|_r < \alpha$ and the solutions of (2.16) are uniformly bounded. $\qquad \square$

The proof in case $\alpha < \rho$ is trivial [228] and we omit the details in this book.

Theorem 2.15. *If in Theorem* 2.14 *condition* (2.22) *is replaced by the condition*

$$D^+_{(2.16)}V(t, x(t)) \leq -c(\|x(t)\|), \ t \neq \tau_k(x(t)), \ k = 1, 2 \ldots, \tag{2.27}$$

where $t \in [t_0, \infty)$, $x \in S^c_\rho \cap \Omega_P$, $c \in K$, *then the solutions of system* (2.16) *are uniformly ultimately bounded.*

Proof. Let $\alpha > 0$ and assume, without loss of generality, that $\alpha \geq \rho$. Choose $\beta = \beta(\alpha) > 0$ so that
$$\beta > \max\{\alpha, \ a^{-1}(b(\alpha))\}.$$

Let $t_0 \in \mathbb{R}$ and $\varphi_0 \in PC[[-r, 0], \mathbb{R}^n]$. Consider the solution $x(t) = x(t; t_0, \varphi_0)$ of (2.16) with $\|\varphi_0\|_r < \alpha$. Since all conditions of Theorem 2.14 are satisfied the solutions of (2.16) are uniformly bounded and for $t \geq t_0$ the following inequalities are valid:

$$\|x(t; t_0, \varphi_0)\| < \beta, \quad \text{and} \quad V(t, x(t)) < a(\beta). \tag{2.28}$$

Now let $B = a^{-1}(b(\rho))$ and $\beta > B$. Let the function $P : \mathbb{R}_+ \to \mathbb{R}_+$ be a continuous and non-decreasing on $\mathbb{R}_+$, and $P(u) > u$ as $u > 0$. We set

$$\eta = \inf\{P(u) - u : a(B) \le u \le a(\beta)\}.$$

Then

$$P(u) > u + \eta \quad \text{as} \quad a(B) \le u \le a(\beta), \tag{2.29}$$

and we choose the integer ν such that

$$a(B) + \nu\eta > a(\beta). \tag{2.30}$$

Let us denote

$$\xi_k = t_0 + k\frac{\eta}{c(\rho)}, \quad k = 0, 1, 2, \dots, \nu.$$

We want to prove

$$V(t, x(t)) < a(B) + (\nu - k)\eta, \quad t \ge \xi_k \tag{2.31}$$

for all $k = 0, 1, 2, \dots, \nu$.

Indeed, using (2.28) and (2.30) we obtain

$$V(t, x(t; t_0, \varphi_0)) < a(\beta) < a(B) + \nu\eta, \quad t \ge t_0 = \xi_0$$

which means the validity of (2.31) for $k = 0$.

Assume (2.31) to be fulfilled for some integer $k, 0 < k < \nu$, i.e.

$$V(s, x(s)) < a(B) + (\nu - k)\eta, \quad s \ge \xi_k. \tag{2.32}$$

Suppose now that

$$V(t, x(t)) \ge a(B) + (\nu - k - 1)\eta, \quad \xi_k \le t \le \xi_{k+1}. \tag{2.33}$$

Then (2.28), (2.29) and (2.33) imply

$$a(B) \le V(t, x(t)) < a(\beta), \quad \xi_k \le t \le \xi_{k+1}$$

and

$$P(V(t, x(t))) > V(t, x(t)) + \eta \ge a(B) + (\nu - k)\eta$$
$$> V(s, x(s)), \quad \xi_k \le s \le t \le \xi_{k+1}.$$

Therefore $x(\cdot) \in \Omega_P$ as $\xi_k \le s \le t \le \xi_{k+1}$. Then (2.21) and (2.33) yield

$$b(\|x(t)\|) \ge V(t, x(t)) \ge a(B) + (\nu - k - 1)\eta \ge a(B),$$

i.e. $x(t) \in S_\rho^c$, $\xi_k \le t \le \xi_{k+1}$. Then by (2.27) we obtain

$$V(\xi_{k+1}, x(\xi_{k+1})) \le V(\xi_k + 0, x(\xi_k + 0)) - \int_{\xi_k}^{\xi_{k+1}} c(\|x(s)\|) \, ds$$

$$< a(B) + (v-k)\eta - c(\rho)[\xi_{k+1} - \xi_k] = a(B) + (v-k-1)\eta$$

$$\le V(\xi_{k+1}, x(\xi_{k+1})),$$

which is a contradiction. Therefore there exists t^*, $\xi_k \le t^* \le \xi_{k+1}$ such that

$$V(t^*, x(t^*)) < a(B) + (v-k-1)\eta$$

and condition (2.19) implies

$$V(t^* + 0, x(t^* + 0)) < a(B) + (v-k-1)\eta.$$

We shall prove that

$$V(t, x(t)) < a(B) + (v-k-1)\eta, \quad t \ge t^*.$$

Supposing the opposite, we set

$$\mu = \inf\{t \ge t^* : V(t, x(t)) \ge a(B) + (v-k-1)\eta\}.$$

We consider two cases:

(A) $\mu \ne t_k$, $k = 1, 2, \ldots$. Then for $h > 0$ sufficiently close to zero, we have

$$V(\mu + h, x(\mu + h)) \ge a(B) + (v-k-1)\eta,$$

whence

$$D_{(2.16)}^+ V(\mu, x(\mu)) \ge 0.$$

On the other hand, we can prove, as above, that $x(\cdot) \in \Omega_P$ as $t^* \le s \le t \le \mu$ and therefore

$$D_{(2.16)}^+ V(\mu, x(\mu)) \le 0.$$

(B) $\mu = t_j$ for some $j \in \{1, 2, \ldots, k, \ldots\}$. We can obtain a contradiction by the analogous arguments, as in the proof of Theorem 2.14.

The contradiction we have already obtained yields

$$V(t, x(t)) < a(B) + (v-k-1)\eta, \quad t \ge \xi_{k+1}.$$

It follows that (2.31) holds for all $k = 0, 1, 2, \ldots, v$.

Let $T = T(\alpha) = v\frac{\eta}{c(\rho)}$. Then (2.31) implies

$$V(t, x(t)) < a(B) \quad \text{as} \quad t \ge t_0 + T$$

or

$$\|x(t)\| < B \quad \text{as} \quad t \geq t_0 + T$$

for any solution $x(t) = x(t; t_0, \varphi_0)$ of (2.16) with $\|\varphi_0\|_r < \alpha$ and the solutions of (2.16) are uniformly ultimately bounded.   □

Theorem 2.16. *If in Theorem* 2.15 *condition* (2.27) *is replaced by the condition*

$$D^+_{(2.16)}V(t, x(t)) \leq M - c(\|x(t)\|), \; t \neq \tau_k(x(t)), \; k = 1, 2, \ldots, \qquad (2.34)$$

where $t \in [t_0, \infty)$, $x \in S^c_\rho \cap \Omega_P$, $c \in K$ *and* $M = \text{const} > 0$, *then the solutions of system* (2.16) *are uniformly ultimately bounded.*

Proof. First, we shall prove the uniform boundedness of the solutions of (2.16).

Let $\rho > 0$ be sufficiently large so that $M - c(\rho) < 0$. Let $\alpha > \max\{\rho, c^{-1}(M)\}$ be given. Choose $\beta = \beta(\alpha) > 0$ so that $\beta = \max\{\alpha, a^{-1}(b(\alpha))\}$.

Let $t_0 \in \mathbb{R}$ and $\varphi_0 \in PC[[-r, 0], \mathbb{R}^n]$. Consider the solution $x(t) = x(t; t_0, \varphi_0)$ of (2.16) with $\|\varphi_0\|_r < \alpha$. Obviously,

$$\|x(t_0 + 0; t_0, \varphi_0)\| = \|\varphi_0(0)\| \leq \|\varphi_0\|_r < \alpha < \beta.$$

We claim that

$$\|x(t)\| < \beta, \quad t > t_0. \qquad (2.35)$$

If it is not true, then there exists some solution $x(t) = x(t; t_0, \varphi_0)$ of (2.16) with $\|\varphi_0\|_r < \alpha$ and a $t^* > t_0$ such that $\|x(t^*; t_0, \varphi_0)\| \geq \beta$. Thus, there exist $s_1, s_2, t_0 \leq s_1 < s_2 \leq t^*$, such that

$$\|x(s_1 + 0)\| \geq \alpha, \; \|x(s_1)\| \leq \alpha, \; \|x(s_2 + 0)\| \geq \beta$$

and

$$x(t) \in S_\beta \cap \overline{S}^c_\alpha, \; t \in [s_1, s_2).$$

As in the proof of Theorem 2.14, we can show that

$$V(s_1 + 0, x(s_1 + 0)) < a(\beta).$$

Next, we want to show that

$$V(t + 0, x(t + 0)) < a(\beta), \; t \in [s_1, s_2]. \qquad (2.36)$$

Suppose that this is not true and let

$$\mu = \inf\{s_2 \geq t > s_1 : V(t + 0, x(t + 0)) \geq a(\beta)\}.$$

We discuss two possibilities:

(A) $\mu \neq t_k$, $k = 1, 2, \ldots$. Since $V(t, x(t))$ is continuous at μ, we have $V(\mu + 0, x(\mu + 0)) = V(\mu, x(\mu)) = a(\beta)$ and

$$D_{(2.16)}^+ V(\mu, x(\mu)) \geq 0. \tag{2.37}$$

It is clear, from the choice of μ, that

$$P(V(\mu, x(\mu))) > V(\mu, x(\mu)) \geq V(s, x(s)), \quad s_1 \leq s \leq \mu.$$

From (2.21) we have

$$b(\|x(\mu)\|) \geq V(\mu, x(\mu)) > b(\alpha),$$

and hence

$$\|x(\mu)\| \geq \alpha. \tag{2.38}$$

Thus, we get using (2.34) and (2.38)

$$D_{(2.16)}^+ V(\mu, x(\mu)) \leq M - c(\|x(\mu)\|) \leq M - c(\alpha)$$

$$< M - c(\max\{\rho, c^{-1}(M)\}) = M - \max\{M, c(\rho)\} \leq 0,$$

which contradicts (2.37).

(B) $\mu = t_j$ for some $j \in \{1, 2, \ldots, k, \ldots\}$. We can obtain a contradiction by the analogous arguments, as in the proof of Theorem 2.14.

On the other hand, using (2.21) we get

$$V(s_2 + 0, x(s_2 + 0)) \geq a(\|x(s_2 + 0)\|) \geq a(\beta),$$

which contradicts the fact that $V(t + 0, x(t + 0)) < a(\beta)$, $t \in [s_1, s_2]$. Thus

$$\|x(t)\| < \beta, \quad t \geq t_0$$

for any solution $x(t) = x(t; t_0, \varphi_0)$ of (2.16) with $\|\varphi_0\|_r < \alpha$ and the system (2.16) is uniformly bounded.

The uniform boundedness of the solutions of (2.16) means that there exists a positive number B such that for each $t_0 \in \mathbb{R}_+$

$$\|\varphi_0\|_r < \rho \quad \text{implies} \quad \|x(t; t_0, \varphi_0)\| < B, \quad t \geq t_0.$$

Let $B = a^{-1}(b(\rho))$ and $\beta > B$. We set η such that (2.29) be true and choose the integer ν such that (2.30) be true.

Let us denote

$$\xi_k = t_0 + k \frac{\eta}{c(\rho) - M}, \quad k = 0, 1, 2, \ldots, \nu.$$

As in the proof of Theorem 2.15, we can prove that

$$\|x(t)\| < B \quad \text{as} \quad t \geq t_0 + T$$

for any solution $x(t) = x(t; t_0, \varphi_0)$ of (2.16) with $\|\varphi_0\|_r < \alpha$, where $T = T(\alpha) = \nu \frac{\eta}{c(\rho) - M}$. Therefore, the solutions of (2.16) are uniformly ultimately bounded. $\qquad \square$

Example 2.17. Let $t_0 \geq 0$ and $\tau_0(x) \equiv t_0$ for $x \in \mathbb{R}$. Consider the nonlinear equation

$$\begin{cases} \dot{x}(t) = -3\xi x^4(t) + \displaystyle\int_{t_0}^t p(t,s)x^3(s)\,ds + q(t), \; t \neq \tau_k(x(t)) \\[2mm] \Delta x(t) = -\beta_k x(t), \; t = \tau_k(x(t)), k = 1, 2, \ldots, \end{cases} \tag{2.39}$$

where $\xi > 0$; $0 \leq \beta_k \leq 2$, $k = 1, 2, \ldots$; $q \in C[\mathbb{R}_+, \mathbb{R}]$, $|q(t)| \leq M$ for some constant $M > 0$; $p \in C[\mathbb{R}_+ \times \mathbb{R}_+, \mathbb{R}_+]$.

Assume that the functions τ_k are such that the conditions H2.9, H2.10, H2.11 and H2.13 for system (2.39) are fulfilled and that there exists a constant $\mu > 1$ such that

$$\int_{t_0}^t p(t,s)\,ds \leq \frac{\xi}{\mu^3}. \tag{2.40}$$

Then (2.40) is a sufficient condition for uniform ultimate boundedness of the solutions of (2.39). In fact, we can choose $a(u) = b(u) = u$, $c(u) = 2\xi u^4$. Let $V(t,x) = |x|$, $S_\rho^c = \{x \in \mathbb{R} : |x| \geq 1\}$ and denote

$$P(u) = \mu.u.$$

Thus, using (2.40) we have

$$V_{(2.39)}^+(t, x(t)) \leq -3\xi|x^4(t)| + \int_{t_0}^t p(t,s)|x^3(s)|\,ds + |q(t)|$$

$$\leq -3\xi|x(t)|^4 + \mu^3|x(t)|^3 \int_{t_0}^t p(t,s)\,ds + |q(t)|$$

$$\leq -3\xi|x(t)|^4 + \mu^3|x(t)|^4.\frac{\xi}{\mu^3} + M$$

$$= M - 2\xi|x(t)|^4,$$

whenever $|x| \geq 1$, $t \neq \tau_k(x(t)), k = 1, 2, \ldots$ and $P(V(t, x(t))) = \mu|x(t)| > |x(s)| = V(s, x(s))$ for $t_0 \leq s \leq t$.

For $t = \tau_k(x(t)), k = 1, 2, \ldots$, we have

$$V(t + 0, x(t) - \beta_k x(t)) = |(1 - \beta_k)x(t)| \leq |x(t)| = V(t, x(t)).$$

Then all conditions of Theorem 2.16 are satisfied. Hence, the solutions of (2.39) are uniformly ultimately bounded.

Now, we shall give the results on the boundedness for impulsive systems of functional differential equations with fixed moments of impulsive perturbations.
Consider the system

$$\begin{cases} \dot{x}(t) = f(t, x_t), \; t \neq t_k \\[2mm] \Delta x(t) = I_k(x(t)), \; t = t_k, \; k = 1, 2, \ldots, \end{cases} \tag{2.41}$$

where $f : [t_0, \infty) \times PC[[-r, 0], \mathbb{R}^n] \to \mathbb{R}^n$; $I_k : \mathbb{R}^n \to \mathbb{R}^n$, $k = 1, 2, \ldots$; $t_0 < t_1 < t_2 < \cdots < t_k < t_{k+1} < \cdots$ and $\lim_{k \to \infty} t_k = \infty$.

Let $\varphi_0 \in PC[[-r, 0], \mathbb{R}^n]$. Denote by $x(t) = x(t; t_0, \varphi_0)$ the solution of system (2.41), satisfying the initial conditions (2.17).

Since the system (2.41) is a particular case of (2.16), then the following theorems follow directly from Theorems 2.13, 2.14 and 2.15.

Theorem 2.18. *Assume that:*

(1) *Conditions H1.12, H1.13, H2.6–H2.8 and H2.12 hold.*

(2) *There exists a function $V \in V_0$ such that H2.5 and (2.18) hold,*

$$V(t + 0, x + I_k(x)) \leq V(t, x), \ x \in \mathbb{R}^n, \ t = t_k, \ k = 1, 2, \ldots, \qquad (2.42)$$

and the inequality

$$D^+_{(2.41)} V(t, x(t)) \leq 0, \ t \neq t_k, \ k = 1, 2, \ldots$$

is valid for $t \in [t_0, \infty)$, $x \in \Omega_P$.

Then the solutions of system (2.41) are equi-bounded.

Theorem 2.19. *Assume that:*

(1) *Condition (1) of Theorem 2.18 holds.*

(2) *For $\rho > 0$, there exists $V \in V_0$ such that (2.21) and (2.42) hold, and the inequality*

$$D^+_{(2.41)} V(t, x(t)) \leq 0, \ t \neq t_k, \ k = 1, 2, \ldots \qquad (2.43)$$

is valid for $t \in [t_0, \infty)$, $x \in S^c_\rho \cap \Omega_P$.

Then the solutions of system (2.41) are uniformly bounded.

Theorem 2.20. *If in Theorem 2.19 condition (2.43) is replaced by the condition*

$$D^+_{(2.41)} V(t, x(t)) \leq -c(\|x(t)\|), \ t \neq t_k, \ k = 1, 2, \ldots,$$

where $t \in [t_0, \infty)$, $x \in S^c_\rho \cap \Omega_P$, $c \in K$, then the solutions of system (2.41) are uniformly ultimately bounded.

Now, we consider a quasilinear system of functional differential equation with impulse effects at fixed moments

$$\begin{cases} \dot{x}(t) = Ax(t) + Bx(t - h(t)) + f_1(x(t), x(t - h(t))), \ t \neq t_k \\ \Delta x(t) = I_k(x(t)), \ t = t_k, \ k = 1, 2, \ldots, \end{cases} \qquad (2.44)$$

where $t \geq t_0$; A and B are constant matrices of type $(n \times n)$; $f_1 : \mathbb{R}^n \times \mathbb{R}^n \to \mathbb{R}^n$; $I_k : \mathbb{R}^n \to \mathbb{R}^n$; $h \in C[[t_0, \infty), \mathbb{R}_+]$; $r = \sup_{t \geq t_0} h(t)$; $t_0 < t_1 < t_2 < \cdots < t_k < t_{k+1} < \cdots$ and $\lim_{k \to \infty} t_k = \infty$.

Let $\varphi_0 \in PC[[-r, 0], \mathbb{R}^n]$. Denote by $x(t) = x(t; t_0, \varphi_0)$ the solution of system (2.44), satisfying the initial conditions (2.17).

Introduce the notations:

- $\lambda_{\max}(A)$ and $\lambda_{\min}(A)$ are the largest and the smallest eigenvalues of the symmetric matrix $A = (a_{ij})$, respectively;

- $\|A\| = [\lambda_{\max}(A^T A)]^{\frac{1}{2}}$ is the norm of the matrix A;

- $\nu[a, b)$ is the number of the points t_k, contained in the interval $[a, b)$.

We need the following conditions in our subsequent analysis.

H2.14. The system
$$\dot{x}(t) = Ax(t)$$

has an asymptotically stable zero solution.

H2.15. $f_1 \in C[\mathbb{R}^n \times \mathbb{R}^n, \mathbb{R}^n]$.

H2.16. There exists a constant $\eta > 0$ such that
$$\|f_1(x, \tilde{x})\| < \eta \|\tilde{x}\|, \quad x, \tilde{x} \in \mathbb{R}^n.$$

H2.17. There exists a constant $M > 0$ such that
$$\nu[t, t + h(t)) < M, \quad t > t_0 - r.$$

H2.18. $I_k \in C[\mathbb{R}^n, \mathbb{R}^n]$, $k = 1, 2, \ldots$.

H2.19. There exists a constant $a > 0$ such that
$$\|I_k(x)\| \leq a\|x\|, \quad x \in \mathbb{R}^n, \quad k = 1, 2, \ldots.$$

In the sequel, we shall use the following lemmas.

Lemma 2.21 ([31]). *Assume that:*

(1) *Conditions H1.12 and H1.13 hold.*

(2) *$q : [t_0, \infty) \to \mathbb{R}$ is a piecewise continuous function with points of discontinuity of the first kind t_k, $t_{k+1}, \ldots$ at which it is continuous from the left.*

(3) *$p : [t_0, \infty) \to \mathbb{R}$ is a continuous function.*

(4) *For $t \geq t_0$, $\beta_k \geq 0$, $k = 1, 2, \ldots$ and $c_0 \geq 0$ the following inequality is valid:*

$$q(t) \leq c_0 + \int_{t_0}^{t} p(s)q(s)\,ds + \sum_{t_0 < t_k < t} \beta_k q(t_k).$$

Then

$$q(t) \leq c_0 \prod_{t_0 < t_k < t} (1 + \beta_k) \exp\left[\int_{t_0}^{t} p(s)\,ds\right], \quad t \geq t_0.$$

Lemma 2.22 ([150]). *Let the condition H2.14 hold and E be the identity operator in $\mathbb{R}^n$. Then the Lyapunov matrix equation*

$$A^T D + DA = -E \tag{2.45}$$

has a unique solution D, which is a symmetric positive definite matrix.

Lemma 2.23. *Assume that:*

(1) *D is a constant symmetric positive definite matrix of type $(n \times n)$.*

(2) *The function $V : \mathbb{R}^n \to \mathbb{R}_+$ is given by $V(x) = x^T D x$.*

Then for each $x \in \mathbb{R}^n$ the following inequalities are valid:

$$\lambda_{\min}(D)\|x\|^2 \leq V(x) \leq \lambda_{\max}(D)\|x\|^2. \tag{2.46}$$

Lemma 2.23 is an immediate corollary of the Lagrange theorem for the quadratic form $V(x) = x^T D x$.

Lemma 2.24. *Let the conditions H1.12, H1.13, H2.14–H2.19 hold.*
Then for $t_0 < t \leq t_0 + r$ for the solution $x(t) = x(t; t_0, \varphi_0)$ of problem (2.44), (2.17) the following inequality is valid:

$$\|x(t)\| < [1 + (\|B\| + \eta)r]\|\varphi_0\|_r \exp[\|A\|r + M \ln(1 + a)]. \tag{2.47}$$

Proof. The solution $x(t) = x(t; t_0, \varphi_0)$ of problem (2.44), (2.17) satisfies the equation

$$x(t) = x(t_0 + 0) + \int_{t_0}^{t} [Ax(s) + Bx(s - h(s)) + f_1(x(s), x(s - h(s)))]\,ds$$

$$+ \sum_{t_0 \leq t_k < t} I_k(x(t_k)).$$

From H2.16, it follows for $t_0 < t \leq t_0 + r$

$$\|x(t)\| < \|x(t_0 + 0)\| + \int_{t_0}^{t} \|A\|\,\|x(s)\|\,ds + \|\varphi_0\|_r(\|B\| + \eta)r + \sum_{t_0 \leq t_k < t} \|I_k(x(t_k))\|.$$

Thus, conditions H2.18 and H2.19 imply that

$$\|x(t)\| < \|x(t_0+0)\| + \int_{t_0}^{t} \|A\| \|x(s)\| \, ds + (\|B\| + \eta)\|\varphi_0\|_r r + \sum_{t_0 \le t_k < t} a\|x(t_k)\|.$$

Hence, Lemma 2.21 yields the estimate

$$\|x(t)\| < (\|x(t_0+0)\| + (\|B\| + \eta)\|\varphi_0\|_r r)(1+a)^{v[t_0,t)} e^{\|A\|r}.$$

Since the moments t_k satisfy condition H2.17, we have

$$v[t_0,t) \le v[t_0, t_0+r) < M, \ t_0 < t \le t_0 + r.$$

Therefore, the inequality

$$\|x(t)\| < [1 + (\|B\| + \eta)r]\|\varphi_0\|_r \exp[\|A\|r + M \ln(1+a)],$$

holds for $t_0 < t \le t_0 + r$ and the proof is complete.   □

Introduce the following condition:

H2.20. The inequality

$$(x + I_k(x))^T D(x + I_k(x)) \le x^T Dx$$

is valid for $x \in \mathbb{R}^n$, $k = 1, 2, \ldots$, where D is the solution of equation (2.45).

Introduce the following notation:

$$\alpha = \left\{ (1 + (\|B\| + \eta)r)\|\varphi_0\|_r \exp[\|A\|r + M \ln(1+a)] \right\}^2 \lambda_{\max}(D). \qquad (2.48)$$

Theorem 2.25. *Assume that:*

(1) *Conditions H1.12, H1.13, H2.14–H2.20 hold.*

(2) *The function $V . \ \mathbb{R}^n \to \mathbb{R}_+$ is given by $V(x) = x^T Dx$, where D is the solution of equation (2.45).*

(3) *The function f_1 and the matrix B are such that*

$$2\phi(D)\|D\| (\|B\| + \eta) < 1,$$

where $\phi(D) = \sqrt{\dfrac{\lambda_{\max}(D)}{\lambda_{\min}(D)}}.$

Then the solutions of system (2.44) are equi-bounded.

Proof. By Lemma 2.24 for $t_0 < t \leq t_0 + r$ the solution $x(t)$ of (2.44), (2.17) satisfies the inequality (2.47).

Let $\alpha > 0$ is the constant defined by (2.48). Introduce the notations:

$$v^\alpha = \{x \in \mathbb{R}^n : V(x) < \alpha\} \text{ and } \partial v^\alpha = \{x \in \mathbb{R}^n : V(x) = \alpha\}.$$

Then, from (2.47) it follows that $x(t) = x(t; t_0, \varphi_0) \in v^\alpha$ for $t_0 - r \leq t \leq t_0 + r$.

We shall prove that $x(t; t_0, \varphi_0) \in v^\alpha$ for $t > t_0 + r$, too. Suppose that this is not true.

Note that, from condition H2.20 and from $V(x(t_k)) < \alpha$, it follows that $V(x(t_k + 0)) \leq V(x(t_k)) < \alpha$, i.e. $x(t)$ cannot leave v^α by jump.

Now the assumption that $x(t; t_0, \varphi_0) \in v^\alpha$ for $t > t_0 + r$ is not true implies the existence of $T > t_0 + r$, $T \neq t_k$, $k = 1, 2, \ldots$ such that $x(t) \in v^\alpha$ for $t_0 - r \leq t < T$ and $x(T) \in \partial v^\alpha$.

Consider the upper right-hand derivative of Lyapunov function $V(x) = x^T D x$ with respect to system (2.44). For $t \neq t_k$, $k = 1, 2, \ldots$ we derive the estimate

$$D^+_{(2.44)} V(x(t)) \leq -\|x(t)\|^2 + 2\|DB\| \, \|x(t)\| \, \|x(t - h(t))\|$$
$$+ 2\|D\| \, \|x(t)\|\eta\|x(t - h(t))\|. \tag{2.49}$$

From inequalities (2.46), we deduce the inequalities

$$\sqrt{\frac{V(x(t))}{\lambda_{\max}(D)}} \leq x(t) \leq \sqrt{\frac{V(x(t))}{\lambda_{\min}(D)}}, \quad t \in \mathbb{R}. \tag{2.50}$$

For $t = T$ from (2.49) and (2.50), we derive the estimate

$$D^+_{(2.44)} V(x(T)) \leq [-\|x(T)\| + 2\|D\|(\|B\| + \eta)\|x(T - h(T))\|]\|x(T)\|$$

$$\leq \left[-\frac{1}{\sqrt{\lambda_{\max}(D)}} + 2\|D\|\frac{\|B\| + \eta}{\sqrt{\lambda_{\min}(D)}} \right] \sqrt{\alpha}\|x(T)\|.$$

Since by condition (3) of Theorem 2.25

$$-\frac{1}{\sqrt{\lambda_{\max}(D)}} + 2\|D\|\frac{\|B\| + \eta}{\sqrt{\lambda_{\min}(D)}} < 0,$$

then, from the above estimate, we obtain $D^+_{(2.44)} V(x(T)) < 0$.

Hence the assumption that $x(t) \in v^\alpha$ for $t_0 - r \leq t < T$ and $x(T) \in \partial v^\alpha$ will not be true, i.e. $x(t) \in v^\alpha$ for all $t \geq t_0 - r$.

Then from (2.46), we obtain that for $t \geq t_0$ for the solution $x(t)$ of problem (2.44), (2.17) the following estimate

$$\|x(t)\| < \left(\phi(D) + \frac{r}{2\|D\|} \right) \|\varphi_0\|_r \, \exp[\|A\|r + M \ln(1 + a)]$$

is valid.

Thus, we have $\|x(t)\| < \sqrt{\frac{\alpha}{\lambda_{\min}(D)}}$ for all $t \geq t_0$. Therefore, the solutions of (2.44) are equi-bounded. □

2.3   Global stability of the solutions

In this section we shall present the main results on the global stability of the zero solution of system (2.16). Here, the results from the previous section will be used.

We shall use the following definitions of global stability of the zero solution of (2.16).

Definition 2.26. The zero solution $x(t) \equiv 0$ of system (2.16) is said to be:

(a) *stable*, if

$$(\forall t_0 \in \mathbb{R})(\forall \varepsilon > 0)(\exists \delta = \delta(t_0, \varepsilon) > 0)$$

$$(\forall \varphi_0 \in PC[[-r, 0], \mathbb{R}^n] : \|\varphi_0\|_r < \delta)$$

$$(\forall t \geq t_0) : \|x(t; t_0, \varphi_0)\| < \varepsilon;$$

(b) *uniformly stable*, if the number δ in (a) is independent of $t_0 \in \mathbb{R}$;

(c) *globally equi-attractive*, if

$$(\forall t_0 \in \mathbb{R})(\forall \alpha > 0)(\forall \varepsilon > 0)(\exists \gamma = \gamma(t_0, \alpha, \varepsilon) > 0)$$

$$(\forall \varphi_0 \in PC[[-r, 0], \mathbb{R}^n] : \|\varphi_0\|_r < \alpha)(\forall t \geq t_0 + \gamma) : \|x(t; t_0, \varphi_0)\| < \varepsilon;$$

(d) *uniformly globally attractive*, if the number γ in (c) is independent of $t_0 \in \mathbb{R}$;

(e) *globally equi-asymptotically stable*, if it is stable and globally equi-attractive;

(f) *uniformly globally asymptotically stable*, if it is uniformly stable, uniformly globally attractive and the solutions of system (2.16) are uniformly bounded;

(g) *globally exponentially stable*, if

$$(\exists c > 0)(\forall \alpha > 0)(\exists \gamma = \gamma(\alpha) > 0)(\forall t_0 \in \mathbb{R})$$

$$(\forall \varphi_0 \in PC[[-r, 0], \mathbb{R}^n] : \|\varphi_0\|_r < \alpha)(\forall t \geq t_0) :$$

$$\|x(t; t_0, \varphi_0)\| \leq \gamma(\alpha)\|\varphi_0\|_r \exp[-c(t - t_0)].$$

In the next theorems, we shall use Lyapunov functions of the class V_0, whose derivatives are estimated by the elements of set Ω_P for $\Omega \equiv \mathbb{R}^n$.

Theorem 2.27. *Assume that:*

(1) *Conditions H2.2, H2.3, H2.6–H2.11, H2.13 and H2.18 hold.*

(2) *There exists a function $V \in V_0$ such that H2.5 holds,*

$$a(\|x\|) \leq V(t, x), \ a \in K, \ (t, x) \in [t_0, \infty) \times \mathbb{R}^n, \tag{2.51}$$

$$V(t + 0, x + I_k(x)) \leq V(t, x), \ (t, x) \in \sigma_k, \ k = 1, 2, \ldots, \tag{2.52}$$

and the inequality

$$D^+_{(2.16)} V(t, x(t)) \leq -c V(t, x(t)), \ t \neq \tau_k(x(t)), \ k = 1, 2, \ldots \tag{2.53}$$

is valid for $t \in [t_0, \infty), \ x \in \Omega_P$ and $c \in \mathbb{R}_+$.

Then the zero solution of system (2.16) is globally equi-asymptotically stable.

Proof. Let $\varepsilon > 0$. From the properties of the function V, it follows that there exists a constant $\delta = \delta(t_0, \varepsilon) > 0$ such that if $x \in \mathbb{R}^n : \|x\| < \delta$, then $\sup_{\|x\| < \delta} V(t_0 + 0, x) < a(\varepsilon)$.

Let $\varphi_0 \in PC[[-r, 0], \mathbb{R}^n] : \|\varphi_0\|_r < \delta$ and $x(t) = x(t; t_0, \varphi_0)$ be the solution of problem (2.16), (2.17). By Theorem 1.16, $J^+(t_0, \varphi_0) = [t_0, \infty)$.

Since all conditions of Corollary 1.25 are met, then

$$V(t, x(t; t_0, \varphi_0)) \leq V(t_0 + 0, \varphi_0(0)), \ t \in [t_0, \infty). \tag{2.54}$$

On the other hand $\|\varphi_0(0)\| \leq \|\varphi_0\|_r < \delta$ and hence $V(t_0 + 0, \varphi_0(0)) < a(\varepsilon)$.

From (2.51), (2.52) and the last inequality, there follow the inequalities

$$a(\|x(t; t_0, \varphi_0)\|) \leq V(t, x(t; t_0, \varphi_0)) \leq V(t_0 + 0, \varphi_0(0)) < a(\varepsilon),$$

which imply that $\|x(t; t_0, \varphi_0)\| < \varepsilon$ for $t \geq t_0$. This implies that the zero solution of system (2.16) is stable.

Now we shall prove that it is globally equi-attractive.

Let $\alpha = \text{const} > 0$ and $\varphi_0 \in PC[[-r, 0], \mathbb{R}^n] : \|\varphi_0\|_r < \alpha$.

From conditions (2.53) and (2.52), it follows that for $t \geq t_0$ the following inequality is valid

$$V(t, x(t; t_0, \varphi_0)) \leq V(t_0 + 0, \varphi_0(0)) \exp[-c(t - t_0)]. \tag{2.55}$$

Let $N(t_0, \alpha) = \sup\{V(t_0 + 0, x) : \|x\| < \alpha\}$ and $\gamma = \gamma(t_0, \alpha, \varepsilon) > \frac{1}{c} \ln \frac{N(t_0, \alpha)}{a(\varepsilon)}$.

Then for $t \geq t_0 + \gamma$ from (2.55), it follows that

$$V(t, x(t; t_0, \varphi_0)) < a(\varepsilon).$$

From the last inequality and (2.51) we have

$$\|x(t; t_0, \varphi_0)\| < \varepsilon,$$

which means that the zero solution of system (2.16) is globally equi-attractive. $\qquad \square$

Theorem 2.28. *Assume that:*

(1) *Condition* (1) *of Theorem* 2.27 *holds.*

(2) *There exists a function* $V \in V_0$ *such that* H2.5 *and* (2.52) *hold,*

$$a(\|x\|) \le V(t, x) \le h(t)b(\|x\|), \ (t, x) \in [t_0, \infty) \times \mathbb{R}^n, \tag{2.56}$$

where $a, b \in K$, $h : [t_0, \infty) \to [1, \infty)$, *and the inequality*

$$D^+_{(2.16)}V(t, x(t)) \le -g(t)c(\|x(t)\|), \ t \ne \tau_k(x(t)), \ k = 1, 2, \ldots \tag{2.57}$$

is valid for $t \in [t_0, \infty)$, $x \in \Omega_P$, $c \in K$, $g : [t_0, \infty) \to (0, \infty)$.

(3) $\displaystyle\int_{t_0}^{\infty} g(s)c\left[b^{-1}\left(\frac{\eta}{h(s)}\right)\right] ds = \infty$ *for each sufficiently small value of* $\eta > 0$.

Then the zero solution of system (2.16) *is globally equi-asymptotically stable.*

Proof. We can prove the stability of the zero solution of system (2.16) by the analogous arguments, as in the proof of Theorem 2.27.

Now we shall prove that the zero solution of (2.16) is globally equi-attractive.

Let $\alpha > 0$ be arbitrary, $\varepsilon > 0$ be given and $\eta = \frac{a(\varepsilon)}{2}$. Let the number $\gamma = \gamma(t_0, \alpha, \varepsilon) > 0$ be chosen so that

$$\int_{t_0}^{t_0+\gamma} g(s)c\left[b^{-1}\left(\frac{\eta}{h(s)}\right)\right] ds > h(t_0)b(\alpha). \tag{2.58}$$

(This is possible in view of condition (3) of Theorem 2.28.)

Let $\varphi_0 \in PC[[-r, 0], \mathbb{R}^n] : \|\varphi_0\|_r < \alpha$ and $x(t) = x(t; t_0, \varphi_0)$ be the solution of problem (2.16), (2.17). If we assume that for any $t \in [t_0, t_0 + \gamma]$ the following inequality holds

$$\|x(t; t_0, \varphi_0)\| \ge b^{-1}\left(\frac{\eta}{h(t)}\right), \tag{2.59}$$

then by (2.58) and (2.59), it follows that

$$V(t, x(t; t_0, \varphi_0)) \le V(t_0 + 0, \varphi_0(0)) - \int_{t_0}^{t} g(s)c(\|x(s; t_0, \varphi_0)\|)\, ds$$

$$\le V(t_0 + 0, \varphi_0(0)) - \int_{t_0}^{t} g(s)c\left[b^{-1}\left(\frac{\eta}{h(s)}\right)\right] ds, \ t \in [t_0, t_0 + \gamma].$$

From the above inequalities, (2.56) and (2.58) for $t = t_0 + \gamma$, we obtain

$$V(t, x(t; t_0, \varphi_0)) \le h(t_0)b(\alpha) - \int_{t_0}^{t} g(s)c\left[b^{-1}\left(\frac{\eta}{h(s)}\right)\right] ds < 0,$$

which contradicts (2.56). The contradiction obtained shows that there exists $t^* \in [t_0, t_0 + \gamma]$, such that

$$\|x(t^*; t_0, \varphi_0)\| < b^{-1}\left(\frac{\eta}{h(t^*)}\right).$$

Then for $t \geq t^*$ (hence for any $t \geq t_0 + \gamma$ as well) the following inequalities are valid

$$a(\|x(t;t_0,\varphi_0)\|) \leq V(t,x(t;t_0,\varphi_0)) \leq V(t^*,x(t^*;t_0,\varphi_0))$$
$$\leq h(t^*)b(\|x(t^*;t_0,\varphi_0)\|) < \eta < a(\varepsilon).$$

Therefore, $\|x(t;t_0,\varphi_0)\| < \varepsilon$ for $t \geq t_0 + \gamma$, i.e. the zero solution of (2.16) is globally equi-attractive. □

Theorem 2.29. *Assume that:*

(1) *Condition* (1) *of Theorem* 2.27 *holds.*

(2) *There exists a function* $V \in V_0$ *such that H2.5 and* (2.52) *hold,*

$$a(\|x\|) \leq V(t,x) \leq b(\|x\|), \ a,b \in K, \ (t,x) \in [t_0,\infty) \times \mathbb{R}^n, \qquad (2.60)$$

where $a(u) \to \infty$ *as* $u \to \infty$, *and the inequality*

$$D^+_{(2.16)} V(t,x(t)) \leq -c(\|x(t)\|), \ t \neq \tau_k(x(t)), \ k = 1,2,\dots \qquad (2.61)$$

is valid for $t \in [t_0,\infty)$, $x \in \Omega_P$, $c \in K$.

Then the zero solution of system (2.16) *is uniformly globally asymptotically stable.*

Proof. First, we shall show that the zero solution of system (2.16) is uniformly stable. For an arbitrary $\varepsilon > 0$ choose the positive number $\delta = \delta(\varepsilon)$ so that $b(\delta) < a(\varepsilon)$. Let $\varphi_0 \in PC[[-r,0],\mathbb{R}^n]$: $\|\varphi_0\|_r < \delta$ and $x(t) = x(t;t_0,\varphi_0)$ be the solution of problem (2.16), (2.17). Then by (2.60), (2.61) and (2.52) for any $t \in J^+(t_0,\varphi_0)$, the following inequalities are valid:

$$a(\|x(t;t_0,\varphi_0)\|) \leq V(t,x(t;t_0,\varphi_0)) \leq V(t_0+0,\varphi_0(0))$$
$$\leq b(\|\varphi_0(0)\|) \leq b(\|\varphi_0\|_r) < b(\delta) < a(\varepsilon).$$

Since by Theorem 1.16 $J^+(t_0,\varphi_0) = [t_0,\infty)$, then $\|x(t;t_0,\varphi_0)\| < \varepsilon$ for $t \geq t_0$. Thus, it is proved that the zero solution of system (2.16) is uniformly stable.

Now, we shall prove that the solutions of system (2.16) are uniformly bounded. Let $\alpha > 0$ and $\varphi_0 \in PC[[-r,0],\mathbb{R}^n]$: $\|\varphi_0\|_r < \alpha$. Since for the function $a \in K$ we have $a(u) \to \infty$ as $u \to \infty$, then we can choose $\beta = \beta(\alpha) > 0$ so that $a(\beta) > b(\alpha)$. Since the conditions of Corollary 1.25 are met, then

$$V(t,x(t;t_0,\varphi_0)) \leq V(t_0+0,\varphi_0(0)), \ t \in [t_0,\infty).$$

From the above inequality, (2.60) and (2.61) we have

$$a(\|x(t;t_0,\varphi_0)\|) \leq V(t,x(t;t_0,\varphi_0)) \leq V(t_0+0,\varphi_0(0))$$
$$\leq b(\|\varphi_0(0)\|) \leq b(\|\varphi_0\|_r) < b(\alpha) < a(\beta),$$

for $t \geq t_0$.

Therefore, $\|x(t; t_0, \varphi_0)\| < \beta$ for $t \geq t_0$. This implies that the solutions of system (2.16) are uniformly bounded.

Finally, we shall prove that the zero solution of system (2.16) is uniformly globally attractive.

Let $\alpha > 0$ be arbitrary, $\varepsilon > 0$ be given. Let the number $\eta = \eta(\varepsilon) > 0$ be chosen so that $b(\eta) > a(\varepsilon)$ and let $\gamma = \gamma(\alpha, \varepsilon) > 0$ be such that $\gamma > \frac{b(\alpha)}{c(\eta)}$.

Let $\varphi_0 \in PC[[-r, 0], \mathbb{R}^n] : \|\varphi_0\|_r < \alpha$ and $x(t) = x(t; t_0, \varphi_0)$ be the solution of problem (2.16), (2.17). If we assume that for any $t \in [t_0, t_0 + \gamma]$ the inequality $\|x(t; t_0, \varphi_0)\| \geq \eta$ holds, then by (2.61) and (2.52) it follows that

$$V(t, x(t; t_0, \varphi_0)) \leq V(t_0 + 0, \varphi_0(0)) - \int_{t_0}^{t} c(\|x(s; t_0, \varphi_0)\|)\, ds$$

$$\leq b(\alpha) - c(\eta)\gamma < 0,$$

which contradicts (2.60). The contradiction obtained shows that there exists $t^* \in [t_0, t_0 + \gamma]$, such that

$$\|x(t^*; t_0, \varphi_0)\| < \eta.$$

Then for $t \geq t^*$ (hence for any $t \geq t_0 + \gamma$ as well) the following inequalities are valid:

$$a(\|x(t; t_0, \varphi_0)\|) \leq V(t, x(t; t_0, \varphi_0)) \leq V(t^*, x(t^*; t_0, \varphi_0))$$

$$\leq b(\|x(t^*; t_0, \varphi_0)\|) < b(\eta) < a(\varepsilon).$$

Therefore $\|x(t; t_0, \varphi_0)\| < \varepsilon$ for $t \geq t_0 + \gamma$, i.e. the zero solution of (2.16) is uniformly globally attractive. $\square$

Corollary 2.30. *If in Theorem 2.29 condition (2.61) is replaced by the condition*

$$D^{+}_{(2.16)} V(t, x(t)) \leq -c V(t, x(t)), \quad t \neq \tau_k(x(t)), \quad k = 1, 2, \ldots, \tag{2.62}$$

where $t \in [t_0, \infty)$, $x \in \Omega_P$, $c = \text{const} > 0$, then the zero solution of system (2.16) is uniformly globally asymptotically stable.

This follows immediately from Theorem 2.29. However, the proof can be carried out using the fact that

$$V(t, x(t; t_0, \varphi_0)) \leq V(t_0 + 0, \varphi_0(0)) \exp[-c(t - t_0)]$$

for $t \geq t_0$ which is obtained from (2.52) and (2.62).

Theorem 2.31. *Assume that:*

(1) *Condition (1) of Theorem 2.27 holds.*

(2) *There exists a function $V \in V_0$ such that H2.5, (2.52) and (2.62) hold, and for any $\alpha > 0$ there exists $\gamma(\alpha) > 0$ such that*

$$\|x\| \leq V(t, x) \leq \gamma(\alpha)\|x\|, \quad (t, x) \in [t_0, \infty) \times \mathbb{R}^n. \tag{2.63}$$

Then the zero solution of system (2.16) is globally exponentially stable.

Proof. Let $\alpha > 0$ be arbitrary. Let $\varphi_0 \in PC[[-r, 0], \mathbb{R}^n] : \|\varphi_0\|_r < \alpha$ and $x(t) = x(t; t_0, \varphi_0)$ be the solution of problem (2.16), (2.17). From (2.52) and (2.62) we have

$$V(t, x(t; t_0, \varphi_0)) \leq V(t_0 + 0, \varphi_0(0)) \exp[-c(t - t_0)], \quad t \geq t_0.$$

From the above inequality and (2.63), we obtain

$$\|x(t; t_0, \varphi_0)\| \leq V(t, x(t; t_0, \varphi_0)) \leq V(t_0 + 0, \varphi_0(0)) \exp[-c(t - t_0)]$$
$$\leq \gamma(\alpha)\|\varphi_0(0)\| \exp[-c(t - t_0)] \leq \gamma(\alpha)\|\varphi_0\|_r \exp[-c(t - t_0)], \quad t \geq t_0,$$

which implies that the zero solution of system (2.16) is globally exponentially stable.

$\square$

Consider impulsive systems of functional differential equations with fixed moments of impulsive perturbations of type (2.41).

Since the system (2.41) is a particular case of (2.16), then the following theorems follow directly from the Theorems 2.27–2.31.

Theorem 2.32. *Assume that:*

(1) *Conditions H1.12, H1.13, H2.2, H2.3, H2.6, H2.7, H2.8 and H2.18 hold.*
(2) *There exists a function $V \in V_0$ such that H2.5 and (2.51) hold,*

$$V(t + 0, x + I_k(x)) \leq V(t, x), \ x \in \mathbb{R}^n, \ t = t_k, \ k = 1, 2, \ldots, \qquad (2.64)$$

and the inequality

$$D^+_{(2.41)} V(t, x(t)) \leq -c V(t, x(t)), \ t \neq t_k, \ k = 1, 2, \ldots$$

is valid for $t \in [t_0, \infty)$, $x \in \Omega_P$ and $c \in \mathbb{R}_+$.

Then the zero solution of system (2.41) is globally equi-asymptotically stable.

Theorem 2.33. *Assume that:*

(1) *Condition (1) of Theorem 2.32 holds.*
(2) *There exists a function $V \in V_0$ such that H2.5, (2.56) and (2.64) hold, and the inequality*

$$D^+_{(2.41)} V(t, x(t)) \leq -g(t)c(\|x(t)\|), \ t \neq t_k, \ k = 1, 2, \ldots$$

is valid for $t \in [t_0, \infty)$, $x \in \Omega_P$, $c \in K$, $g : [t_0, \infty) \rightarrow (0, \infty)$.

(3) $\displaystyle\int_{t_0}^{\infty} g(s)c\left[b^{-1}\left(\frac{\eta}{h(s)}\right)\right] ds = \infty$ *for each sufficiently small value of $\eta > 0$.*

Then the zero solution of system (2.41) is globally equi-asymptotically stable.

Theorem 2.34. *Assume that:*

(1) *Condition* (1) *of Theorem* 2.32 *holds.*

(2) *There exists a function* $V \in V_0$ *such that H2.5,* (2.60) *and* (2.64) *hold, and the inequality*

$$D^+_{(2.41)}V(t, x(t)) \leq -c(\|x(t)\|), \ t \neq \tau_k(x(t)), \ k = 1, 2, \ldots$$

is valid for $t \in [t_0, \infty), \ x \in \Omega_P, \ c \in K.$

Then the zero solution of system (2.41) *is uniformly globally asymptotically stable.*

Theorem 2.35. *Assume that:*

(1) *Condition* (1) *of Theorem* 2.32 *holds.*

(2) *There exists a function* $V \in V_0$ *such that H2.5,* (2.63) *and* (2.64) *hold, and the inequality*

$$D^+_{(2.41)}V(t, x(t)) \leq -cV(t, x(t)), \ t \neq t_k, \ k = 1, 2, \ldots$$

is valid for $t \in [t_0, \infty), \ x \in \Omega_P, \ c > 0.$

Then the zero solution of system (2.41) *is globally exponentially stable.*

Remark 2.36. All theorems in this section are true if instead of the set Ω_P we use the set Ω_1 for $\Omega \equiv \mathbb{R}^n$.

Example 2.37. Let $x \subset \mathbb{R}, r > 0, \sigma \in C[\mathbb{R}_+, \mathbb{R}_+], t - \sigma(t) \to \infty$ as $t \to \infty$.
Consider the following impulsive equation:

$$\begin{cases} \dot{x}(t) = -\alpha(t)x(t) - \beta(t)x(t-r) + g(t)x(t-\sigma(t)), \ t \neq t_k, \ t \geq 0 \\ \Delta x(t_k) = c_k x(t_k), \ t_k > 0, \ k = 1, 2, \ldots, \end{cases} \quad (2.65)$$

where $\beta \in C[\mathbb{R}_+, (0, \infty)]; \ \alpha, \ g \in C[\mathbb{R}_+, \mathbb{R}_+]; \ -1 < c_k \leq 0, \ k = 1, 2, \ldots;$
$0 < t_1 < t_2 < \cdots < t_k < t_{k+1} < \cdots$ and $\lim_{k \to \infty} t_k = \infty.$

Let $t_{-1} = \min\{-r, \inf_{t \geq 0}\{t - \sigma(t)\}\} < 0$ and $\varphi_0 \in C[[t_{-1}, 0], \mathbb{R}].$
Define the function $V(t, x) = \frac{1}{2}x^2$. Then the set

$$\Omega_1 = \{x \in PC[\mathbb{R}_+, \mathbb{R}] : x^2(s) \leq x^2(t), \ t + t_{-1} \leq s \leq t\}.$$

If there exists a constant $c > 0$ such that $|\beta(t)| + g(t) \leq \alpha(t) - c$ for $t \geq 0$, then
for $t \geq 0, \ x \in \Omega_1$ we have

$$D^+_{(2.65)}V(t, x(t)) = [-\alpha(t)x(t) - \beta(t)x(t-r) + g(t)x(t-\sigma(t))]x(t)$$
$$\leq [-\alpha(t)x(t) + |\beta(t)|x(t-r) + g(t)x(t-\sigma(t))]x(t)$$
$$\leq -2cV(t, x(t)), \ t \neq t_k.$$

Also,

$$V(t_k + 0, x(t_k) + c_k x(t_k)) = \frac{1}{2}(1 + c_k)^2 x^2(t_k) \le V(t_k, x(t_k)), \ k = 1, 2, \dots .$$

Thus, all conditions of Theorem 2.32 are satisfied and the zero solution of (2.65) is globally equi-asymptotically stable.

Example 2.38. Consider the following system:

$$\begin{cases} \dot{x}(t) = -y(t)\sin(x(t-1)) - 4x(t) + y(t-1), & t \ge 0, \ t \ne t_k \\ \dot{y}(t) = x(t)\sin(x(t-1)) - 3y(t), & t \ge 0, \ t \ne t_k \\ x(t_k + 0) = \left(1 - \frac{2}{k^2}\right)x(t_k), & t_k > 0, \ k = 1, 2, \dots \\ y(t_k + 0) = \left(1 - \frac{3}{k^2}\right)y(t_k), & t_k > 0, \ k = 1, 2, \dots, \end{cases} \tag{2.66}$$

where $x, \ y \in \mathbb{R}; 0 < t_1 < t_2 < \cdots < t_k < t_{k+1} < \cdots$ and $\lim_{k \to \infty} t_k = \infty$.

Let $\varphi_0 \in C[[-1, 0], \mathbb{R}^2]$. Define the function $V(t, x, y) = x^2 + y^2$. Then the set
$$\Omega_1 = \{(x, y) \in PC[\mathbb{R}_+, \mathbb{R}^2] : \ x^2(s) + y^2(s) \le x^2(t) + y^2(t), \ t - 1 \le s \le t\}.$$
For $t \ne t_k, k = 1, 2, \dots$ and $(x, y) \in \Omega_1$, we have
$$D^+_{(2.66)} V(t, x(t), y(t)) = 2x(t)\dot{x}(t) + 2y(t)\dot{y}(t) = 2x(t)y(t-1) - 8x^2(t) - 6y^2(t)$$
$$\le x^2(t) + y^2(t-1) - 8x^2(t) - 6y^2(t) \le -5V(t, x(t), y(t)).$$

Also, for $t = t_k, k = 1, 2, \dots$, we have
$$V(t + 0, x(t + 0), y(t + 0)) = x^2(t + 0) + y^2(t + 0)$$
$$= \left(1 - \frac{2}{k^2}\right)^2 x^2(t) + \left(1 - \frac{3}{k^2}\right)^2 y^2(t)$$
$$\le V(t, x(t), y(t)).$$

Thus, all conditions of Theorem 2.35 are satisfied and the zero solution of (2.66) is globally exponentially stable.

Example 2.39. Consider the following equation:

$$\begin{cases} \dot{x}(t) = a(t)\dfrac{x(t)}{2x^2(t) + 1} + \dfrac{e^{-x^2(t)}}{2x^2(t) + 1} \displaystyle\int_0^t b(t, s)e^{x^2(s)}x(s)\,ds, \ t \ne t_k \\ \Delta x(t_k) = \alpha_k x(t_k), \quad t_k > 0, \ k = 1, 2, \dots, \end{cases} \tag{2.67}$$

where $t \ge 0; x \in \mathbb{R}; a \in C[\mathbb{R}_+, \mathbb{R}]; b \in C[\mathbb{R}_+ \times \mathbb{R}_+, \mathbb{R}], b(t, s) = b(t-s); \alpha_k \in \mathbb{R}, k = 1, 2, \dots; 0 < t_1 < t_2 < \cdots < t_k < t_{k+1} < \cdots$ and $\lim_{k \to \infty} t_k = \infty$.

Let $\varphi \in PC[[0, t], \mathbb{R}]$, $t \geq 0$ and $x(t) = x(t; 0, \varphi)$ be the solution of (2.67), satisfying the initial condition $x(s) = \varphi(s)$, $s \in [0, t]$.

We shall prove that the conditions

(1) $\sup\limits_{t \geq 0} \int_0^t |b(t, s)| \, ds < \infty,$

(2) $a(t) + \sup\limits_{t \geq 0} \int_0^t |b(t, s)| \, ds < -c,$

(3) $-1 < \alpha_k \leq 0, \quad k = 1, 2, \ldots$

are sufficient for global uniform asymptotic stability of the zero solution of (2.67).
Define the function $V(t, x) = |x|e^{x^2}$. Then the set

$$\Omega_1 = \{x \in PC[\mathbb{R}_+, \mathbb{R}] : |x(s)|e^{x^2(s)} \leq |x(t)|e^{x^2(t)}, \ 0 \leq s \leq t\}.$$

For $t \neq t_k$, $k = 1, 2, \ldots$, we have

$D^+_{(2.67)}V(t, x(t))$

$$= e^{x^2(t)}\left[a(t)\frac{|x(t)|}{2x^2(t) + 1} + \frac{e^{-x^2(t)}}{2x^2(t) + 1}\int_0^t |b(t, s)|e^{x^2(s)}|x(s)| \, ds\right]$$

$$+ 2x(t)|x(t)|e^{x^2(t)}\left[a(t)\frac{x(t)}{2x^2(t) + 1} + \frac{e^{-x^2(t)}}{2x^2(t) + 1}\int_0^t b(t, s)e^{x^2(s)}x(s) \, ds\right]$$

$$\leq a(t)|x(t)|e^{x^2(t)} + \int_0^t |b(t, s)|e^{x^2(s)}|x(s)| \, ds.$$

For $t \neq t_k$, $k = 1, 2, \ldots$ and $x \in \Omega_1$ from (1) and (2) we obtain

$$D^+_{(2.67)}V(t, x(t)) \leq -cV(t, x(t)).$$

Also, for $t = t_k$, $k = 1, 2, \ldots$, we have

$$V(t_k + 0, x(t_k) + \alpha_k x(t_k)) = |(1 + \alpha_k)x(t_k)|e^{[(1+\alpha_k)x(t_k)]^2}$$

$$= |1 + \alpha_k| \, |x(t_k)|e^{(1+\alpha_k)^2 x^2(t_k)} \leq V(t_k, x(t_k)).$$

Thus, all conditions of Theorem 2.34 are satisfied and the zero solution of (2.67) is uniformly globally asymptotically stable.

Example 2.40. Consider the following equation:

$$\begin{cases} \dot{x}(t) = x(t) \ln \sqrt{2} + \dfrac{1}{10} (\ln 2) \displaystyle\int_{-\infty}^0 e^s x(t + s) \, ds, \ t \neq t_k \\[2mm] \Delta x(t_k) = \alpha_k x(t_k), \quad t_k > 0, \ k = 1, 2, \ldots, \end{cases} \qquad (2.68)$$

where $t \geq 0$; $x \in \mathbb{R}$; $-1 < \alpha_k \leq 0$, $k = 1, 2, \ldots$; $0 < t_1 < t_2 < \cdots < t_k < t_{k+1} <$ $\cdots$ and $\lim_{k \to \infty} t_k = \infty$.

Let $\varphi \in PC[(-\infty, 0], \mathbb{R}]$ and $x(t) = x(t; 0, \varphi)$ be the solution of (2.68), satisfying the initial condition $x(s) = \varphi(s)$, $s \in (-\infty, 0]$.

Define the function $V(t, x) = x^2$ and let $P(u) = 4u^2$, $u \geq 0$. Then the set

$$\Omega_P = \{x \in PC[\mathbb{R}_+, \mathbb{R}] : V(t+s, x(t+s)) \leq P(V(t, x(t))), -\infty < s \leq 0\}$$

$$= \{x \in PC[\mathbb{R}_+, \mathbb{R}] : x^2(t+s) \leq 4x^2(t), -\infty < s \leq 0\}.$$

For $t \neq t_k$, $k = 1, 2, \ldots$, we have

$$D^+_{(2.68)} V(t, x(t)) \leq 2x^2(t) \ln \sqrt{2} + \frac{1}{5} x(t)(\ln 2) \int_{-\infty}^{0} e^s |x(t+s)| \, ds.$$

For $t \neq t_k$, $k = 1, 2, \ldots$ and $x \in \Omega_P$, we obtain

$$D^+_{(2.68)} V(t, x(t)) \leq 2x^2(t) \ln \sqrt{2} + \frac{1}{5} x(t)(\ln 2) \int_{-\infty}^{0} e^s 2|x(t)| \, ds$$

$$= 2x^2(t) \left[\ln \sqrt{2} + \frac{1}{5}(\ln 2) \int_{-\infty}^{0} e^s \, ds \right] = \frac{7}{5}(\ln 2) V(t, x(t)).$$

Also, for $t = t_k$, $k = 1, 2, \ldots$, we have

$$V(t_k + 0, x(t_k) + \alpha_k x(t_k)) = (x(t_k) + \alpha_k x(t_k))^2$$

$$= (1 + \alpha_k)^2 x^2(t_k) \leq V(t_k, x(t_k)).$$

Thus, all conditions of Theorem 2.32 are satisfied and the zero solution of (2.68) is globally equi-asymptotically stable.

Notes and comments

Piecewise continuous Lyapunov functions which are applied in Chapter 2 were introduced by Bainov and Simeonov [30]. Moreover, the technique of investigation here essentially depends on the choice of minimal subsets of a suitable space of functions, by the elements of which the derivatives of Lyapunov functions are estimated [94, 98, 99, 131–134, 141, 148, 149, 169, 184, 229, 231].

The results exposed in Section 2.1 are of Bainov and Stamova [52] and Stamova and Stamov [207]. In [148], [149] and [155] the problem of stability for impulsive functional differential equations with infinite delays is considered and in [38] theorems on stability of impulsive differential-difference equations are proved. Similar results are given in [141], [184], [213] and [229].

The exposition of the problem of boundedness of impulsive functional differential equations in Section 2.2 is based on the works of Stamova [197] and [204]. Similar results for impulsive integro-differential systems with fixed moments of impulse effects are given in [87].

The results of the Section 2.3 are new. Close to them are the results of Bainov and Stamova [41] and [48].

Similar results for the impulsive functional differential equations with infinite delays are given by Luo and Shen in [148] and [149] and by Martynyuk, Shen and Stavroulakis in [155].

Chapter 3

Extensions of stability and boundedness theory

In the present chapter, we shall discuss some extensions of Lyapunov stability and boundedness theory for systems of impulsive functional differential equations.

Section 3.1 will deal with *stability, boundedness and global stability of sets* with respect to impulsive functional differential equations. Such results are generalizations of the stability and boundedness results given in Chapter 2.

In Section 3.2, we shall use vector Lyapunov functions and the comparison principle and we shall give sufficient conditions for *conditional stability* of the zero solution of systems under consideration.

Section 3.3 will investigate *parametric stability* properties of impulsive functional differential equations with fixed moments of impulse effect. The obtained results are parallel to results of Siljak, Ikeda and Ohta [186].

Section 3.4 is devoted to the development of Lyapunov–Razumikhin method in studying of *eventual stability and eventual boundedness* for impulsive functional differential equations with variable impulsive perturbations. The results for systems with fixed moments of impulse effect will be also given.

In Section 3.5, we shall continue to use Lyapunov direct method and provide several *practical stability* results. The advantage of using vector Lyapunov functions is demonstrated.

In Section 3.6, we shall consider a scalar comparison equation and we shall analyze conditions for *Lipschitz stability* of the solutions of impulsive functional differential equations.

Section 3.7 will deal with *stability in terms of two measures* which unify various stability concepts for the systems under consideration.

Finally, in Section 3.8, the results on *boundedness in terms of two measures* will be given.

3.1   Stability and boundedness of sets

The notion of stability of sets, which includes as a special case stability in the sense of Lyapunov, is one of the most important notions in the stability theory. The stability of sets with respect to systems of ordinary differential equations without impulses has been considered by Yoshizawa in [223]. We refer to [99, 131–134] for the results on stability and boundedness of sets for functional differential equations, and to [30, 129, 131, 132] for impulsive differential equations.

In this section, we shall discuss problems related to stability, boundedness and global stability of sets of a sufficiently general type contained in some domain (an open connected set) with respect to impulsive functional differential equations with variable impulsive perturbations. The results for the systems with impulse effect at fixed moments will also be considered.

Stability of sets

Let $t_0 \in \mathbb{R}$, $r = \mathrm{const} > 0$, $\Omega \subseteq \mathbb{R}^n$, $\Omega \neq \emptyset$ and $= \|x\| = \sqrt{x_1^2 + x_2^2 + \cdots + x_n^2}$ define the norm of $x \in \mathbb{R}^n$. Consider the following system of impulsive functional differential equations with variable impulsive perturbations:

$$\begin{cases} \dot{x}(t) = f(t, x_t), \ t \neq \tau_k(x(t)) \\ \Delta x(t) = I_k(x(t)), \ t = \tau_k(x(t)), \ k = 1, 2, \ldots, \end{cases} \tag{3.1}$$

where $f : [t_0, \infty) \times PC[[-r, 0], \Omega] \to \mathbb{R}^n$; $\tau_k : \Omega \to (t_0, \infty)$, $I_k : \Omega \to \mathbb{R}^n$, $k = 1, 2, \ldots$; $\Delta x(t) = x(t + 0) - x(t - 0)$; and for $t \geq t_0$, $x_t \in PC[[-r, 0], \Omega]$ is defined by $x_t(s) = x(t + s)$, $-r \leq s \leq 0$.

Let $\varphi_0 \in PC[[-r, 0], \Omega]$. Denote by $x(t) = x(t; t_0, \varphi_0)$ the solution of system (3.1), satisfying the initial conditions

$$\begin{cases} x(t; t_0, \varphi_0) = \varphi_0(t - t_0), \ t_0 - r \leq t \leq t_0 \\ x(t_0 + 0; t_0, \varphi_0) = \varphi_0(0), \end{cases} \tag{3.2}$$

and by $J^+(t_0, \varphi_0)$ the maximal interval of type $[t_0, \beta)$ in which the solution $x(t; t_0, \varphi_0)$ is defined.

Let $\tau_0(x) \equiv t_0$ for $x \in \Omega$. Let $M \subset [t_0 - r, \infty) \times \Omega$. We shall use the following notations:

$$M(t) = \{x \in \Omega : (t, x) \in M, t \in [t_0, \infty)\};$$

$$M_0(t) = \{x \in \Omega : (t, x) \in M, t \in [t_0 - r, t_0]\};$$

$$d(x, M(t)) = \inf_{y \in M(t)} \|x - y\| \text{ is the distance between } x \in \Omega \text{ and } M(t);$$

$$M(t, \varepsilon) = \{x \in \Omega : d(x, M(t)) < \varepsilon\} \ (\varepsilon > 0) \text{ is an } \varepsilon\text{-neighbourhood of } M(t);$$

$$d_0(\varphi, M_0(t)) = \sup_{t \in [t_0 - r, t_0]} d(\varphi(t - t_0), M_0(t)), \ \varphi \in PC[[-r, 0], \Omega];$$

$M_0(t, \varepsilon) = \{\varphi \in PC[[-r, 0], \Omega] : d_0(\varphi, M_0(t)) < \varepsilon\}$ is an ε-neighbourhood of $M_0(t)$;

$$\overline{S_\alpha} = \{x \in \mathbb{R}^n : \|x\| \leq \alpha\}; \quad \overline{S_\alpha}(PC_0) = \{\varphi \in PC[[-r, 0], \mathbb{R}^n] : \|\varphi\|_r \leq \alpha\}.$$

We shall give the following definitions of stability of the set M with respect to system (3.1).

Definition 3.1. The set M is said to be:

(a) *stable* with respect to system (3.1), if

$$(\forall t_0 \in \mathbb{R})(\forall \alpha > 0)(\forall \varepsilon > 0)(\exists \delta = \delta(t_0, \alpha, \varepsilon) > 0)$$

$$(\forall \varphi_0 \in \overline{S_\alpha}(PC_0) \cap M_0(t, \delta))(\forall t \geq t_0) : \ x(t; t_0, \varphi_0) \in M(t, \varepsilon);$$

(b) *uniformly stable* with respect to system (3.1), if the number δ from (a) depends only on ε;

(c) *attractive* with respect to system (3.1), if

$$(\forall t_0 \in \mathbb{R})(\forall \alpha > 0)(\exists \eta > 0)(\forall \varepsilon > 0)(\forall \varphi_0 \in \overline{S_\alpha}(PC_0) \cap M_0(t, \eta))$$

$$(\exists \sigma = \sigma > 0)(\forall t \geq t_0 + \sigma) : \ x(t; t_0, \varphi_0) \in M(t, \varepsilon);$$

(d) *asymptotically stable* with respect to system (3.1), if it is stable and attractive;

(e) *unstable* with respect to system (3.1), if (a) fails to hold.

Introduce the following conditions:

H3.1. $M(t) \neq \emptyset$ for $t \in [t_0, \infty)$.

H3.2. $M_0(t) \neq \emptyset$ for $t \in [t_0 - r, t_0]$.

H3.3. For any compact subset F of $[t_0, \infty) \times \Omega$ there exists a constant $K_1 > 0$ depending on F such that if (t, x), $(t', x) \in F$, then the following inequality is valid:

$$|d(x, M(t)) - d(x, M(t'))| \leq K_1 |t - t'|.$$

In the further considerations, we shall use piecewise continuous auxiliary functions $V : [t_0, \infty) \times \Omega \to \mathbb{R}_+$, which belong to the class V_0 and are such that the following condition holds

H3.4. $V(t, x) = 0$ for $(t, x) \in M$, $t \geq t_0$ and $V(t, x) > 0$ for $(t, x) \in \{[t_0, \infty) \times \Omega\} \setminus M$.

Theorem 3.2. *Assume that:*

(1) *Conditions H1.1, H1.2, H1.3, H1.5, H1.6, H1.9, H1.10, H2.1, H3.1–H3.3 hold.*

(2) *There exists a function $V \in V_0$ such that H3.4 holds,*

$$a(d(x, M(t))) \leq V(t, x), \ a \in K, \ (t, x) \in [t_0, \infty) \times \Omega, \tag{3.3}$$

$$V(t + 0, x + I_k(x)) \leq V(t, x), \ (t, x) \in \sigma_k, \ k = 1, 2, \ldots, \tag{3.4}$$

and the inequality

$$D^+_{(3.1)} V(t, x(t)) \leq 0, \ t \neq \tau_k(x(t)), \ k = 1, 2, \ldots \tag{3.5}$$

is valid for $t \in [t_0, \infty)$, $x \in \Omega_1$.

Then the set M is stable with respect to system (3.1).

Proof. Let $t_0 \in \mathbb{R}$, $\varepsilon > 0$, $\alpha > 0$. From the properties of the function V, it follows that there exists a constant $\delta = \delta(t_0, \alpha, \varepsilon) > 0$ such that if $x \in \overline{S_\alpha} \cap M(t_0 + 0, \delta)$, then $V(t_0 + 0, x) < a(\varepsilon)$.

Let $\varphi_0 \in \overline{S_\alpha}(PC_0) \cap M_0(t, \delta)$. Then $d(\varphi_0(0), M(t_0 + 0)) < \delta$, i.e. $\varphi_0(0) \in \overline{S_\alpha} \cap M(t_0 + 0, \delta)$, hence $V(t_0 + 0, \varphi_0(0)) < a(\varepsilon)$.

Let $x(t) = x(t; t_0, \varphi_0)$ be the solution of problem (3.1), (3.2). By Theorem 1.16, it follows that $J^+(t_0, \varphi_0) = [t_0, \infty)$. Since the conditions of Corollary 1.25 are met, then

$$V(t, x(t; t_0, \varphi_0)) \leq V(t_0 + 0, \varphi_0(0)), \ t \in [t_0, \infty). \tag{3.6}$$

From (3.3), (3.4) and (3.6), there follow the inequalities:

$$a(d(x(t; t_0, \varphi_0), M(t))) \leq V(t, x(t; t_0, \varphi_0))$$
$$\leq V(t_0 + 0, \varphi_0(0)) < a(\varepsilon), \ t \in [t_0, \infty).$$

Hence, $x(t; t_0, \varphi_0) \in M(t, \varepsilon)$ for $t \geq t_0$, i.e. the set M is stable with respect to system (3.1). $\square$

Theorem 3.3. *Let the conditions of Theorem 3.2 hold, and let a function $b \in K$ exist such that*

$$V(t, x) \leq b(d(x, M(t))), \ (t, x) \in [t_0, \infty) \times \Omega. \tag{3.7}$$

Then the set M is uniformly stable with respect to system (3.1).

Proof. Let $\varepsilon > 0$. Choose $\delta = \delta(\varepsilon) > 0$ so that $b(\delta) < a(\varepsilon)$. Let $\varphi_0 \in \overline{S_\alpha}(PC_0) \cap M_0(t, \delta)$. Using successively (3.3), (3.6) and (3.7), we obtain

$$a(d(x(t; t_0, \varphi_0), M(t))) \leq V(t, x(t; t_0, \varphi_0))$$
$$\leq V(t_0 + 0, \varphi_0(0)) \leq b(d(\varphi_0(0), M_0(t_0)))$$
$$\leq b(d_0(\varphi_0, M_0(t))) < b(\delta) < a(\varepsilon),$$

for $t \in [t_0, \infty)$. Hence, $x(t; t_0, \varphi_0) \in M(t, \varepsilon)$ for $t \geq t_0$.

This proves that the set M is uniformly stable with respect to system (3.1). $\square$

Theorem 3.4. *If in Theorem* 3.2 *condition* (3.5) *is replaced by the condition*

$$D^+_{(3.1)}V(t, x(t)) \leq -cV(t, x(t)), \ t \neq \tau_k(x(t)), \ k = 1, 2, \ldots, \qquad (3.8)$$

where $t \in [t_0, \infty)$, $x \in \Omega_1$, $c = \text{const} > 0$, *then the set* M *is asymptotically stable with respect to system* (3.1).

Proof. Since the conditions of Theorem 3.2 are met, then the set M is stable with respect to system (3.1). We shall show that it is an attractive set with respect to system (3.1).

Let $t_0 \in \mathbb{R}$, $\varepsilon > 0$, $\alpha > 0$. From (3.8) and (3.4), we have

$$V(t, x(t; t_0, \varphi_0)) \leq V(t_0 + 0, \varphi_0(0)) \exp[-c(t - t_0)], \quad t \geq t_0. \qquad (3.9)$$

Let $\eta = \text{const} > 0 : \varphi_0 \in \overline{S_\alpha}(PC_0) \cap M_0(t, \eta)$ and $x(t) = x(t; t_0, \varphi_0)$ be the solution of problem (3.1), (3.2). We set $N = N(t_0, \eta, \varepsilon) = \sup\{V(t_0 + 0, x) : x \in \overline{S_\alpha} \cap M(t_0 + 0, \eta)\}$.

Choose $\sigma > 0$ so that

$$\sigma > \frac{1}{c} \ln \frac{N(t_0, \eta, \alpha)}{a(\varepsilon)}.$$

Then, from (3.3) and (3.9) for $t \geq t_0 + \sigma$, the following inequalities hold:

$$a(d(x(t; t_0, \varphi_0), M(t))) \leq V(t, x(t; t_0, \varphi_0))$$
$$\leq V(t_0 + 0, \varphi_0(0)) \exp[-c(t - t_0)] < a(\varepsilon).$$

Hence, $x(t; t_0, \varphi_0) \in M(t, \varepsilon)$ for $t \geq t_0 + \sigma$ and the set M is an attractive set with respect to system (3.1). □

Consider the system of impulsive functional differential equations with fixed moments of impulsive perturbations

$$\begin{cases} \dot{x}(t) = f(t, x_t), \ t \neq t_k \\ \Delta x(t) = I_k(x(t)), \ t = t_k, \ k = 1, 2, \ldots, \end{cases} \qquad (3.10)$$

where $f : [t_0, \infty) \times PC[[-r, 0], \Omega] \to \mathbb{R}^n$; $I_k : \Omega \to \mathbb{R}^n$, $k = 1, 2, \ldots$; $t_0 < t_1 < t_2 < \cdots < t_k < t_{k+1} < \cdots$ and $\lim_{k \to \infty} t_k = \infty$.

Let $\varphi_0 \in PC[[-r, 0], \Omega]$. Denote by $x(t) = x(t; t_0, \varphi_0)$ the solution of system (3.10), satisfying the initial conditions (3.2).

Since (3.10) is a special case of (3.1), the following theorems follow directly from Theorems 3.2 and 3.3.

Theorem 3.5. *Assume that:*

(1) *Conditions H1.1, H1.2, H1.3, H1.9, H1.10, H1.12, H1.13, H3.1–H3.3 hold.*

(2) *There exists a function $V \in V_0$ such that H3.4 and (3.3) hold,*

$$V(t+0, x + I_k(x)) \leq V(t, x), \quad x \in \Omega, \ t = t_k, \ k = 1, 2, \ldots,$$

and the inequality

$$D^+_{(3.10)} V(t, x(t)) \leq 0, \ t \neq t_k, \ k = 1, 2, \ldots$$

is valid for each $t \in [t_0, \infty)$, $x \in \Omega_1$.

Then the set M is stable with respect to system (3.10).

Theorem 3.6. *Let the conditions of Theorem 3.5 hold, and let a function $b \in K$ exist such that*

$$V(t, x) \leq b(d(x, M(t))), \quad (t, x) \in [t_0, \infty) \times \Omega.$$

Then the set M is uniformly stable with respect to system (3.10).

We shall next consider a result which gives asymptotic stability of the set M with respect to system (3.10). We shall use two Lyapunov like functions.

Theorem 3.7. *Assume that:*

(1) *Condition (1) of Theorem 3.5 holds.*

(2) *There exist functions V, $W \in V_0$ such that H3.4 holds,*

$$a(d(x, M(t))) \leq V(t, x), \ a \in K, \ (t, x) \in [t_0, \infty) \times \Omega, \tag{3.11}$$

$$b(d(x, M(t))) \leq W(t, x), \ b \in K, \ (t, x) \in [t_0, \infty) \times \Omega, \tag{3.12}$$

$$\sup\{D^+_{(3.10)} W(t, x) : t \geq t_0, t \neq t_k\} \leq N_1 < \infty, \tag{3.13}$$

$$V(t+0, x + I_k(x)) \leq V(t, x), \ x \in \Omega, \ t = t_k, \ k = 1, 2, \ldots, \tag{3.14}$$

$$W(t+0, x + I_k(x)) \leq W(t, x), \ x \in \Omega, \ t = t_k, \ k = 1, 2, \ldots,$$

and the inequality

$$D^+_{(3.10)} V(t, x(t)) \leq -c(W(t, x(t))), \ t \neq t_k, \ k = 1, 2, \ldots \tag{3.15}$$

is valid for each $t \in [t_0, \infty)$, $x \in \Omega_1$ and $c \in K$.

Then the set M is asymptotically stable with respect to system (3.10).

Proof. From Theorem 3.5, it follows that the set M is a stable set of system (3.10). Let $t_0 \in \mathbb{R}$, $\alpha = \text{const} > 0 : M(t, \alpha) \subset \Omega$ for $t \geq t_0$. For an arbitrary $t \geq t_0$, we put

$$V_{t,\alpha}^{-1} = \{x \in \Omega : V(t + 0, x) \leq a(\alpha)\}.$$

From (3.11), we have that for each $t \geq t_0$ the following inclusions are valid:

$$V_{t,\alpha}^{-1} \subset M(t, \alpha) \subset \Omega.$$

From (3.14) and (3.15), we obtain that if $\varphi_0 \in PC[[-r, 0], \Omega]: \varphi_0(0) \in V_{t_0,\alpha}^{-1}$, then $x(t; t_0, \varphi_0) \in V_{t,\alpha}^{-1}$ for $t \in [t_0, \infty)$.

Let $\varphi_0 \in PC[[-r, 0], \Omega]: \varphi_0(0) \in V_{t_0,\alpha}^{-1}$. We shall prove that

$$\lim_{t \to \infty} d(x(t; t_0, \varphi_0), M(t)) = 0.$$

Suppose that this is not true. Then there exist $\varphi_0 \in PC[[-r, 0], \Omega] : \varphi_0(0) \in V_{t_0,\alpha}^{-1}$, $\beta > 0$, $l > 0$, and a sequence $\{\xi_k\}_{k=1}^{\infty} \subset (t_0, \infty)$ such that for $k = 1, 2, \ldots$ the following inequalities are valid:

$$\xi_{k+1} - \xi_k \geq \beta$$

and

$$d(x(\xi_k; t_0, \varphi_0), M(\xi_k)) \geq l.$$

From the last inequality and (3.12), we have

$$W(\xi_k, x(\xi_k; t_0, \varphi_0)) \geq b(l), \quad k = 1, 2, \ldots . \tag{3.16}$$

Choose the constant $\gamma : 0 < \gamma < \min\{\beta, \frac{b(l)}{2N_1}\}$ and from (3.13) and (3.16), we obtain

$$W(t, x(t; t_0, \varphi_0)) = W(\xi_k, x(\xi_k; t_0, \varphi_0)) + \int_{\xi_k}^{t} D_{(3.10)}^{+} W(s, x(s; t_0, \varphi_0)) \, ds$$

$$\geq b(l) - N_1(\xi_k - t) \geq b(l) - N_1\gamma > \frac{b(l)}{2}$$

for $t \in [\xi_k - \gamma, \xi_k]$. From the above estimate and from (3.14) and (3.15), we conclude that

$$V(\xi_q, x(\xi_q; t_0, \varphi_0)) \leq V(t_0 + 0, \varphi_0(0)) + \int_{t_0}^{\xi_q} D_{(3.10)}^{+} V(s, x(s; t_0, \varphi_0)) \, ds$$

$$\leq V(t_0 + 0, \varphi_0(0)) - \int_{t_0}^{\xi_q} c(W(s, x(s; t_0, \varphi_0))) \, ds$$

$$\leq V(t_0 + 0, \varphi_0(0)) - \sum_{k=1}^{q} \int_{\xi_k - \gamma}^{\xi_k} c(W(s, x(s; t_0, \varphi_0))) \, ds$$

$$\leq V(t_0 + 0, \varphi_0(0)) - c\left(\frac{b(l)}{2}\right)\gamma q \to -\infty \quad \text{as } q \to \infty,$$

which contradicts (3.11).

Consequently, $\lim_{t\to\infty} d(x(t;t_0,\varphi_0), M(t)) = 0$ and since $V_{t_0,\alpha}^{-1}$ is a neighbourhood of the origin which is contained in $M(t_0,\alpha)$, then the set M is an attractive set of system (3.10). $\quad\square$

In Theorem 3.7 two auxiliary functions of class V_0 were used. The function $W(t, x)$ may have a special form. In the case when $W(t, x) = d(x, M(t))$, we deduce the following corollary of Theorem 3.7.

Corollary 3.8. *Assume that:*

(1) *Condition* (1) *of Theorem 3.5 holds.*

(2) *There exists a function* $V \in V_0$ *such that* H3.4, (3.11) *and* (3.14) *hold,*

$$d(x + I_k(x), M(t)) \le d(x, M(t)), \quad x \in \Omega, \, t = t_k, \, k = 1,2,\ldots,$$

and the inequality

$$D^+_{(3.10)}V(t, x(t)) \le -c(d(x(t), M(t))), \, t \ne t_k, \, k = 1,2,\ldots$$

is valid for each $t \in [t_0, \infty)$, $x \in \Omega_1$ *and* $c \in K$.

Then the set M *is asymptotically stable with respect to system* (3.10).

In the case when $W(t, x) = V(t, x)$, we deduce the following corollary of Theorem 3.7.

Corollary 3.9. *Assume that:*

(1) *Condition* (1) *of Theorem 3.5 holds.*

(2) *There exists a function* $V \in V_0$ *such that* H3.4, (3.11) *and* (3.14) *hold, and the inequality*

$$D^+_{(3.10)}V(t, x(t)) \le -c(V(t, x(t))), \, t \ne t_k, \, k = 1,2,\ldots$$

is valid for each $t \in [t_0, \infty)$, $x \in \Omega_1$ *and* $c \in K$.

Then the set M *is asymptotically stable with respect to system* (3.10).

Example 3.10. Consider a system of four biological spaces with impulse effects at fixed moments of time $t_1, t_2, \ldots$ such that $0 < t_1 < t_2 < \cdots$ and $\lim_{k \to \infty} t_k = \infty$:

$$
\begin{cases}
\dot{x}_1(t) = -a_1 x_1(t) + \lambda_1(t) \int_{-r}^{0} x_1(t+s)ds + b_1 x_1(t) x_2^2(t) x_3^2(t), \ t \neq t_k \\[2mm]
\dot{x}_2(t) = -a_2 x_2(t) + \lambda_2(t) \int_{-r}^{0} x_2(t+s)ds + b_2 x_2(t) x_1^2(t) x_3^2(t), \ t \neq t_k \\[2mm]
\dot{x}_3(t) = -a_3 x_3(t) + \lambda_3(t) \int_{-r}^{0} x_3(t+s)ds + x_1(t-r) x_2(t-r) x_4(t), \ t \neq t_k \\[2mm]
\dot{x}_4(t) = -a_4 x_4(t) + \lambda_4(t) \int_{-r}^{0} x_4(t+s)ds - x_1(t-r) x_2(t-r) x_3(t), \ t \neq t_k \\[2mm]
\Delta x_1(t_k) = c_{1k} x_1(t_k), \quad \Delta x_2(t_k) = c_{2k} x_2(t_k) \\[1mm]
\Delta x_3(t_k) = c_{3k} x_3(t_k), \quad \Delta x_4(t_k) = c_{4k} x_4(t_k)
\end{cases}
$$

$$(3.17)$$

where $t \geq 0$; $r > 0$; $b_1, b_2 \in \mathbb{R}$; $x_i \in \mathbb{R}_+$, $a_i > 0$, $\lambda_i : \mathbb{R}_+ \to \mathbb{R}_+$, $-1 < c_{ik} \leq 0$ for $i = 1, 2, 3, 4$ and $k = 1, 2, \ldots$.

Let $M = \{(t, 0, 0, 0, 0) : t \in [-r, \infty)\}$.

Consider the function $V(t, x_1, x_2, x_3, x_4) = x_1^2 + x_2^2 + x_3^2 + x_4^2$. Then the set

$$
\Omega_1 = \big\{ \mathrm{col}(x_1(t), x_2(t), x_3(t), x_4(t)) \in PC[\mathbb{R}_+, \mathbb{R}_+^4] :
$$

$$
x_1^2(s) + x_2^2(s) + x_3^2(s) + x_4^2(s) \leq x_1^2(t) + x_2^2(t) + x_3^2(t) + x_4^2(t),
$$

$$
t - r \leq s \leq t \big\}.
$$

For $t = t_k, k = 1, 2, \ldots$, we have

$$
V(t+0, x_1(t) + c_{1k} x_1(t), x_2(t) + c_{2k} x_2(t), x_3(t) + c_{3k} x_3(t), x_4(t) + c_{4k} x_4(t))
$$

$$
= (1 + c_{1k})^2 x_1^2(t) + (1 + c_{2k})^2 x_2^2(t) + (1 + c_{3k})^2 x_3^2(t) + (1 + c_{4k})^2 x_4^2(t)
$$

$$
\leq V(t, x_1(t), x_2(t), x_3(t), x_4(t)).
$$

Moreover, if $b_1 + b_2 \leq 0$, $\int_r^\infty \lambda(t)dt < \infty$, $\lambda(t) = \max_i \lambda_i(t)$, $a = \min_i a_i$, $i = 1, 2, 3, 4$ and $a - \lambda(t)r > c \geq 0$, then for $t \geq 0$, $t \neq t_k$ and $x = \mathrm{col}(x_1, x_2, x_3, x_4) \in \Omega_1$, we have

$$
D_{(3.17)}^+ V(t, x_1(t), x_2(t), x_3(t), x_4(t))
$$

$$
= 2x_1(t)\Big[-a_1 x_1(t) + \lambda_1(t) \int_{-r}^{0} x_1(t+s)ds + b_1 x_1(t) x_2^2(t) x_3^2(t) \Big]
$$

$$
+ 2x_2(t)\Big[-a_2 x_2(t) + \lambda_2(t) \int_{-r}^{0} x_2(t+s)ds + b_2 x_2(t) x_1^2(t) x_3^2(t) \Big]
$$

$$+ 2x_3(t)\left[-a_3x_3(t) + \lambda_3(t)\int_{-r}^0 x_3(t+s)ds + x_1(t-r)x_2(t-r)x_4(t)\right]$$

$$+ 2x_4(t)\left[-a_4x_4(t) + \lambda_4(t)\int_{-r}^0 x_4(t+s)ds - x_1(t-r)x_2(t-r)x_3(t)\right]$$

$$\leq -2aV(t,x(t)) + 2\lambda_1(t)\int_{-r}^0 x_1(t+s)x_1(t)ds + 2\lambda_2(t)\int_{-r}^0 x_2(t+s)x_2(t)\,ds$$

$$+ 2\lambda_3(t)\int_{-r}^0 x_3(t+s)x_3(t)ds + 2\lambda_4(t)\int_{-r}^0 x_4(t+s)x_4(t)ds$$

$$+ 2(b_1 + b_2)x_1^2(t)x_2^2(t)x_3^2(t)$$

$$\leq -2aV(t,x(t)) + \lambda(t)\int_{-r}^0 \left[V(t+s,x(t+s)) + V(t,x(t))\right]ds$$

$$\leq -2[a - \lambda(t)r]V(t,x(t)) \leq -2cV(t,x(t)).$$

Thus, all conditions of Corollary 3.9 are satisfied and the set M is an asymptotically stable set with respect to system (3.17).

Boundedness with respect to sets

In this part of Section 3.1, we shall apply the direct method of Lyapunov for investigation of boundedness of the solutions of system of the type (3.1) for $\Omega \equiv \mathbb{R}^n$, i.e. we shall consider the system

$$\begin{cases} \dot{x}(t) = f(t,x_t), \ t \neq \tau_k(x(t)) \\ \Delta x(t) = I_k(x(t)), \ t = \tau_k(x(t)), \ k = 1,2,\ldots, \end{cases} \tag{3.18}$$

where $f : [t_0, \infty) \times PC[[-r,0], \mathbb{R}^n] \to \mathbb{R}^n$; $\tau_k : \mathbb{R}^n \to (t_0, \infty)$, $I_k : \mathbb{R}^n \to \mathbb{R}^n$, $k = 1, 2, \ldots$; $\Delta x(t) = x(t+0) - x(t-0)$; and for $t \geq t_0$, $x_t \in PC[[-r,0], \mathbb{R}^n]$ is defined by $x_t(s) = x(t+s)$, $-r \leq s \leq 0$.

Let $\varphi_0 \in PC[[-r,0], \mathbb{R}^n]$. Denote by $x(t) = x(t; t_0, \varphi_0)$ the solution of (3.18), satisfying the initial conditions

$$\begin{cases} x(t; t_0, \varphi_0) = \varphi_0(t - t_0), \ t_0 - r \leq t \leq t_0 \\ x(t_0 + 0; t_0, \varphi_0) = \varphi_0(0), \end{cases} \tag{3.19}$$

and by $J^+(t_0, \varphi_0)$ the maximal interval of type $[t_0, \beta)$ in which the solution $x(t; t_0, \varphi_0)$ is defined.

Let $M \subset [t_0 - r, \infty) \times \mathbb{R}^n$. We shall use the notations of the first part of this section for $\Omega \equiv \mathbb{R}^n$.

Definition 3.11. We say that the solutions of system (3.18) are:

(a) *equi-M-bounded*, if

$$(\forall t_0 \in \mathbb{R})(\forall \eta > 0)(\forall \alpha > 0)(\exists \beta = \beta(t_0, \eta, \alpha) > 0)$$

$$(\forall \varphi_0 \in \overline{S_\alpha(PC_0)} \cap \overline{M_0(t, \eta)})(\forall t \geq t_0): \ x(t; t_0, \varphi_0) \in M(t, \beta);$$

(b) *t- (or α-) uniformly M-bounded*, if the number β from (a) is independent of t_0 (or of α);

(c) *uniformly M-bounded*, if the number β from (a) depends only on η.

   In proof of the main results, we shall use piecewise continuous auxiliary functions $V : [t_0, \infty) \times \mathbb{R}^n \to \mathbb{R}_+$, which belong to the class V_0 and are such that condition H3.4 holds for $\Omega \equiv \mathbb{R}^n$.

Theorem 3.12. *Assume that:*

(1) *Conditions H2.6–H2.13 and H3.1–H3.3 for $\Omega \equiv \mathbb{R}^n$ hold.*

(2) *There exists a function $V \in V_0$ such that H3.4 holds,*

$$V(t, x) \geq a(d(x, M(t))), \ a \in K, \ (t, x) \in [t_0, \infty) \times \mathbb{R}^n, \tag{3.20}$$

where $a(u) \to \infty$ as $u \to \infty$,

$$V(t + 0, x + I_k(x)) \leq V(t, x), \ (t, x) \in \sigma_k, \ k = 1, 2, \ldots,$$

and the inequality

$$D^+_{(3.18)}V(t, x(t)) \leq 0, \ t \neq \tau_k(x(t)), \ k = 1, 2, \ldots$$

is valid for $t \in [t_0, \infty), \ x \in \Omega_1$.

Then the solutions of system (3.18) are equi-M-bounded.

Proof. Let $\alpha > 0, \ \eta > 0$ and $t_0 \in \mathbb{R}$. From the properties of the function V, it follows that there exists a number $k = k(t_0, \eta, \alpha) > 0$ such that if $x \in \overline{S_\alpha} \cap \overline{M(t_0 + 0, \eta)}$, then $V(t_0 + 0, x) \leq k$.
   From the condition $a(u) \to \infty$ as $u \to \infty$, it follows that there exists a number $\beta = \beta(t_0, \eta, \alpha) > 0$ such that $a(\beta) > k$.
   Let $\varphi_0 \in \overline{S_\alpha(PC_0)} \cap \overline{M_0(t, \eta)}$. Then $\varphi_0(0) \in \overline{S_\alpha} \cap \overline{M(t_0 + 0, \eta)}$, hence $V(t_0 + 0, \varphi_0(0)) \leq k$.
   Let $x(t) = x(t; t_0, \varphi_0)$ be the solution of problem (3.18), (3.19). Since the conditions of Corollary 1.25 are met, then

$$V(t, x(t; t_0, \varphi_0)) \leq V(t_0 + 0, \varphi_0(0)), \ t \in [t_0, \infty). \tag{3.21}$$

From (3.20) and (3.21), we obtain

$$a(d(x(t), M(t))) \leq V(t, x(t)) \leq V(t_0 + 0, \varphi_0(0)) \leq k < a(\beta)$$

for $t \in [t_0, \infty)$. This shows that $d(x(t), M(t)) < \beta$ for $t \geq t_0$, hence the solutions of (3.18) are equi-M-bounded.   □

Theorem 3.13. *Let the conditions of Theorem 3.12 hold, and let a function $b \in K$ exist such that*

$$V(t, x) \leq b(d(x, M(t))), \quad (t, x) \in [t_0, \infty) \times \mathbb{R}^n. \tag{3.22}$$

Then the solutions of system (3.18) are uniformly M-bounded.

Proof. Let $\eta > 0$. Choose the number $\beta = \beta(\eta) > 0$ so that $b(\eta) < a(\beta)$, $\beta > \eta$.
   Let $\alpha > 0$, $t_0 \in \mathbb{R}$ and $\varphi_0 \in \overline{S_\alpha}(PC_0) \cap \overline{M_0(t, \eta)}$. Using successively (3.20), (3.21) and (3.22), we obtain

$$a(d(x(t), M(t))) \leq V(t, x(t)) \leq V(t_0 + 0, \varphi_0(0))$$
$$\leq b(d(\varphi_0(0), M_0(t_0))) \leq b(d_0(\varphi_0, M_0(t))) \leq b(\eta) < a(\beta),$$

for $t \in [t_0, \infty)$. Hence, $x(t; t_0, \varphi_0) \in M(t, \beta)$ for $t \geq t_0$.   □

   In analogous way, the following two theorems are proved which supply sufficient conditions for t- (respectively for α-) uniform M-boundedness of the solutions of system (3.18).

Theorem 3.14. *Let the conditions of Theorem 3.12 hold, condition (3.20) being replaced by the condition*

$$a(d(x, M(t))) \leq V(t, x) \leq b(d(x, M(t)), \|x\|) \text{ for } (t, x) \in [t_0, \infty) \times \mathbb{R}^n,$$

where $a \in K$, $a(u) \to \infty$ as $u \to \infty$ and the function $b(\cdot, s) \in K$ for each $s \geq 0$ fixed.
   Then the solutions of system (3.18) are t-uniformly M-bounded.

Theorem 3.15. *Let the conditions of Theorem 3.12 hold, condition (3.20) being replaced by the condition*

$$a(d(x, M(t))) \leq V(t, x) \leq b(t, d(x, M(t))) \text{ for } (t, x) \in [t_0, \infty) \times \mathbb{R}^n,$$

where $a \in K$, $a(u) \to \infty$ as $u \to \infty$ and the function $b(t, \cdot) \in K$ for any $t \geq t_0$ fixed.
   Then the solutions of system (3.18) are α-uniformly M-bounded.

Example 3.16. Let $\tau_0(x, y) \equiv 0$ for $x, y \in \mathbb{R}$. Consider the system:

$$
\begin{cases}
\dot{x}(t) = c(t)y(t) + d(t)x(t)g(\rho^2(t - r)), \ t \neq \tau_k(x(t), y(t)), \ t \geq 0 \\
\dot{y}(t) = -c(t)x(t) + \beta(t)y(t)g(\rho^2(t - r)), \ t \neq \tau_k(x(t), y(t)), \ t \geq 0 \\
\Delta x(t) = c_k x(t), \ \Delta y(t) = d_k y(t), \ t = \tau_k(x(t), y(t)), \ k = 1, 2, \ldots,
\end{cases}
\tag{3.23}
$$

where $x, y \in \mathbb{R}$; $r > 0$; the functions $c(t)$, $d(t)$ and $\beta(t)$ are continuous in $(0, \infty)$; $g(u)$ is a non-negative continuous function; $d(t) \leq 0$; $\beta(t) \leq 0$; $-1 < c_k \leq 0$, $-1 < d_k \leq 0$, $k = 1, 2, \ldots$; and $\rho^2(s) = x^2(s) + y^2(s)$.

Assume that the functions τ_k are such that the conditions H2.9, H2.10, H2.11 and H2.13 for system (3.23) are fulfilled.

Let $M = \{(t, 0, 0) : t \in [-r, \infty)\}$.

The function $V(t, x, y) = x^2 + y^2 = \rho^2$ satisfies the condition of Theorem 3.13. In fact,

$$
\Omega_1 = \left\{ \mathrm{col}(x(t), y(t)) \in PC[\mathbb{R}_+, \mathbb{R}^2] : \rho^2(s) \leq \rho^2(t), \ t - r \leq s \leq t \right\}.
$$

Then for $t = \tau_k(x(t), y(t))$, $k = 1, 2, \ldots$, we have

$$
V(t + 0, x(t) + c_k x(t), y(t) + d_k y(t)) = (1 + c_k)^2 x^2(t) + (1 + d_k)^2 y^2(t)
$$
$$
\leq V(t, x(t), y(t)).
$$

Also, for $t \geq 0$, $t \neq \tau_k(x(t), y(t))$, $k = 1, 2, \ldots$ and $(x, y) \in \Omega_1$, we have

$$
D^+_{(3.23)}V(t, x(t), y(t)) = 2d(t)x^2(t)g(\rho^2(t - r)) + 2\beta(t)y^2(t)g(\rho^2(t - r))
$$
$$
\leq 2d(t)x^2(t)g(\rho^2(t)) + 2\beta(t)y^2(t)g(\rho^2(t)).
$$

Since $g(u) \geq 0$, $d(t) \leq 0$ and $\beta(t) \leq 0$ it follows that

$$
D^+_{(3.23)}V(t, x(t), y(t)) \leq 0, \ t \neq \tau_k(x(t), y(t)), k = 1, 2, \ldots.
$$

Therefore, all conditions of Theorem 3.13 are satisfied and the solutions of system (3.23) are uniformly M-bounded.

As consequences of Theorem 3.12 and Theorem 3.13 we obtain the next two results for the system of impulsive functional differential equations with fixed moments of impulsive perturbations

$$
\begin{cases}
\dot{x}(t) = f(t, x_t), \ t \neq t_k \\
\Delta x(t) = I_k(x(t)), \ t = t_k, \ k = 1, 2, \ldots,
\end{cases}
\tag{3.24}
$$

where $f : [t_0, \infty) \times PC[[-r, 0], \mathbb{R}^n] \to \mathbb{R}^n$; $I_k : \mathbb{R}^n \to \mathbb{R}^n$, $k = 1, 2, \ldots$; $t_0 < t_1 < t_2 < \cdots < t_k < t_{k+1} < \cdots$ and $\lim_{k \to \infty} t_k = \infty$.

Let $\varphi_0 \in PC[[-r, 0], \mathbb{R}^n]$. Denote by $x(t) = x(t; t_0, \varphi_0)$ the solution of system (3.24), satisfying the initial conditions (3.19).

Theorem 3.17. *Assume that:*

(1) *Conditions H1.12, H1.13, H2.6, H2.7, H2.8, H2.12 and H3.1–H3.3 for $\Omega \equiv \mathbb{R}^n$ hold.*

(2) *There exists a function $V \in V_0$ such that H3.4 and (3.20) hold,*

$$V(t + 0, x + I_k(x)) \leq V(t, x), \quad x \in \mathbb{R}^n, \, t = t_k, \, k = 1, 2, \ldots,$$

and the inequality

$$D^+_{(3.24)} V(t, x(t)) \leq 0, \, t \neq t_k, \, k = 1, 2, \ldots$$

is valid for $t \in [t_0, \infty), \, x \in \Omega_1$.

Then the solutions of system (3.24) are equi-M-bounded.

Theorem 3.18. *Let the conditions of Theorem 3.17 hold, and let a function $b \in K$ exist such that*

$$V(t, x) \leq b(d(x, M(t))), \, (t, x) \in [t_0, \infty) \times \mathbb{R}^n.$$

Then the solutions of system (3.24) are uniformly M-bounded.

Global stability of sets

Let $M \subset [t_0 - r, \infty) \times \mathbb{R}^n$. We shall use the notations of the first part of this section for $\Omega = \mathbb{R}^n$ as well as the following definition:

Definition 3.19. The set M is said to be:

(a) *stable* with respect to system (3.18), if

$$(\forall t_0 \in \mathbb{R})(\forall \alpha > 0)(\forall \varepsilon > 0)(\exists \delta = \delta(t_0, \alpha, \varepsilon) > 0)$$

$$(\forall \varphi_0 \in \overline{S_\alpha}(PC_0) \cap M_0(t, \delta))(\forall t \geq t_0) : \quad x(t; t_0, \varphi_0) \in M(t, \varepsilon);$$

(b) *uniformly stable* with respect to system (3.18), if the number δ from point (a) depends only on ε;

(c) *uniformly globally attractive* with respect to system (3.18), if

$$(\forall \eta > 0)(\forall \varepsilon > 0)(\exists \sigma = \sigma(\eta, \varepsilon) > 0)$$

$$(\forall t_0 \in \mathbb{R})(\forall \alpha > 0)(\forall \varphi_0 \in \overline{S_\alpha}(PC_0) \cap \overline{M_0(t, \eta)})$$

$$(\forall t \geq t_0 + \sigma) : \quad x(t; t_0, \varphi_0) \in M(t, \varepsilon);$$

(d) *uniformly globally asymptotically stable* with respect to system (3.18), if M is a uniformly stable and uniformly globally attractive set of system (3.18), and if the solutions of system (3.18) are uniformly M-bounded.

Definition 3.20. Let $\lambda : [t_0, \infty) \to \mathbb{R}_+$ be a measurable function. Then, we say that $\lambda(t)$ is *integrally positive*, if

$$\int_J \lambda(t)dt = \infty$$

whenever $J = \bigcup_{k=1}^{\infty} [\alpha_k, \beta_k]$, $\alpha_k < \beta_k < \alpha_{k+1}$, and $\beta_k - \alpha_k \geq \theta > 0$, $k = 1, 2, \ldots$.

Theorem 3.21. *Assume that:*

(1) *Conditions H2.6–H2.13, and H3.1–H3.3 for $\Omega \equiv \mathbb{R}^n$ hold.*

(2) *There exists a function $V \in V_0$ such that H3.4 holds,*

$$a(d(x, M(t))) \leq V(t, x) \leq b(d(x, M(t))), \quad a, b \in K, \quad (t, x) \in [t_0, \infty) \times \mathbb{R}^n,$$
$$(3.25)$$

where $a(u) \to \infty$ as $u \to \infty$,

$$V(t + 0, x + I_k(x)) \leq V(t, x), \quad (t, x) \in \sigma_k, \quad k = 1, 2, \ldots, \quad (3.26)$$

and the inequality

$$D^+_{(3.18)} V(t, x(t)) \leq -p(t)c(d(x(t), M(t))), \quad t \neq \tau_k(x(t)), \quad k = 1, 2, \ldots \quad (3.27)$$

is valid for $t \in [t_0, \infty)$, $x \in \Omega_1$, $p : [t_0, \infty) \to (0, \infty)$, $c \in K$.

(3) $\int_0^{\infty} p(s)c[b^{-1}(\eta)]\, ds = \infty$ *for each sufficiently small value of $\eta > 0$.*

Then the set M is uniformly globally asymptotically stable with respect to system (3.18).

Proof. Let $\varepsilon > 0$. Choose $\delta = \delta(\varepsilon) > 0$, $\delta < \varepsilon$ so that $b(\delta) < a(\varepsilon)$.

Let $\alpha > 0$ be arbitrary, $\varphi_0 \in \overline{S_\alpha}(PC_0) \cap M_0(t, \delta)$ and $x(t) = x(t; t_0, \varphi_0)$.

From condition (2) of Theorem 3.21, it follows that for $t \in J^+(t_0, \varphi_0)$ the following inequalities are valid:

$$a(d(x(t; t_0, \varphi_0), M(t))) \leq V(t, x(t)) \leq V(t_0 + 0, \varphi_0(0))$$
$$\leq b(d(\varphi_0(0), M_0(t_0))) \leq b(d_0(\varphi_0, M_0(t))) < b(\delta) < a(\varepsilon).$$

Since $J^+(t_0, \varphi_0) = [t_0, \infty)$, then $x(t) \in M(t, \varepsilon)$ for all $t \geq t_0$. Thus, it is proved that the set M is uniformly stable with respect to system (3.18).

Now let $\eta > 0$ and $\varepsilon > 0$ be given and let the number $\sigma = \sigma(\eta, \varepsilon) > 0$ be chosen so that

$$\int_{t_0}^{t_0+\sigma} p(s)c\left[b^{-1}\left(\frac{a(\varepsilon)}{2}\right)\right] ds > b(\eta). \quad (3.28)$$

(This is possible in view of condition (3) of Theorem 3.21.)

Let $\alpha > 0$ be arbitrary, $\varphi_0 \in \overline{S_\alpha}(PC_0) \cap \overline{M_0(t, \eta)}$ and $x(t) = x(t; t_0, \varphi_0)$.

Assume that for any $t \in [t_0, t_0 + \sigma]$ the following inequality holds:

$$d(x(t), M(t)) \geq b^{-1}\left(\frac{a(\varepsilon)}{2}\right). \qquad (3.29)$$

Then, by (3.27), it follows that

$$\int_{t_0}^{t_0+\sigma} D^+_{(3.18)} V(s, x(s)) \, ds \leq - \int_{t_0}^{t_0+\sigma} p(s)c\left[b^{-1}\left(\frac{a(\varepsilon)}{2}\right)\right] ds < -b(\eta). \qquad (3.30)$$

On the other hand, if $t_0 + \sigma \in (t_r, t_{r+1}]$ for some r, then we obtain

$$\int_{t_0}^{t_0+\sigma} D^+_{(3.18)} V(s, x(s)) \, ds$$

$$= \sum_{k=1}^{r} \int_{t_{k-1}}^{t_k} D^+_{(3.18)} V(s, x(s)) \, ds + \int_{t_r}^{t_0+\sigma} D^+_{(3.18)} V(s, x(s)) \, ds$$

$$= \sum_{k=1}^{r} [V(t_k, x(t_k)) - V(t_{k-1} + 0, x(t_{k-1} + 0))] + V(t_0 + \sigma, x(t_0 + \sigma))$$

$$- V(t_r + 0, x(t_r + 0))$$

$$\geq V(t_0 + \sigma, x(t_0 + \sigma)) - V(t_0 + 0, \varphi_0(0)),$$

whence, in view of (3.30), it follows that $V(t_0 + \sigma, x(t_0 + \sigma)) < 0$, which contradicts (3.25).

The contradiction obtained shows that there exists $t^* \in [t_0, t_0 + \sigma]$, such that

$$d(x(t^*), M(t^*)) < b^{-1}\left(\frac{a(\varepsilon)}{2}\right). \qquad (3.31)$$

Then for $t \geq t^*$ (hence for any $t \geq t_0 + \sigma$ as well) the following inequalities are valid:

$$a(d(x(t), M(t))) \leq V(t, x(t)) \leq V(t^*, x(t^*))$$

$$\leq b(d(x(t^*), M(t^*))) < \frac{a(\varepsilon)}{2} < a(\varepsilon).$$

Hence, $x(t) \in M(t, \varepsilon)$ for $t \geq t_0 + \sigma$, i.e. the set M is uniformly globally attractive with respect to system (3.18).

Finally, since the conditions of Theorem 3.13 are met, then the solutions of system (3.18) are uniformly M-bounded. $\square$

Theorem 3.22. *Assume that:*

(1) *Condition (1) of Theorem 3.21 holds.*

(2) *There exists a function $V \in V_0$ such that* H3.4, *(3.25) and (3.26) hold, and the inequality*

$$D^+_{(3.18)} V(t, x(t)) \leq -\lambda(t) c(d(x(t), M(t))), \; t \neq \tau_k(x(t)), \; k = 1, 2, \ldots \quad (3.32)$$

is valid for $t \in [t_0, \infty)$, $x \in \Omega_1$, $c \in K$, where $\lambda(t)$ is an integrally positive function.

Then the set M is uniformly globally asymptotically stable with respect to system (3.18).

Proof. The fact that the set M is uniformly stable with respect to system (3.18) is proved as in the proof of Theorem 3.21, and the uniform M-boundedness of the solutions of system (3.18) follows from Theorem 3.13.

Now, we shall prove that the set M is uniformly globally attractive with respect to the system (3.18).

Let again $\varepsilon > 0$ and $\eta > 0$ be given. Choose the number $\delta = \delta(\varepsilon) > 0$ so that $b(\delta) < a(\varepsilon)$.

We shall prove that there exists $\sigma = \sigma(\varepsilon, \eta) > 0$ such that for any solution $x(t) = x(t; t_0, \varphi_0)$ of system (3.18) for which $t_0 \in \mathbb{R}$, $\varphi_0 \in \overline{S_\alpha}(PC_0) \cap M_0(t, \eta)$ ($\alpha > 0$ arbitrary) and for any $t^* \in [t_0, t_0 + \sigma]$ the following inequality is valid:

$$d(x(t^*), M(t^*)) < \delta(\varepsilon). \quad (3.33)$$

Suppose that this is not true. Then, for any $\sigma > 0$ there exists a solution $x(t) = x(t; t_0, \varphi_0)$ of system (3.18) for which $t_0 \in \mathbb{R}$, $\varphi_0 \in \overline{S_\alpha}(PC_0) \cap \overline{M_0(t, \eta)}$, $\alpha > 0$, such that

$$d(x(t), M(t)) \geq \delta(\varepsilon), \; t \in [t_0, t_0 + \sigma]. \quad (3.34)$$

From (3.32) and (3.26), it follows that

$$V(t, x(t)) - V(t_0 + 0, \varphi_0(0)) \leq \int_{t_0}^{t} D^+_{(3.18)} V(s, x(s)) \, ds$$

$$\leq -\int_{t_0}^{t} \lambda(s) c(d(x(s), M(s))) \, ds, \; t \geq t_0. \quad (3.35)$$

From the properties of the function $V(t, x(t))$ in the interval $[t_0, \infty)$, it follows that there exists the finite limit

$$\lim_{t \to \infty} V(t, x(t)) = v_0 \geq 0. \quad (3.36)$$

Then from (3.32), (3.34), (3.35) and (3.36), it follows that

$$\int_{t_0}^{\infty} \lambda(t) c(d(x(t), M(t))) \, dt \leq b(\eta) - v_0.$$

From the integral positivity of the function $\lambda(t)$, it follows that the number σ can be chosen so that

$$\int_{t_0}^{t_0+\sigma} \lambda(t)\,dt > \frac{b(\eta) - v_0 + 1}{c(\delta(\varepsilon))}.$$

Then, we obtain

$$b(\eta) - v_0 \geq \int_{t_0}^{\infty} \lambda(t)c(d(x(t), M(t)))\,dt$$

$$\geq \int_{t_0}^{t_0+\sigma} \lambda(t)c(d(x(t), M(t)))\,dt$$

$$\geq c(\delta(\varepsilon)) \int_{t_0}^{t_0+\sigma} \lambda(t)\,dt > b(\eta) - v_0 + 1.$$

The contradiction obtained shows that there exists a positive constant $\sigma = \sigma(\varepsilon, \eta)$ such that for any solution $x(t) = x(t; t_0, \varphi_0)$ of system (3.18) for which $t_0 \in \mathbb{R}$, $\varphi_0 \in \overline{S_\alpha}(PC_0) \cap \overline{M_0(t, \eta)}$, $\alpha > 0$, there exists $t^* \in [t_0, t_0 + \sigma]$ such that inequality (3.33) holds.

Then for $t \geq t^*$ (hence for any $t \geq t_0 + \sigma$ as well) the following inequalities are valid:

$$a(d(x(t), M(t))) \leq V(t, x(t)) \leq V(t^*, x(t^*))$$
$$\leq b(d(x(t^*), M(t^*))) < b(\delta) < a(\varepsilon),$$

which proves that the set M is uniformly globally attractive with respect to system (3.18).   □

We shall use Theorem 3.22 to prove the global uniform asymptotic stability of a set with respect to the system

$$\begin{cases} \dot{x}(t) = \begin{cases} A(t)x(t) + B(t)x(t - h(t)), \ x(t) > 0, \ t \neq \tau_k(x(t)) \\ 0, \quad x(t) \leq 0, \ t \neq \tau_k(x(t)) \end{cases} \\ \Delta x(t) = \begin{cases} C_k x(t), \ x(t) > 0, \ t = \tau_k(x(t)) \\ 0, \quad x(t) \leq 0, \ t = \tau_k(x(t)), \end{cases} \end{cases} \quad (3.37)$$

where $t \geq t_0$; $x \in PC[[t_0, \infty), \mathbb{R}^n]$; $A(t)$ and $B(t)$ are $(n \times n)$-matrix-valued functions, $C_k, k = 1, 2, \ldots$ are $(n \times n)$-matrices; $h \in C[[t_0, \infty), \mathbb{R}_+]$.

Such systems seem to have application, among other things, in the study of active suspension height control. In the interest of improving the overall performance of automotive vehicles, in recent years, suspensions, incorporating active components have been developed. The designs may cover a spectrum of performance capabilities, but the active components alter only the vertical force reactions of the suspensions, not the kinematics. The conventional passive suspensions consist of usual components with spring and damping properties, which are time-invariant. The interest in active or semi-active suspensions derives from the potential for improvements to vehicle ridden performance with no compromise or enhancement in handling. The full active suspensions incorporate actuators to generate the desired forces in the suspension. The actuators are normally hydraulic cylinders [205].

Let $\tau = \inf_{t \geq t_0}(t - h(t))$ and $\varphi_1 \in C[[\tau, t_0], \mathbb{R}^n]$. Denote by $x(t) = x(t; t_0, \varphi_1)$ the solution of system (3.37), satisfying the initial conditions

$$x(s; t_0, \varphi_1) = \varphi_1(s), \quad \tau \leq s \leq t_0,$$

$$x(t_0 + 0) = \varphi_1(t_0).$$

Theorem 3.23. *Assume that:*

(1) *Conditions H2.9, H2.10, H2.11 and H2.13 hold for the system (3.37).*

(2) *The matrix functions $A(t)$ and $B(t)$ are continuous for $t \in [t_0, \infty)$.*

(3) *$t - h(t) \to \infty$ as $t \to \infty$.*

(4) *For each $k = 1, 2, \ldots$ the elements of the matrix C_k are non-negative.*

(5) *There exists a continuous real $(n \times n)$-matrix $D(t)$, $t \in [t_0, \infty)$, which is symmetric, positive definite, differentiable for $t \neq \tau_k(x(t))$, $k = 1, 2, \ldots$ and such that for each $k = 1, 2, \ldots$*

$$x^T[A^T(t)D(t) + D(t)A(t) + \dot{D}(t)]x \leq -c(t)\|x\|^2, \quad x \in \mathbb{R}^n, \quad t \neq \tau_k(x(t)), \tag{3.38}$$

$$x^T[C_k^T D(t) + D(t)C_k + C_k^T D(t)C_k]x \leq 0, \quad x \in \mathbb{R}^n, \quad t = \tau_k(x(t)), \tag{3.39}$$

where $c(t) > 0$ is a continuous function.

(6) *There exists an integrally positive function $\lambda(t)$ such that for $t \geq t_0$*

$$d(t) = c(t) - \max\{\alpha(t)\lambda(t), \beta(t)\lambda(t)\} \geq 0, \tag{3.40}$$

$$\frac{2\beta^{1/2}(t)}{\alpha^{1/2}(t - h(t))}\|D(t)B(t)\| \leq d(t), \tag{3.41}$$

where $\alpha(t)$ and $\beta(t)$ are, respectively, the smallest and the greatest eigenvalues of matrix $D(t)$.

Then the set $M = [\tau - t_0, \infty) \times \{x \in \mathbb{R}^n : x \leq 0\}$ is uniformly globally asymptotically stable with respect to system (3.37).

Proof. Consider the function

$$V(t, x) = \begin{cases} x^T D(t)x, & x > 0 \\ 0, & x \le 0. \end{cases}$$

From the condition that $D(t)$ is a real symmetric matrix, it follows that for $x \in \mathbb{R}^n$, $x \ne 0$ the following inequalities hold:

$$\alpha(t)\|x\|^2 \le x^T D(t)x \le \beta(t)\|x\|^2. \tag{3.42}$$

From the last inequalities, it follows that condition (3.25) is satisfied.

For the chosen function $V(t, x)$ the set Ω_1 is

$$\Omega_1 = \left\{ x \in PC[[t_0, \infty), \mathbb{R}^n] : x^T(s)D(s)x(s) \le x^T(t)D(t)x(t), \ \tau \le s \le t \right\}.$$

For $t \ge t_0$ and $x \in \Omega_1$ the following inequalities are valid:

$$\alpha(t - h(t))\|x(t - h(t))\|^2 \le x^T(t - h(t))D(t - h(t))x(t - h(t))$$
$$\le x^T(t)D(t)x(t) \le \beta(t)\|x(t)\|^2,$$

from which we obtain the estimate

$$\|x(t - h(t))\| \le \frac{\beta^{1/2}(t)}{\alpha^{1/2}(t - h(t))}\|x(t)\|. \tag{3.43}$$

Let $t \ne \tau_k(x(t))$, $k = 1, 2, \ldots$, and $x \in \Omega_1$. From (3.38), (3.40)–(3.43), we have

$$D^+_{(3.37)}V(t, x(t)) - \begin{cases} -c(t)\|x(t)\|^2 + 2\|D(t)B(t)\|\|x(t)\|\|x(t - h(t))\|, & x(t) > 0, \\ 0, & x(t) \le 0 \end{cases}$$

$$\le \begin{cases} -[c(t) - d(t)]\|x(t)\|^2, & x(t) > 0, \\ 0, & x(t) \le 0 \end{cases}$$

$$\le -\lambda(t)V(t, x(t)).$$

Let $t = \tau_k(x(t))$, $k = 1, 2, \ldots$. Then from (3.39), we have

$$V(t + 0, x(t) + C_k x(t)) = \begin{cases} (x^T(t) + x^T(t)C_k^T)D(t)(x(t) + C_k x(t)), & x(t) > 0, \\ 0, & x(t) \le 0 \end{cases}$$

$$= \begin{cases} x^T(t)D(t)x(t) + x^T(t)[C_k^T D(t) + D(t)C_k + C_k^T D(t)C_k]x(t), & x(t) > 0, \\ 0, & x(t) \le 0 \end{cases}$$

$$\le V(t, x(t)).$$

Thus, we have checked that all the conditions of Theorem 3.22 are satisfied.

Hence, the set $M = [\tau - t_0, \infty) \times \{x \in \mathbb{R}^n : x \le 0\}$ is uniformly globally asymptotically stable with respect to system (3.37). □

As consequences of Theorem 3.21 and Theorem 3.22, we obtain the next two results for the system (3.24).

Theorem 3.24. *Assume that:*

(1) *Conditions H1.12, H1.13, H2.6, H2.7, H2.8, H2.12, and H3.1–H3.3 for $\Omega \equiv \mathbb{R}^n$ hold.*

(2) *There exists a function $V \in V_0$ such that H3.4 and (3.25) hold,*

$$V(t+0, x+I_k(x)) \leq V(t,x), \ x \in \mathbb{R}^n, \ t = t_k, \ k = 1,2,\ldots, \tag{3.44}$$

and the inequality

$$D^+_{(3.24)}V(t,x(t)) \leq -p(t)c(d(x(t), M(t))), \ t \neq t_k, \ k = 1,2,\ldots$$

is valid for $t \in [t_0, \infty)$, $x \in \Omega_1$, $p : [t_0, \infty) \to (0, \infty)$, $c \in K$.

(3) $\displaystyle\int_0^\infty p(s)c[b^{-1}(\eta)]ds = \infty$ *for each sufficiently small value of $\eta > 0$.*

Then the set M is uniformly globally asymptotically stable with respect to system (3.24).

Theorem 3.25. *Assume that:*

(1) *Condition (1) of Theorem 3.24 holds.*

(2) *There exists a function $V \in V_0$ such that H3.4, (3.25) and (3.44) hold, and the inequality*

$$D^+_{(3.24)}V(t,x(t)) \leq -\lambda(t)c(d(x(t), M(t))), \ t \neq t_k, \ k = 1,2,\ldots$$

is valid for $t \in [t_0, \infty)$, $x \in \Omega_1$, $c \in K$, where $\lambda(t)$ is an integrally positive function.

Then the set M is uniformly globally asymptotically stable with respect to system (3.24).

Example 3.26. Consider the system

$$\begin{cases} \dot{x}(t) = a(t)x(t)[x^2(t-r) + y^2(t-r)] - b(t)x(t)[x^2(t) + 2y^2(t)], \ t \neq t_k \\ \dot{y}(t) = a(t)y(t)[x^2(t-r) + y^2(t-r)] - b(t)y^3(t), \ t \neq t_k \\ \Delta x(t_k) = c_k x(t_k), \ \Delta y(t_k) = d_k y(t_k), \end{cases}$$

$$\tag{3.45}$$

where $t \geq 0$; $x, y \in \mathbb{R}$; $r > 0$; $a, b \in C[\mathbb{R}_+, (0, \infty)]$; $-1 < c_k \leq 0$, $-1 < d_k \leq 0$ for $k = 1,2,\ldots$; $0 < t_1 < t_2 < \cdots$ and $\lim_{k\to\infty} t_k = \infty$.
   Let

$$X(t) = \text{col}(x(t), y(t)) = \varphi_1(t), \ t \in [-r, 0], \quad \varphi_1 \in C[[-r, 0], \mathbb{R}^2].$$

Consider the functions $V(t, x, y) = x^2 + y^2$. Then, the set

$$\Omega_1 = \{X = \mathrm{col}(x(t), y(t)) \in PC[\mathbb{R}_+, \mathbb{R}^2] :$$
$$x^2(s) + y^2(s) \le x^2(t) + y^2(t), \ t - r \le s \le t\}.$$

Let $M = \{(t, 0, 0) : t \in [-r, \infty)\}$.
For $t \ge 0$ and $X \in \Omega_1$, we have

$$D^+_{(3.45)} V(t, x(t), y(t))$$
$$= 2\{a(t)[x^2(t) + y^2(t)][x^2(t - r) + y^2(t - r)] - b(t)[x^2(t) + y^2(t)]^2\}$$
$$\le 2(a(t) - b(t))[x^2(t) + y^2(t)]^2, \ t \ne t_k, \ k = 1, 2, \dots .$$

Also, for $k = 1, 2, \dots$, we obtain

$$V(t_k + 0, x(t_k) + c_k x(t_k), y(t_k) + d_k y(t_k))$$
$$= (1 + c_k)^2 x^2(t_k) + (1 + d_k)^2 y^2(t_k) \le V(t_k, x(t_k), y(t_k)).$$

If $a(t) - b(t) = p(t) < 0$ for $t \ge 0$, then all the conditions of Theorem 3.24 are satisfied. Hence, the set M is an uniformly globally asymptotically stable set with respect to system (3.45).

3.2 Conditional stability

Let $t_0 \in \mathbb{R}$, $r > 0$. Let $\|x\| = |x_1| + |x_2| + \cdots + |x_n|$ be the norm of $x \in \mathbb{R}^n$. Consider the system (3.24). Let $\varphi_0 \in PC[[-r, 0], \mathbb{R}^n]$. Denote by $x(t) = x(t; t_0, \varphi_0)$ the solution of system (3.24), satisfying the initial conditions (3.19).

Let $M(n - l)$, $l < n$ be a $(n - l)$-dimensional manifold in $\mathbb{R}^n$, containing the origin. We set

$$M_0(n - l) = \{\varphi : \varphi \in PC[[-r, 0], M(n - l)]\}.$$

We shall give the following definitions of conditional stability of the zero solution of system (3.24) with respect to the manifold $M(n - l)$.

Definition 3.27. The zero solution of system (3.24) is said to be:

(a) *conditionally stable* with respect to the manifold $M(n - l)$, if

$$(\forall t_0 \in \mathbb{R})(\forall \varepsilon > 0)(\exists \delta = \delta(t_0, \varepsilon) > 0)$$

$$(\forall \varphi_0 \in \overline{S}_\delta(PC_0) \cap M_0(n - l))(\forall t \ge t_0) : \ x(t; t_0, \varphi_0) \in S_\varepsilon;$$

(b) *conditionally uniformly stable* with respect to $M(n - l)$, if the function δ in (a) is independent of t_0;

(c) *conditionally globally equi-attractive* with respect to $M(n - l)$, if

$$(\forall t_0 \in \mathbb{R})(\forall \alpha > 0)(\forall \varepsilon > 0)(\exists T = T(t_0, \alpha, \varepsilon) > 0)$$

$$(\forall \varphi_0 \in \overline{S}_\alpha(PC_0) \cap M_0(n - l))(\forall t \geq t_0 + T) : \ x(t; t_0, \varphi_0) \in S_\varepsilon;$$

(d) *conditionally uniformly globally attractive* with respect to $M(n-l)$, if the number T in (c) is independent of t_0;

(e) *conditionally globally equi-asymptotically stable* with respect to $M(n - l)$, if it is conditionally stable and conditionally globally equi-attractive with respect to $M(n - l)$;

(f) *conditionally uniformly globally asymptotically stable* with respect to $M(n-l)$, if it is conditionally uniformly stable and conditionally uniformly globally attractive with respect to $M(n - l)$;

(g) *conditionally unstable* with respect to the manifold $M(n - l)$, if (a) fails to hold.

Remark 3.28. If $M(n - l) = \mathbb{R}^n$, then the definitions (a)–(g) are reduced to the usual definitions of stability by Lyapunov for the zero solution of system (3.24).

Together with the system (3.24), we shall consider the following system of impulsive ordinary differential equations:

$$\begin{cases} \dot{u}(t) = D(t)u(t), \ t \neq t_k, \ t \geq t_0 \\ \Delta u(t_k) = D_k u(t_k), \ k = 1, 2, \ldots, t_k > t_0, \end{cases} \tag{3.46}$$

where $u : \ [t_0, \infty) \to \mathbb{R}_+^m$; $D(t)$ is an $(m \times m)$-matrix valued function; $D_k, \ k = 1, 2, \ldots$ are $(m \times m)$-constant matrices.

Let $u_0 \in \mathbb{R}_+^m$. We denote by $u(t) = u(t; t_0, u_0)$ the solution of system (3.46), which satisfies the initial condition $u(t_0) = u_0$, and by $J^+(t_0, u_0)$ the maximal interval of type $[t_0, \beta)$ in which the solution $u(t; t_0, u_0)$ is defined.

Let $e \in \mathbb{R}_+^m$ be the vector $(1, 1, \ldots, 1)$. We introduce the sets:

$$B(\alpha) = \left\{ u \in \mathbb{R}_+^m : 0 \leq u < \alpha e \right\},$$

$$\overline{B}(\alpha) = \left\{ u \in \mathbb{R}_+^m : 0 \leq u \leq \alpha e \right\}, \ \alpha = \text{const} > 0,$$

$$R(m - l) = \left\{ u = (u_1, \ldots, u_m) \in \mathbb{R}^m : u_1 = u_2 = \cdots = u_l = 0, \right\}, \ l < m.$$

Introduce the following conditions:

H3.5. The matrix-valued $(m \times m)$-function $D(t)$ is continuous for $t \in [t_0, \infty)$.

H3.6. The functions $\psi_k : \mathbb{R}^m_+ \to \mathbb{R}^m_+$, $\psi_k(u) = u + D_k u$, $k = 1, 2, \ldots$, are non-decreasing in $\mathbb{R}^m_+$.

H3.7. $J^+(t_0, u_0) = [t_0, \infty)$.

We shall consider such solutions $u(t)$ of the system (3.46) for which $u(t) \geq 0$. That is why the following definitions on conditional stability of the zero solution of this system will be used.

Definition 3.29. The zero solution of system (3.46) is said to be:

(a) *conditionally stable* with respect to the manifold $R(m - l)$, if

$$(\forall t_0 \in \mathbb{R})(\forall \varepsilon > 0)(\exists \delta = \delta(t_0, \varepsilon) > 0)$$

$$(\forall u_0 \in \overline{B}(\delta) \cap R(m - l))(\forall t \geq t_0) : u^+(t; t_0, u_0) \in B(\varepsilon);$$

(b) *conditionally uniformly stable* with respect to $R(m - l)$, if the function δ from (a) does not depend on t_0;

(c) *conditionally globally equi-attractive* with respect to $R(m - l)$, if

$$(\forall t_0 \in \mathbb{R})(\forall \alpha > 0)(\forall \varepsilon > 0)(\exists T = T(t_0, \alpha, \varepsilon) > 0)$$

$$(\forall u_0 \in \overline{B}(\alpha) \cap R(m - l))(\forall t \geq t_0 + T) : u^+(t; t_0, u_0) \in B(\varepsilon);$$

(d) *conditionally uniformly globally attractive* with respect to $R(m-l)$, if the number T in (c) does not depend on t_0;

(e) *conditionally globally equi-asymptotically stable* with respect to $R(m - l)$, if it is conditionally stable and conditionally globally equi-attractive with respect to $R(m - l)$;

(f) *conditionally uniformly globally asymptotically stable* with respect to $R(m-l)$, if it is conditionally uniformly stable and conditionally uniformly globally attractive with respect to $R(m - l)$;

(g) *conditionally unstable* with respect to the manifold $R(m - l)$, if (a) fails to hold.

In the successive investigations, we shall use piecewise continuous auxiliary vector functions $V : [t_0, \infty) \times \mathbb{R}^n \to \mathbb{R}^m_+$, $V = \mathrm{col}(V_1, \ldots, V_m)$ such that $V_j \in V_0$, $j = 1, 2, \ldots, m$.

Theorem 3.30. *Assume that:*

(1) *Conditions H1.12, H1.13, H2.2, H2.3, H2.6–H2.8, H2.18, H3.5, H3.6 and H3.7 hold.*

(2) *There exists a function $V : [t_0, \infty) \times \mathbb{R}^n \to \mathbb{R}^m_+$, $m \leq n$, $V = \text{col}(V_1, \ldots, V_m)$, $V_j \in V_0$, $j = 1, 2, \ldots, m$ such that $\sup_{[t_0, \infty) \times \mathbb{R}^n} \|V(t, x)\| = K \leq \infty$,*

$$V(t, 0) = 0, \ t \geq t_0,$$

$$a(\|x\|)e \leq V(t, x), \ a \in K, \ (t, x) \in [t_0, \infty) \times \mathbb{R}^n, \tag{3.47}$$

$$V(t + 0, x + I_k(x)) \leq \psi_k(V(t, x)), \ x \in \mathbb{R}^n, \ t = t_k, \ k = 1, 2, \ldots,$$

and the inequality

$$D^+_{(3.24)}V(t, x(t)) \leq D(t)V(t, x(t)), \ t \neq t_k, \ k = 1, 2, \ldots$$

is valid for $t \geq t_0$ and $x \in \Omega_1$.

(3) *The set $M(n - l) = \{x \in \mathbb{R}^n : V_k(t + 0, x) \equiv 0, \ k = 1, 2, \ldots, l\}$ is an $(n - l)$-dimensional manifold in $\mathbb{R}^n$, containing the origin, $l < n$.*

Then:

(1) *If the zero solution of system (3.46) is conditionally stable with respect to the manifold $R(m - l)$, then the zero solution of system (3.24) is conditionally stable with respect to the manifold $M(n - l)$.*

(2) *If the zero solution of system (3.46) is conditionally globally equi-attractive with respect to the manifold $R(m - l)$, then the zero solution of system (3.24) is conditionally globally equi-attractive with respect to the manifold $M(n - l)$.*

Proof of Assertion 1. Let $t_0 \in \mathbb{R}$ and $\varepsilon > 0$ $(a(\varepsilon) < K)$ be given. Let the zero solution of system (3.46) be conditionally stable with respect to $R(m - l)$. Then, there exists a positive function $\delta_1 = \delta_1(t_0, \varepsilon)$ which is continuous in t_0 for given ε and is such that, if $u_0 \in \overline{B}(\delta_1) \cap R(m - l)$, then $u^+(t; t_0, u_0) < a(\varepsilon)e$ for $t \geq t_0$.

It follows, from the properties of the function V, that there exists $\delta = \delta(t_0, \varepsilon) > 0$ such that if $x \in \overline{S}_\delta$ then $V(t_0 + 0, x) \in \overline{B}(\delta_1)$.

Let $\varphi_0 \in \overline{S}_\delta(PC_0) \cap M_0(n - l)$. Then $\varphi_0(0) \in \overline{S}_\delta$ and therefore, $V(t_0 + 0, \varphi_0(0)) \in \overline{B}(\delta_1)$. Moreover, $V_k(t_0 + 0, \varphi_0(0)) = 0$ for $k = 1, 2, \ldots, l$, i.e. $V(t_0 + 0, \varphi_0(0)) \in R(m - l)$. Thus,

$$u^+(t; t_0, V(t_0 + 0, \varphi_0(0))) < a(\varepsilon)e, \ t \geq t_0. \tag{3.48}$$

Let $x(t) = x(t; t_0, \varphi_0)$ be the solution of the initial value problem (3.24), (3.19). Then, the function V satisfies all conditions of Theorem 1.23 for $u_0 = V(t_0 + 0, \varphi_0(0))$ and by (3.47) and (3.48), we arrive at

$$a(\|x(t)\|)e \leq V(t, x(t)) \leq u^+(t; t_0, V(t_0 + 0, \varphi_0(0))) < a(\varepsilon)e$$

for $t \geq t_0$. Hence, $x(t; t_0, \varphi_0) \in S_\varepsilon$ for $t \geq t_0$, i.e. the zero solution of system (3.24) is conditionally stable with respect to the manifold $M(n - l)$.

Proof of Assertion 2. Let $t_0 \in \mathbb{R}$, $\alpha > 0$ and $\varepsilon > 0$ $(a(\varepsilon) < K)$ be given.

It follows, from the properties of the function V, that there exists $\alpha_1 = \alpha_1(t_0, \alpha) > 0$ such that if $x \in \overline{S}_\alpha$, then $V(t_0 + 0, x) \in \overline{B}(\alpha_1)$.

If the zero solution of system (3.46) is conditionally globally equi-attractive with respect to $R(m - l)$, then there exists a number $T = T(t_0, \alpha_1, \varepsilon) > 0$ such that if $u_0 \in \overline{B}(\alpha_1) \cap R(m - l)$, then $u^+(t; t_0, u_0) < a(\varepsilon)e$ for $t \geq t_0 + T$.

Let $\varphi_0 \in \overline{S}_\alpha(PC_0) \cap M_0(n - l)$. Then $\varphi_0(0) \in \overline{S}_\alpha$ and $V(t_0 + 0, \varphi_0(0)) \in \overline{B}(\alpha_1) \cap R(m - l)$. Therefore,

$$u^+(t; t_0, V(t_0 + 0, \varphi_0(0))) < a(\varepsilon)e, \ t \geq t_0 + T. \tag{3.49}$$

If $x(t) = x(t; t_0, \varphi_0)$ is the solution of the initial value problem (3.24), (3.19), then it follows from Theorem 1.23 that

$$V(t, x(t)) \leq u^+(t; t_0, V(t_0 + 0, \varphi_0(0))), \ t \geq t_0.$$

The last inequality, (3.47) and (3.49) imply the inequalities

$$a(\|x(t)\|)e \leq V(t, x(t)) \leq u^+(t; t_0, V(t_0 + 0, \varphi_0(0))) < a(\varepsilon)e$$

for $t \geq t_0 + T$.

Therefore, $\|x(t; t_0, \psi_0)\| < \varepsilon$ for $t \geq t_0 + T$, that leads to the conclusion that the zero solution of system (3.24) is conditionally globally equi-attractive with respect to the manifold $M(n - l)$. □

Corollary 3.31. *Let the conditions of Theorem 3.30 be fulfilled.*

Then conditional global equi-asymptotic stability of the zero solution of system (3.46) with respect to the manifold $R(m - l)$ implies the conditional global equi-asymptotic stability of the zero solution of system (3.24) with respect to the manifold $M(n - l)$.

Theorem 3.32. *Let the conditions of Theorem 3.30 be fulfilled, and let a function $b \in K$ exist such that $V(t, x) \leq b(\|x\|)e$ for $(t, x) \in [t_0, \infty) \times \mathbb{R}^n$.*

Then:

(1) *If the zero solution of system (3.46) is conditionally uniformly stable with respect to the manifold $R(m - l)$, then the zero solution of system (3.24) is conditionally uniformly stable with respect to the manifold $M(n - l)$.*

(2) *If the zero solution of system (3.46) is conditionally uniformly globally attractive with respect to the manifold $R(m - l)$, then the zero solution of system (3.24) is conditionally uniformly globally attractive with respect to the manifold $M(n - l)$.*

The proof of Theorem 3.32 is analogous to the proof of Theorem 3.30. We shall note that in this case the function δ and the number T can be chosen independently of t_0.

Corollary 3.33. *Let the conditions of Theorem 3.32 be satisfied.*

Then conditional uniform global asymptotic stability of the zero solution of system (3.46) *with respect to the manifold* $R(m-l)$ *implies the conditional uniform global asymptotic stability of the zero solution of system* (3.24) *with respect to the manifold* $M(n-l)$.

Example 3.34. We shall apply Theorem 3.30 to the system

$$
\begin{cases}
\dot{x}(t) = (1 + t^2)x(t - r(t)) + (1 - t^2)y(t - r(t)) \\
\quad + (t^2 - 1)z(t - r(t)), \ t \neq t_k \\
\dot{y}(t) = (1 - e^{-t})x(t - r(t)) + (1 + e^{-t})y(t - r(t)) \\
\quad + (e^{-t} - 1)z(t - r(t)), \ t \neq t_k \\
\dot{z}(t) = (t^2 - e^{-t})x(t - r(t)) + (e^{-t} - t^2)y(t - r(t)) \quad (3.50) \\
\quad + (e^{-t} + t^2)z(t - r(t)), \ t \neq t_k \\
\Delta x(t_k) = a_{1k}x(t_k) + b_{1k}[y(t_k) - z(t_k)], \ k = 1, 2, \ldots \\
\Delta y(t_k) = a_{2k}y(t_k) + b_{2k}[z(t_k) - x(t_k)], \ k = 1, 2, \ldots \\
\Delta z(t_k) = a_{3k}z(t_k) + b_{3k}[x(t_k) - y(t_k)], \ k = 1, 2, \ldots,
\end{cases}
$$

where $t \geq 0; 0 \leq r(t) \leq r$;

$$
a_{1k} = \frac{1}{2}\left(\sqrt{1 + d_{1k}} + \sqrt{1 + d_{3k}} - 2\right),
$$

$$
a_{2k} = \frac{1}{2}\left(\sqrt{1 + d_{2k}} + \sqrt{1 + d_{1k}} - 2\right),
$$

$$
a_{3k} = \frac{1}{2}\left(\sqrt{1 + d_{3k}} + \sqrt{1 + d_{2k}} - 2\right);
$$

$$
b_{1k} = \frac{1}{2}\left(\sqrt{1 + d_{1k}} - \sqrt{1 + d_{3k}}\right),
$$

$$
b_{2k} = \frac{1}{2}\left(\sqrt{1 + d_{2k}} - \sqrt{1 + d_{1k}}\right),
$$

$$
b_{3k} = \frac{1}{2}\left(\sqrt{1 + d_{3k}} - \sqrt{1 + d_{2k}}\right);
$$

$-1 < d_{ik} \leq 0, i = 1, 2, 3, k = 1, 2, \ldots; 0 < t_1 < t_2 < \cdots$ and $\lim_{k \to \infty} t_k = \infty$.
Consider the manifold $M(2) = \{\text{col}(x, y, z) \in \mathbb{R}^3 : x + y = z\}$.

We shall use the vector function

$$V(x, y, z) = \left((x + y - z)^2, (-x + y + z)^2, (x - y + z)^2\right)^T.$$

Then, the set

$$\Omega_1 = \Big\{(x, y, z) \in PC[\mathbb{R}_+, \mathbb{R}^3] :$$

$$V(x(s), y(s), z(s)) \leq V(x(t), y(t), z(t)), t - r \leq s \leq t\Big\}.$$

For $t \geq 0$, $t \neq t_k$ and $(x, y, z)^T \in \Omega_1$, we have

$$D^+_{(3.50)} V(x(t), y(t), z(t))$$

$$\leq 2 \begin{pmatrix} 1 & 0 & 0 \\ 0 & e^{-t} & 0 \\ 0 & 0 & t^2 \end{pmatrix} V(x(t), y(t), z(t))$$

$$+ 2 \begin{pmatrix} 1 & 0 & 0 \\ 0 & e^{-t} & 0 \\ 0 & 0 & t^2 \end{pmatrix} V(x(t - r(t)), y(t - r(t)), z(t - r(t)))$$

$$\leq 4 \begin{pmatrix} 1 & 0 & 0 \\ 0 & e^{-t} & 0 \\ 0 & 0 & t^2 \end{pmatrix} V(x(t), y(t), z(t)).$$

Also, for $k = 1, 2, \ldots$

$$V(x(t_k + 0), y(t_k + 0), z(t_k + 0))$$

$$= V(x(t_k), y(t_k), z(t_k)) + \begin{pmatrix} d_{1k} & 0 & 0 \\ 0 & d_{2k} & 0 \\ 0 & 0 & d_{3k} \end{pmatrix} V(x(t_k), y(t_k), z(t_k)).$$

Since the zero solution of the comparison system

$$\begin{cases} \dot{u}_1(t) = 4u_1(t), \ t \neq t_k, \ t \geq 0 \\ \dot{u}_2(t) = 4e^{-t}u_2(t), \ t \neq t_k, \ t \geq 0 \\ \dot{u}_3(t) = 4t^2 u_3(t), \ t \neq t_k, \ t \geq 0 \\ \Delta u_1(t_k) = d_{1k}u_1(t_k), \ \Delta u_2(t_k) = d_{2k}u_2(t_k) \\ \Delta u_3(t_k) = d_{3k}u_3(t_k), \ k = 1, 2, \ldots \end{cases}$$

is conditionally stable with respect to the manifold $R(2) = \{\mathrm{col}(0, u_2, u_3) \in \mathbb{R}^3 : u_2 \geq 0, u_3 \geq 0\}$ [129] and all the conditions of Theorem 3.30 are fulfilled, the zero solution of (3.50) is conditionally stable with respect to the manifold $M(2)$.

3.3   Parametric stability

The feasibility of equilibria and their stability are two basic problems of analysis of a wide variety of dynamic models in diverse fields like population biology, economics, neural networks, and chemical processes. Although in a majority of models the two problems are strongly interdependent, they are always considered separately. The standard approach is first to locate the equilibria, then select one that is of interest, translate it to the origin, and lastly determine its stability properties. The translation of the equilibrium is justified by the fact that a stability analysis can be developed "without loss of generality" for the equilibrium at the origin and then universally used for other equilibria of the model. This approach may break down when parametric uncertainties are present because of modelling inaccuracies or changes in the environment of the model. Each time a parameter is changed, the original equilibrium may either shift to a new location or disappear, thus making the stability analysis of the translated equilibrium at the origin either imprecise or entirely useless.

Uncertain parameters appear, in a general way, throughout the models in population biology, especially those of Lotka–Volterra type. Siljak, in collaboration with Ikeda and Ohta, formulated [186] the concept of *parametric stability*, which addresses simultaneously the twin problem of existence and stability of a moving equilibrium.

The objective of this section is to extend the notion of parametric stability for impulsive functional differential systems.

Let $t_0 \in \mathbb{R}$, $r > 0$, Ω be a bounded domain in $\mathbb{R}^n$ containing the origin and $\|x\| = (\sum_{i=1}^{n} x_i^2)^{\frac{1}{2}}$ be the norm of $x \in \mathbb{R}^n$. Consider the system

$$
\begin{cases}
\dot{x}(t) = f(t, x_t, p), \ t \geq t_0, \ t \neq t_k \\
\Delta x(t) = I_k(x(t), p), \ t = t_k, \ t_k > t_0, \ k = 1, 2, \ldots,
\end{cases}
\tag{3.51}
$$

where $p \in \mathbb{R}^m$ is a constant parameter vector; $f : [t_0, \infty) \times PC[[-r, 0], \Omega] \times \mathbb{R}^m \to \mathbb{R}^n$; $I_k : \Omega \times \mathbb{R}^m \to \Omega$, $k = 1, 2, \ldots$; $t_0 < t_1 < t_2 < \cdots$ and $\lim_{k \to \infty} t_k = \infty$.

Let $\varphi_0 \in PC[[-r, 0], \Omega]$ and $p \in \mathbb{R}^m$ be a fixed parameter. Denote by $x(t; t_0, \varphi_0, p)$ the solution of system (3.51), satisfying the initial conditions

$$
\begin{cases}
x(t; t_0, \varphi_0, p) = \varphi_0(t - t_0), \ t_0 - r \leq t \leq t_0 \\
x(t_0 + 0; t_0, \varphi_0, p) = \varphi_0(0),
\end{cases}
\tag{3.52}
$$

and by $J^+(t_0, \varphi_0, p)$ the maximal interval of type $[t_0, \beta)$ in which the solution $x(t; t_0, \varphi_0, p)$ is defined.

We also assume that for some nominal value p^* of the parameter vector p, there is an *equilibrium* state x^*, that is,

$$
\begin{cases}
f(t, x^*, p^*) = 0, \ t \geq t_0, \ t \neq t_k \\
\Delta x^*(t_k) = x^*(t_k + 0) - x^*(t_k) = 0, \ t_k > t_0, k = 1, 2, \ldots,
\end{cases}
\tag{3.53}
$$

and x^* is stable. Suppose that the parameter vector p is changed from p^* to another value. The question arises: Does a new equilibrium x^ε of (3.51) exist there? If x^ε exists, is it stable as x^* was, or is its stability destroyed by the change of p?

Consider the equilibrium $x^\varepsilon : \mathbb{R}^m \to \Omega$ as a function $x^\varepsilon(p)$ and introduce the following definitions of parametric stability.

Definition 3.35. The system (3.51) is said to be *parametrically stable* at $p^* \in \mathbb{R}^m$, if there exists a neighborhood $N(p^*)$ such that for any $p \in N(p^*)$:

(i) there exists an equilibrium $x^\varepsilon(p) \in \Omega$;

(ii) $(\forall t_0 \in \mathbb{R})(\forall \varepsilon > 0)(\exists \delta = \delta(t_0, \varepsilon, p) > 0)$
$(\forall \varphi_0 \in PC[[-r, 0], \Omega] : \|\varphi_0 - x^\varepsilon(p)\|_r < \delta)$
$(\forall t \geq t_0) : \|x(t; t_0, \varphi_0, p) - x^\varepsilon(p)\| < \varepsilon.$

Remark 3.36. If the system (3.51) is not stable in the above sense, we say it is *parametrically unstable* at p^*. This means that for any neighborhood $N(p^*)$, there exists a $p \in N(p^*)$ for which either there is no equilibrium $x^\varepsilon(p)$ of (3.51), or there is an equilibrium $x^\varepsilon(p)$, which is unstable in the sense of Lyapunov.

Definition 3.37. The system (3.51) is said to be *parametrically uniformly stable* at $p^* \in \mathbb{R}^m$, if the number δ from Definition 3.35 is independent of $t_0 \in \mathbb{R}$.

Definition 3.38. The system (3.51) is said to be *parametrically uniformly asymptotically stable* at $p^* \in \mathbb{R}^m$, if there exists a neighborhood $N(p^*)$ such that for any $p \in N(p^*)$:

(i) it is parametrically uniformly stable at p^*;

(ii) for all $p \in N(p^*)$, there exists a number $\mu = \mu(p) > 0$ such that $\|\varphi_0 - x^\varepsilon(p)\|_r < \mu$ implies

$$\lim_{t \to \infty} \|x(t; t_0, \varphi_0, p) - x^\varepsilon(p)\| = 0.$$

We introduce the following conditions:

H3.8. $f \in C[[t_0, \infty) \times PC[[-r, 0], \Omega] \times \mathbb{R}^m, \mathbb{R}^n].$

H3.9. The function $f(t, \phi, p)$ is Lipschitz continuous with respect to ϕ in $PC[[-r, 0], \Omega]$ and $p \in \mathbb{R}^m$ uniformly on $t \in [t_0, \infty)$.

H3.10. There exists a constant $P > 0$ such that for all $(t, \phi, p) \in [t_0, \infty) \times PC[[-r, 0], \Omega] \times \mathbb{R}^m$

$$\|f(t, \phi, p)\| \leq P < \infty.$$

H3.11. $I_k \in C[\Omega \times \mathbb{R}^m, \mathbb{R}^n], \ k = 1, 2, \ldots.$

H3.12. The functions $(I + I_k)$ map $\Omega \times \mathbb{R}^m$ into $\Omega \times \mathbb{R}^m$, $k = 1, 2, \ldots$ where I is the identity in $\Omega \times \mathbb{R}^m$.

In this section, we shall use the Lyapunov functions $V : [t_0, \infty) \times \Omega \to \mathbb{R}_+$, which belong to the class V_0, and satisfy the condition

H3.13. $V(t, x^\varepsilon(p)) = 0$, $t \in [t_0, \infty)$, $p \in N(p^*)$.

Note that if the hypothesis H1.12, H1.13, H3.8, H3.10, H3.11 and H3.12 are met, then by Theorem 1.17, $J^+(t_0, \varphi_0, p) = [t_0, \infty)$.

In the proof of the main results, we shall use the following lemma.

Lemma 3.39. *Assume that:*

(1) *Conditions* H1.12, H1.13, H3.8, H3.10, H3.11 *and* H3.12 *hold.*

(2) *The solution* $x(t) = x(t; t_0, \varphi_0, p)$ *of the initial value problem* (3.51), (3.52) *is such that* $x \in PC[(t_0 - r, \infty), \Omega] \cap PC^1[[t_0, \infty), \Omega]$.

(3) *The function* $F : [t_0, \infty) \times \mathbb{R}_+ \times \mathbb{R}_+ \to \mathbb{R}$ *is continuous in each of the sets* $(t_{k-1}, t_k] \times \mathbb{R}_+ \times \mathbb{R}_+$, $k = 1, 2, \ldots$ *and* $F(t, u, \mu)$ *is non-decreasing in* u *for each* $t \in [t_0, \infty)$ *and* $\mu \in \mathbb{R}_+$, *where* $\mu = \mu(p)$ *is a parameter.*

(4) $\psi_k(u, \mu) \in C[\mathbb{R}_+ \times \mathbb{R}_+, \mathbb{R}_+]$, $k = 1, 2, \ldots$ *are non-decreasing with respect to* u.

(5) *The maximal solution* $R(t; t_0, u_0, \mu)$ *of the problem*

$$\begin{cases} \dot{u} = F(t, u, \mu), \ t \geq t_0, \ t \neq t_k \\ u(t_0) = u_0 \geq 0 \\ u(t_k + 0) = \psi_k(u(t_k), \mu), \ t_k > t_0, \ k = 1, 2, \ldots \end{cases}$$

is defined in the interval $[t_0, \infty)$.

(6) *There exists a function* $V \in V_0$ *such that* $V(t_0 + 0, \varphi_0(0)) \leq u_0$,

$$V(t + 0, x + I_k(x, p)) \leq \psi_k(V(t, x), \mu), \ p \in \mathbb{R}^m, \ \mu \in \mathbb{R}_+, \ x \in \Omega,$$

$t = t_k, k = 1, 2, \ldots$, *and the inequality*

$$D^+_{(3.51)} V(t, x(t)) \leq F(t, V(t, x(t)), \mu), \ t \neq t_k, \ k = 1, 2, \ldots$$

is valid for each $t \in [t_0, \infty)$ *and* $x \in \Omega_1$.

Then

$$V(t, x(t; t_0, \varphi_0, p)) \leq R(t; t_0, u_0, \mu), \ t \in [t_0, \infty).$$

The proof of Lemma 3.39 is similar to that of Theorem 1.23. We omit it here.

In the case when $F(t, u, \mu) \equiv 0$ for $t \in [t_0, \infty)$, $u \in \mathbb{R}_+$, $\mu \in \mathbb{R}_+$ and $\psi_k(u, \mu) \equiv u$ for $u \in \mathbb{R}_+$, $\mu \in \mathbb{R}_+$, $k = 1, 2, \ldots$, we deduce the following corollary from Lemma 3.39.

Corollary 3.40. *Assume that:*

(1) *Conditions* H1.12, H1.13, H3.8–H3.12 *hold.*

(2) *The condition* (2) *of Lemma 3.39 is satisfied.*

(3) *There exists a function* $V \in V_0$ *such that*

$$D^+_{(3.51)}V(t, x(t)) \leq 0, \ x \in \Omega_1, \ t \geq t_0, \ t \neq t_k, \ k = 1, 2, \ldots,$$

$$V(t + 0, x + I_k(x, p)) \leq V(t, x), \ p \in \mathbb{R}^m, \ x \in \Omega, \ t = t_k, \ k = 1, 2, \ldots.$$

Then

$$V(t, x(t; t_0, \varphi_0, p)) \leq V(t_0 + 0, \varphi_0(0)), \ t \in [t_0, \infty).$$

The following theorem provides sufficient conditions for requirement (i) of Definition 3.35.

Theorem 3.41. *Assume that:*

(1) *Conditions* H1.12, H1.13, H3.8–H3.12 *hold.*

(2) *For some nominal value* p^* *of the parameter vector* p, *there is an equilibrium state* x^* *which satisfies* (3.53).

(3) $\det D_x f(t, x^*, p^*) \neq 0, \ t \neq t_k, \ k = 1, 2, \ldots.$

Then there exists a neighbourhood $N(p^*)$ *of* p^* *such that for any* $p \in N(p^*)$, *the system*

$$\begin{cases} f(t, x_t, p) = 0, \ t \geq t_0, \ t \neq t_k \\ \Delta x(t_k) = x(t_k + 0) - x(t_k) - 0, \ t_k > t_0, \ k = 1, 2, \ldots \end{cases} \tag{3.54}$$

has a solution $x^{\varepsilon}(p) \in \Omega$.

The proof of Theorem 3.41 is similar to the proof of Assertion 1 in Theorem 1.17 for the existence of a solution of a system of functional differential equations. A theorem for the existence of an equilibrium $x^{\varepsilon}(p) \in \Omega$ in the continuous case is being used [186].

Theorem 3.42. *Assume that:*

(1) *The conditions of Theorem 3.41 hold.*

(2) *There exists a function* $V \in V_0$ *such that* H3.13 *holds,*

$$a(\|x - x^{\varepsilon}(p)\|) \leq V(t, x), \ a \in K, \ (t, x) \in [t_0, \infty) \times \Omega, \tag{3.55}$$

$$V(t + 0, x + I_k(x, p)) \leq V(t, x), \ x \in \Omega, \ p \in N(p^*), \ t = t_k, \ k = 1, 2, \ldots, \tag{3.56}$$

and the inequality

$$D^+_{(3.51)}V(t, x(t)) \leq 0, \ t \neq t_k, \ k = 1, 2, \ldots$$

is valid for $t \in [t_0, \infty), \ x \in \Omega_1$.

Then the system (3.51) *is parametrically stable at* p^*.

Proof. Let $\varepsilon > 0$. From the properties of the function V, it follows that there exists a constant $\delta = \delta(t_0, \varepsilon, p) > 0$ such that if $x \in \Omega :\ \|x - x^\varepsilon(p)\| < \delta$, then $\sup_{\|x - x^\varepsilon(p)\| < \delta} V(t_0 + 0, x) < a(\varepsilon)$.

Let $\varphi_0 \in PC[[-r, 0], \Omega] :\ \|\varphi_0 - x^\varepsilon(p)\|_r < \delta$. Then $\|\varphi_0(0) - x^\varepsilon(p)\| \leq \|\varphi_0 - x^\varepsilon(p)\|_r < \delta$ hence

$$V(t_0 + 0, \varphi_0(0)) < a(\varepsilon). \tag{3.57}$$

Let $x(t) = x(t; t_0, \varphi_0, p)$ be the solution of problem (3.51), (3.52). Since the conditions of Corollary 3.40 are met, then

$$V(t, x(t; t_0, \varphi_0, p)) \leq V(t_0 + 0, \varphi_0(0)), \quad t \in [t_0, \infty).$$

From the last inequality and (3.55)–(3.57), there follow the inequalities

$$a(\|x(t; t_0, \varphi_0, p) - x^\varepsilon(p)\|) \leq V(t, x(t; t_0, \varphi_0, p)) \leq V(t_0 + 0, \varphi_0(0)) < a(\varepsilon),$$

which imply that $\|x(t; t_0, \varphi_0, p) - x^\varepsilon(p)\| < \varepsilon$ for $t \geq t_0$. This implies that the system (3.51) is parametrically stable at p^*.   □

Theorem 3.43. *Let the conditions of Theorem* 3.42 *hold, and let a function* $b \in K$ *exist such that*

$$V(t, x) \leq b(\|x - x^\varepsilon(p)\|), \quad (t, x) \in [t_0, \infty) \times \Omega.$$

Then the system (3.51) *is parametrically uniformly stable at* p^*.

Theorem 3.44. *Assume that:*

(1) *The conditions of Theorem* 3.41 *hold.*

(2) *There exists a function* $V \in V_0$ *such that* H3.13 *and* (3.56) *hold,*

$$a(\|x - x^\varepsilon(p)\|) \leq V(t, x) \leq b(\|x - x^\varepsilon(p)\|), \quad a, b \in K, \ (t, x) \in [t_0, \infty) \times \Omega,$$

and the inequality

$$D^+_{(3.51)} V(t, x(t)) \leq -c(\|(x(t) - x^\varepsilon(p)\|), \ t \neq t_k, \ k = 1, 2, \ldots \tag{3.58}$$

is valid for $c \in K, \ t \in [t_0, \infty)$ *and* $x \in \Omega_1$.

Then the system (3.51) *is parametrically uniformly asymptotically stable at* p^*.

Theorem 3.43 and Theorem 3.44 are similar to Theorem 2.3 and Theorem 2.4, respectively.

Corollary 3.45. *If in Theorem* 3.44 *condition* (3.58) *is replaced by the condition*

$$D^+_{(3.51)} V(t, x(t)) \leq -c V(t, x(t)), \ t \in [t_0, \infty), \ t \neq t_k, \ k = 1, 2, \ldots,$$

where $x \in \Omega_1, \ c = \text{const} > 0$, *then the system* (3.51) *is parametrically uniformly asymptotically stable.*

In the next examples we study the parametric stability of two impulsive Lotka–Volterra models.

Example 3.46. To illustrate the idea of the parametric stability, let us first consider an impulsive Lotka–Volterra model of two interacting species:

$$
\begin{cases}
\dot{x}_1(t) = \dfrac{r_1}{K_1} x_1(t)\big(K_1 - x_1(t) - e_{12}\alpha_{12}x_2(t - \tau_2(t))\big), \quad t \neq t_k \\[2mm]
\dot{x}_2(t) = \dfrac{r_2}{K_2} x_2(t)\big(K_2 - x_2(t) - e_{21}\alpha_{21}x_1(t - \tau_1(t))\big), \quad t \neq t_k \\[2mm]
x_1(t_k + 0) = (d_{1k} + 1)x_1(t_k) - d_{1k}\dfrac{K_1 - K_2 e_{12}\alpha_{12}}{1 - e_{12}e_{21}\alpha_{12}\alpha_{21}}, \quad k = 1, 2, \ldots \\[2mm]
x_2(t_k + 0) = (d_{2k} + 1)x_2(t_k) - d_{2k}\dfrac{K_2 - K_1 e_{21}\alpha_{21}}{1 - e_{12}e_{21}\alpha_{12}\alpha_{21}}, \quad k = 1, 2, \ldots,
\end{cases}
\tag{3.59}
$$

where $x_1(t)$ and $x_2(t)$ are populations of the two species at time t; r_1 and r_2 are intrinsic growth rates; K_1 and K_2 are the carrying capacities of the environment; α_{12} and α_{21} are inter-specific coefficients, and $0 \leq \tau_i(t) \leq t_0, i = 1, 2, t \geq 0$. All parameters r_1, r_2, K_1, K_2 and α_{12} and α_{21} are positive numbers. The uncertain parameters are e_{12} and e_{21}, which can take values from the interval $[0, 1]$ and represent the interaction strength between the species. The values $x_i(t_k)$ and $x_i(t_k + 0)$ are the population numbers of ith species before and after the impulsive effect at the time t_k, respectively; and $d_{ik} > -1$ for all $i = 1, 2$ and $k = 1, 2, \ldots$.

It is easy to show that for (3.59) there exists an equilibrium x^ε at

$$
\begin{cases}
x_1^\varepsilon = \dfrac{K_1 - K_2 e_{12}\alpha_{12}}{1 - e_{12}e_{21}\alpha_{12}\alpha_{21}} \\[3mm]
x_2^\varepsilon = \dfrac{K_2 - K_1 e_{21}\alpha_{21}}{1 - e_{12}e_{21}\alpha_{12}\alpha_{21}},
\end{cases}
\tag{3.60}
$$

which is positive for all permissible values of e_{12} and e_{21} whenever the carrying capacity ratio K_1/K_2 satisfies the condition

$$
e_{12}\alpha_{12} < \frac{K_1}{K_2} < \frac{1}{e_{21}\alpha_{21}}.
\tag{3.61}
$$

It is known [2, 3, 4, 186] that under the conditions (3.61) for any closed interval contained in $t \in (t_{k-1}, t_k], k = 1, 2, \ldots$, there exist positive numbers r_* and r^* such that for $i = 1, 2$,

$$
r_* \leq \frac{r_i x_i}{K_i} \leq r^*.
\tag{3.62}
$$

Theorem 3.47. *Assume that:*

(1) *Conditions (3.61) and (3.62) hold.*

(2) $0 < t_1 < t_2 < \cdots$ and $\lim_{k \to \infty} t_k = \infty$.

(3) $-1 < d_{1k} \leq 0, -1 < d_{2k} \leq 0, k = 1, 2, \ldots$.

(4) *There exists a constant $c > 0$ such that*

$$r_* > c + (\alpha_{12} + \alpha_{21})r^*.$$

Then the system (3.59) is parametrically uniformly asymptotically stable for all permissible values of e_{12} and e_{21}.

Proof. Choose

$$V(t, x_1, x_2) = \left(x_1 - x_1^\varepsilon\right)^2 + \left(x_2 - x_2^\varepsilon\right)^2 = d^2.$$

For $t \geq 0$ and $t \neq t_k$, we have

$$D^+_{(3.59)} V(t, x_1(t), x_2(t))$$

$$= 2(x_1(t) - x_1^\varepsilon)\frac{r_1 x_1(t)}{K_1}\left[K_1 - x_1(t) - e_{12}\alpha_{12}x_2(t - \tau_2(t))\right]$$

$$+ 2(x_2(t) - x_2^\varepsilon)\frac{r_2 x_2(t)}{K_2}\left[K_2 - x_2(t) - e_{21}\alpha_{21}x_1(t - \tau_1(t))\right].$$

Since $(x_1^\varepsilon, x_2^\varepsilon)$ is an equilibrium of (3.59), from (3.62) we obtain

$$D^+_{(3.59)} V(t, x_1(t), x_2(t))$$

$$\leq -2r_*(x_1(t) - x_1^\varepsilon)^2 + 2r^*|\alpha_{12}| \, |x_1(t) - x_1^\varepsilon| \, |x_2(t - \tau_2(t)) - x_2^\varepsilon|$$

$$-2r_*(x_2(t) - x_2^\varepsilon)^2 + 2r^*|\alpha_{21}| \, |x_1(t - \tau_1(t)) - x_1^\varepsilon| \, |x_2(t) - x_2^\varepsilon|,$$

for $t \neq t_k$, $k = 1, 2, \ldots$.

Using the inequality $2|a| \, |b| \leq a^2 + b^2$, we get

$$D^+_{(3.59)} V(t, x_1(t), x_2(t))$$

$$\leq -2r_*(x_1(t) - x_1^\varepsilon)^2 + r^*|\alpha_{12}|\left((x_1(t) - x_1^\varepsilon)^2 + (x_2(t - \tau_2(t)) - x_2^\varepsilon)^2\right)$$

$$-2r_*(x_2(t) - x_2^\varepsilon)^2 + r^*|\alpha_{21}|\left((x_2(t) - x_2^\varepsilon)^2 + (x_1(t - \tau_1(t)) - x_1^\varepsilon)^2\right),$$

for $t \neq t_k$, $k = 1, 2, \ldots$.

The set Ω_1 is

$$\Omega_1 = \left\{\text{col}(x_1(t), x_2(t)) \in PC[\mathbb{R}_+, \mathbb{R}_+^2] : d^2(s) \leq d^2(t), t - \tau_0 \leq s \leq t\right\}.$$

Then, for $(x_1, x_2) \in \Omega_1$, we have

$$D^+_{(3.59)} V(t, x_1(t), x_2(t))$$

$$\leq 2\left[-r_* + (\alpha_{12} + \alpha_{21})r^* \right]\left((x_1(t) - x_1^\varepsilon)^2 + (x_2(t) - x_2^\varepsilon)^2\right)$$

$$< -2c V(t, x_1(t), x_2(t)), \ t \neq t_k, \ k = 1, 2, \ldots.$$

Also,

$$V(t_k + 0, x_1(t_k + 0), x_2(t_k + 0))$$

$$= \left(x_1(t_k + 0) - x_1^\varepsilon\right)^2 + \left(x_2(t_k + 0) - x_2^\varepsilon\right)^2$$

$$= (1 + d_{1k})^2\left(x_1(t_k) - x_1^\varepsilon\right)^2 + (1 + d_{2k})^2\left(x_2(t_k) - x_2^\varepsilon\right)^2$$

$$\leq V(t_k, x_1(t_k), x_2(t_k)), \quad k = 1, 2, \ldots.$$

Since all conditions of Theorem 3.44 are satisfied, the system (3.59) is parametrically uniformly asymptotically stable for all permissible values of e_{12} and e_{21}. We can therefore conclude that the equilibrium x^ε is uniformly asymptotically stable at $e_{12} = e_{21} = 1$, i.e. for $e_{ij} \in [0, 1]$, $i, j = 1, 2, i \neq j$ it remains stable. ☐

Example 3.48. In this example, we shall consider uncertain parameters in a nonlinear model, which appear in a general way throughout the model. In the context of example 3.46, this means that the carrying capacities and intrinsic growth rates can be considered as uncertain parameters as well. In this example we consider a very general class of impulsive Lotka–Volterra models of n species represented by

$$S^n : \begin{cases} \dot{x}_i(t) = x_i(t)\left[g_i(x_i(t), p) + h_i(x_t, p)\right], \ t \geq 0, \ t \neq t_k \\ \\ \Delta x_i(t_k) = c_{ik}(x_i(t_k), p), \ t_k > 0, k = 1, 2, \ldots, \end{cases}$$

where the state $x_i(t) \in \mathbb{R}$; the functions $g_i : \mathbb{R} \times \mathbb{R}^m \to \mathbb{R}$ and $h_i : PC[[-r, 0], \Omega] \times \mathbb{R}^m \to \mathbb{R}$; $c_{ik} \in \mathbb{R}$ for all $i = 1, 2, \ldots, n$ and $k = 1, 2, \ldots$ and $0 < t_1 < t_2 < \cdots$, $\lim_{k \to \infty} t_k = \infty$.

A compact notation of S^n is

$$S^n : \begin{cases} \dot{x}(t) = X(t)\left[g(x(t), p) + h(x_t, p)\right], \ t \geq 0, \ t \neq t_k \\ \\ \Delta x(t_k) = C_k(x(t_k), p), \ t_k > 0, k = 1, 2, \ldots, \end{cases}$$

where the state $x(t) \in \mathbb{R}^n$; $X = \text{diag}\{x_1, x_2, \ldots, x_n\}$; the functions $g : \mathbb{R}^n \times \mathbb{R}^m \to \mathbb{R}^n$ and $h : PC[[-r, 0], \Omega] \times \mathbb{R}^m \to \mathbb{R}^n$ are given as $g(x, p) = \text{col}(g_1(x_1, p), g_2(x_2, p), \ldots, g_n(x_n, p)), h(x_t, p) = \text{col}(h_1(x_t, p), h_2(x_t, p), \ldots, h_n(x_t, p))$ and $C_k(x(t_k), p) = \text{col}(c_{1k}(x_1(t_k), p), c_{2k}(x_2(t_k), p), \ldots, c_{nk}(x_n(t_k), p))$.

We are interested only in the equilibrium of S^n located in the $\mathbb{R}_+^n$. The following theorem is a direct consequence of Theorem 3.41.

Theorem 3.49. *Assume that:*

(1) $0 < t_1 < t_2 < \cdots$ *and* $\lim_{k \to \infty} t_k = \infty$.

(2) $C_k \in C[\mathbb{R}_+^n \times \mathbb{R}^m, \mathbb{R}^n]$, $k = 1, 2, \ldots$.

(3) $g \in C[\mathbb{R}_+^n \times \mathbb{R}^m, \mathbb{R}_+^n]$.

(4) $h \in C[PC[[-r, 0], \Omega_+] \times \mathbb{R}^m, \mathbb{R}_+^n]$.

(5) *For some nominal value p^* of the parameter vector p, there is an equilibrium state $x^* \in \mathbb{R}_+^n$ which satisfies*

$$\begin{cases} g(x^*, p^*) + h(x^*, p^*) = 0 \\ \Delta x^*(t_k) = 0, \ t_k > 0, k = 1, 2, \ldots. \end{cases}$$

(6) $\det D_x[g(x^*, p^*) + h(x^*, p^*)] \neq 0$.

Then there exists a neighbourhood $N(p^*)$ such that for any $p \in N(p^*)$, the system

$$\begin{cases} g(x(t), p) + h(x_t, p) = 0, \ t \geq 0, \ t \neq t_k \\ \Delta x(t_k) = x(t_k + 0) - x(t_k) = 0, \ t_k > 0, k = 1, 2, \ldots, \end{cases}$$

has a solution $x^\varepsilon(p) \in \mathbb{R}_+^n$.

Once existence of an equilibrium $x^\varepsilon(p)$ is established, we turn our attention to the second part of the parametric problem, which is stability of $x^\varepsilon(p)$.

Definition 3.50. A matrix $A_{n \times n} = (a_{ij})_{n \times n}$, $a_{ii} > 0$, $a_{ij} \leq 0$, $i \neq j$ is said to be an *M-matrix* if, there exists is a positive diagonal matrix $Q_{n \times n}$ such that the matrix $QA + A^T Q$ is positive definite.

Define a Lyapunov function

$$V(t, x) = \sum_{i=1}^n q_i \left(x_i - x_i^\varepsilon(p) - x_i^\varepsilon(p) \ln \frac{x_i}{x_i^\varepsilon(p)} \right),$$

where q_i, $i = 1, 2, \ldots, n$ are all positive numbers. Obviously, the function $V(t, x) \in V_0$ and there exist comparison functions $a, b \in K$ such that

$$a(\|x - x^\varepsilon(p)\|) \leq V(t, x) \leq b(\|x - x^\varepsilon(p)\|), \ t \geq 0. \tag{3.63}$$

Theorem 3.51. *Assume that:*

(1) *The conditions of Theorem 3.49 hold.*

(2) *For any* $p \in N(p^*)$, *there exist positive continuous functions* $c_i \in K$, $i = 1, 2, \ldots, n$, *such that*

$$(x_i - x_i^\varepsilon(p))[g_i(x_i, p) - g_i(x_i^\varepsilon(p), p)] \leq -c_i^2(|x_i - x_i^\varepsilon(p)|).   \quad (3.64)$$

(3) *For any* $p \in N(p^*)$, *there exist positive definite functions* $e_i \in K$, $i = 1, 2, \ldots, n$, *such that*

$$(x_i - x_i^\varepsilon(p))[h_i(x_t, p) - h_i(x^\varepsilon(p), p)] \leq e_i(\|x(s) - x^\varepsilon(p)\|)   \quad (3.65)$$

for all $t - r \leq s \leq t$, $t \in \mathbb{R}_+$ *and all* $i = 1, 2, \ldots, n$.

(4) *For any* $p \in N(p^*)$, *there exists a matrix* $W = (w_{ij})$, *such that*

$$e_i\left(a^{-1}(b(\|x - x^\varepsilon(p)\|))\right) \leq c_i(|x_i - x_i^\varepsilon(p)|) \sum_{j=1}^{n} w_{ij} c_j(|x_j - x_j^\varepsilon(p)|),   \quad (3.66)$$

where w_{ij} *are nonnegative numbers*, $i, j = 1, 2, \ldots, n$.

(5) *The matrix* $(E - W)$ *is an M-matrix.*

(6) *For any* $p \in N(p^*)$ *there exist constants* L_{ik}, $0 < L_{ik} < 1$, $i = 1, 2, \ldots, n$, $k = 1, 2, \ldots$, *such that*

$$L_{ik} x_i \leq x_i + c_{ik}(x_i) \leq x_i + x_i^\varepsilon \ln(L_{ik}).$$

Then the system S^n *is parametrically uniformly asymptotically stable at* p^*.

Proof. With respect to the system S^n for $t \geq 0$ and $t \neq t_k$, we compute

$$D^+_{(S^n)} V(t, x(t)) = \sum_{i=1}^{n} q_i(x_i(t) - x_i^\varepsilon(p))(\dot{x}_i(t)/x_i(t))$$

$$= \sum_{i=1}^{n} q_i(x_i(t) - x_i^\varepsilon(p)) \Big[g_i(x_i(t), p) + h_i(x_t, p) \Big]$$

$$= \sum_{i-1}^{n} q_i(x_i(t) - x_i^\varepsilon(p)) \Big[(g_i(x_i(t), p) - g_i(x_i^\varepsilon(p), p))$$

$$+ (h_i(x_t, p) - h_i(x^\varepsilon(p), p)) \Big].$$

From (3.64) and (3.65), we deduce the inequality

$$D^+_{(S^n)} V(t, x(t)) \leq \sum_{i=1}^{n} q_i \Big[-c_i^2(|x_i(t) - x_i^\varepsilon(p)|) + e_i\left(\|x(s) - x^\varepsilon(p)\|\right) \Big].$$

for all $t - r \leq s \leq t$, $t \in \mathbb{R}_+$.

Consider the set Ω_1, where

$$\Omega_1 = \left\{ x \in PC[(0, \infty), \mathbb{R}_+^n] : V(s, x(s)) \le V(t, x(t)), \; t - r \le s \le t \right\}.$$

For all $x \in \Omega_1$, $t \ne t_k$ from (3.63), we have

$$a(\|x(s) - x^\varepsilon(p)\|) \le V(s, x(s)) \le V(t, x(t))$$
$$\le b(\|x(t) - x^\varepsilon(p)\|), \; t - r \le s \le t$$

and then

$$\|x(s) - x^\varepsilon(p)\| \le a^{-1}(b(\|x(t) - x^\varepsilon(p)\|)).$$

Now, from the last inequality and from (3.66) for $t \ne t_k$, $k = 1, 2, \ldots$, we have

$$D_{(S^n)}^+ V(t, x(t))$$
$$\le -c(\|x(t) - x^\varepsilon(p)\|)[Q(E - W) + (E - W)^T Q]c(\|x(t) - x^\varepsilon(p)\|)/2,$$

where

$$c(\|x - x^\varepsilon(p)\|) = (c_1(|x_1 - x_1^\varepsilon(p)|), \ldots, c_n(|x_n - x_n^\varepsilon(p)|))$$

and

$$Q = \text{diag}\{q_1, q_2, \ldots, q_n\}.$$

Since $(E - W)$ is an M-matrix, the condition (3.58) of Theorem 3.44 is satisfied. Also, from the condition (6) of Theorem 3.51, we have

$$V(t_k + 0, x(t_k) + C_k(x(t_k), p)) - V(t_k, x(t_k))$$
$$= \sum_{i=1}^n q_i \left(x_i(t_k + 0) - x_i(t_k) - x_i^\varepsilon(p) \ln \frac{x_i(t_k + 0)}{x_i(t_k)} \right)$$
$$= \sum_{i=1}^n q_i \left(c_{ik}(x_i(t_k), p) - x_i^\varepsilon(p) \ln \frac{x_i(t_k) + c_{ik}(x_i(t_k), p)}{x_i(t_k)} \right)$$
$$\le \sum_{i=1}^n q_i \left(c_{ik}(x_i(t_k), p) - x_i^\varepsilon(p) \ln L_{ik} \right) \le 0, \quad k = 1, 2, \ldots .$$

Since all conditions of Theorem 3.44 are satisfied, the system S^n is parametrically uniformly asymptotically stable at p^*.   □

3.4   Eventual stability and boundedness

In many real cases, it is obligatory to study the stability of such sets, which are not invariant with respect to a given system of differential equations. This immediately excludes the stability in the sense of Lyapunov. Examples for that can be found when self-controlled systems of management are being studied. For the problem, arisen in this situation, to be solved, a new notion is introduced – *eventual stability* [131, 132, 223]. In this case, the set under consideration, despite not being invariant in the usual sense, is invariant in the asymptotic sense.

In this section, we shall discuss questions related to eventual stability of $x = 0$ and eventual boundedness of the solutions with respect to impulsive functional differential equations with variable impulsive perturbations. The results for the systems with impulse effect at fixed moments will also be considered.

Let $t_0 \in \mathbb{R}$, $r = \text{const} > 0$, Ω be a domain in $\mathbb{R}^n$ containing the origin and $\|x\| = \sqrt{x_1^2 + x_2^2 + \cdots + x_n^2}$ define the norm of $x \in \mathbb{R}^n$.

Eventual stability

Consider the system of impulsive functional differential equations with variable impulsive perturbations (3.1). Let $\varphi_0 \in PC[[-r, 0], \Omega]$. Denote by $x(t) = x(t; t_0, \varphi_0)$ the solution of system (3.1), satisfying the initial conditions (3.2).

We shall use also the following notations:

$$B_\alpha = \{(t, x) \in [t_0, \infty) \times \mathbb{R}^n : \|x\| < \alpha\};$$
$$\overline{B_\alpha} = \{(t, x) \in [t_0, \infty) \times \mathbb{R}^n : \|x\| \le \alpha\}, \quad \alpha > 0.$$

We shall use the following definitions of eventual stability of $x = 0$ for the system (3.1).

Definition 3.52. The set $x(t) \equiv 0$ is said to be:

(1) *eventually stable* of system (3.1), if

$$(\forall \varepsilon > 0)(\exists T = T(\varepsilon) > 0)(\forall t_0 \ge T)(\exists \delta = \delta(t_0, \varepsilon) > 0)$$

$$(\forall \varphi_0 \in PC[[-r, 0], \Omega] : \|\varphi\|_r < \delta)(\forall t \ge t_0) : \|x(t; t_0, \varphi_0)\| < \varepsilon;$$

(2) *uniformly eventually stable* of system (3.1), if the number δ in (a) is independent of $t_0 \in \mathbb{R}$.

In the proofs of our main theorems, we shall use the piecewise continuous Lyapunov functions $V : [t_0, \infty) \times \Omega \to \mathbb{R}_+$, $V \in V_0$ for which the condition H2.5 is true.

Theorem 3.53. *Assume that:*

(1) *Conditions* H1.1, H1.2, H1.3, H1.5, H1.6, H1.9, H1.11, H2.1–H2.4 *hold.*

(2) *There exists a function* $V \in V_0$ *such that* H2.5 *holds,*

$$a(\|x\|) \leq V(t, x), \ a \in K, \ (t, x) \in [t_0, \infty) \times \Omega, \tag{3.67}$$

$$V(t + 0, x + I_k(x)) \leq V(t, x), \ (t, x) \in \sigma_k, \ k = 1, 2, \ldots, \tag{3.68}$$

and the inequality

$$D^+_{(3.1)} V(t, x(t)) \leq p(t)q(t, x(t)), \ t \neq \tau_k(x(t)), \ k = 1, 2, \ldots, \tag{3.69}$$

is valid for $t \in [t_0, \infty)$, $x \in \Omega_1$, $p : [t_0, \infty) \to \mathbb{R}$, $q : [t_0, \infty) \times \Omega \to \mathbb{R}$.

(3) *There exists a number* $\Gamma > 0$ *such that*

$$|q(t, x)| \leq \Gamma, \ (t, x) \in [t_0, \infty) \times \Omega.$$

(4) $\displaystyle\int_{t_0}^{\infty} |p(t)| dt < \infty$.

Then the set $x = 0$ *is an eventually stable set of system* (3.1).

Proof. Let $\varepsilon > 0$ be such that $S_\varepsilon \subset \Omega$ and $\Gamma > 0$. Let the number $T = T(\varepsilon) > 0$ be chosen so that for $t \geq T$

$$\int_t^{\infty} |p(s)| ds < \frac{a(\varepsilon)}{2\Gamma}. \tag{3.70}$$

(This is possible in view of condition (4) of Theorem 3.53.)

Let $t_0 \geq T$. From the properties of the function V, it follows that there exists a constant $\delta = \delta(t_0, \varepsilon) > 0$ such that if $(t_0 + 0, x) \in B_\delta$, then $V(t_0 + 0, x) < \frac{1}{2}a(\varepsilon)$. Let $\varphi_0 \in PC[[-r, 0], \Omega] : \|\varphi_0\|_r < \delta$ and $x(t) = x(t; t_0, \varphi_0)$ be the solution of problem (3.1), (3.2). Then $\|\varphi_0(0)\| \leq \|\varphi_0\|_r < \delta$, $(t_0 + 0, \varphi_0(0)) \in B_\delta$, hence

$$V(t_0 + 0, \varphi_0(0)) < \frac{1}{2}a(\varepsilon). \tag{3.71}$$

From condition (3) of Theorem 3.53, (3.69) and (3.71), we have

$$\int_{t_0}^t D^+_{(3.1)} V(s, x(s)) ds \leq \Gamma \int_{t_0}^t |p(s)| ds < \frac{1}{2}a(\varepsilon), \ t \geq t_0. \tag{3.72}$$

Let $t_1, t_2, \ldots$ ($t_0 < t_1 < t_2 < \cdots$) be the moments in which the integral curve $(t, x(t; t_0, \varphi_0))$ of problem (3.1), (3.2) meets the hypersurfaces σ_k, $k = 1, 2, \ldots$ and let $t_{k+l} < t < t_{k+l+1}$.

Then, we have

$$\int_{t_0}^{t} D_{(3.1)}^{+} V(s, x(s)) \, ds$$

$$= \int_{t_0}^{t_1} D_{(3.1)}^{+} V(s, x(s)) \, ds + \sum_{j=2}^{k+l} \int_{t_{j-1}}^{t_j} D_{(3.1)}^{+} V(s, x(s)) \, ds$$

$$+ \int_{t_{k+l}}^{t} D_{(3.1)}^{+} V(s, x(s)) \, ds$$

$$= V(t_1, x(t_1)) - V(t_0 + 0, \varphi_0(0)) + \sum_{j=2}^{k+l} [V(t_j, x(t_j))$$

$$- V(t_{j-1} + 0, x(t_{j-1}))] + V(t, x(t)) - V(t_{k+l} + 0, x(t_{k+l} + 0))$$

$$\geq V(t, x(t)) - V(t_0 + 0, \varphi_0(0)). \tag{3.73}$$

From (3.67), (3.70)–(3.73), we obtain

$$a(\|x(t; t_0, \varphi_0)\|) \leq V(t_0 + 0, \varphi_0(0)) + \int_{t_0}^{t} D_{(3.1)}^{+} V(s, x(s)) \, ds$$

$$< V(t_0 + 0, \varphi_0(0)) + \frac{1}{2} a(\varepsilon) < a(\varepsilon).$$

Therefore, $\|x(t; t_0, \varphi_0))\| < \varepsilon$ for $t \geq t_0$.   □

Theorem 3.54. *Assume that:*

(1) *Condition (1) of Theorem 3.53 holds.*

(2) *There exists a function $V \in V_0$ such that H2.5 and (3.68) hold,*

$$V(t, x) \geq 0, \quad (t, x) \in [t_0, \infty) \times \Omega, \tag{3.74}$$

$$\|x + I_k(x)\| \leq \|x\|, \quad (t, x) \in \sigma_k, \quad k = 1, 2, \dots, \tag{3.75}$$

and the inequality

$$D_{(3.1)}^{+} V(t, x(t)) \leq |p_1(t)|, \quad t \neq \tau_k(x(t)), \quad k = 1, 2, \dots \tag{3.76}$$

is valid for $t \in [t_0, \infty)$, $x \in \Omega_1$, $p_1 : [t_0, \infty) \to \mathbb{R}$.

(3) *For each $\varepsilon_1, \varepsilon_2, 0 < \varepsilon_1 < \varepsilon_2$ there exist $\delta_1 = \delta_1(\varepsilon_1, \varepsilon_2) > 0$ and $T_1 = T_1(\varepsilon_1, \varepsilon_2) > 0$ such that*

$$D_{(3.1)}^{+} V(t, x(t)) \leq -\delta_1 \|f(t, x_t)\| + |p_2(t)|, \quad t \neq \tau_k(x(t)), \quad k = 1, 2, \dots$$

for $t \geq T_1$, $x \in \Omega_1 \cap (\overline{S}_{\varepsilon_2} \setminus S_{\varepsilon_1})$, $p_2 : [t_0, \infty) \to \mathbb{R}$.

(4) $\displaystyle\int_{t_0}^{\infty} |p_i(t)|\, dt < \infty,\ i = 1, 2.$

Then the set $x = 0$ is an eventually stable set of system (3.1).

Proof. Let $\varepsilon > 0$ be such that $S_\varepsilon \subset \Omega$ and let $\delta_1 = \delta_1(\frac{\varepsilon}{2}, \varepsilon) > 0$, $T_1 = T_1(\frac{\varepsilon}{2}, \varepsilon) > 0$ are the numbers from the condition (3) of Theorem 3.54. Let the number $T_2 = T_2(\varepsilon) > 0$ be chosen so that

$$\int_{t}^{\infty} \Big[|p_1(s)| + |p_2(s)| \Big]\, ds < \frac{\delta_1 \varepsilon}{4},\ t \geq T_2 \tag{3.77}$$

and set $T = T(\varepsilon) = \max\Big(T_1(\frac{\varepsilon}{2}, \varepsilon), T_2(\varepsilon) \Big)$.

Let $t_0 \geq T$. Since $V(t_0 + 0, 0) = 0$, there exists $\delta = \delta(t_0, \varepsilon) > 0$ such that if $(t_0 + 0, x) \in B_\delta$, then

$$V(t_0 + 0, x) < \frac{\delta_1 \varepsilon}{4}.$$

Let $\varphi_0 \in PC[[-r, 0], \Omega] : \|\varphi_0\|_r < \delta$. Then $\|\varphi_0(0)\| \leq \|\varphi_0\|_r < \delta$, $(t_0 + 0, \varphi_0(0)) \in B_\delta$, hence

$$V(t_0 + 0, \varphi_0(0)) < \frac{\delta_1 \varepsilon}{4}. \tag{3.78}$$

Let $x(t) = x(t; t_0, \varphi_0)$ be the solution of problem (3.1), (3.2). We shall prove that $\|x(t; t_0, \varphi_0)\| < \varepsilon$ for $t \geq t_0$.

Suppose that this is not true. Then there exist a solution $x(t; t_0, \varphi_0)$ of (3.1) with $\|\varphi_0\|_r < \delta$ and a $t^* > t_0$ such that $t_k < t^* \leq t_{k+1}$ for some fixed k and

$$\|x(t^*)\| \geq \varepsilon, \quad \text{and}\quad \|x(t; t_0, \varphi_0)\| < \varepsilon,\ t \in [t_0, t_k]. \tag{3.79}$$

From (3.79) and (3.75) for each $k = 1, 2, \ldots$, we have

$$\|x(t_k + 0)\| = \|x(t_k) + I_k(x(t_k))\| \leq \|x(t_k)\| < \varepsilon$$

and hence

$$\|x(t_k + 0)\| < \varepsilon.$$

Set $t'' = \inf\{t \geq t_0 : \|x(t)\| \geq \varepsilon\}$. Since $t'' \neq t_k$, $k = 1, 2, \ldots$, then the function $x(t)$ is continuous at $t = t''$ and

$$\|x(t'')\| = \varepsilon.$$

By the similar arguments, we can prove the existence of t', $t_0 < t' < t''$, $t' \neq t_k$, $k = 1, 2, \ldots$ such that $\|x(t')\| = \frac{\varepsilon}{2}$ and $\frac{\varepsilon}{2} < \|x(t)\| < \varepsilon$ for $t \in (t', t'')$. (It is enough to set $t' = \sup\{t \geq t_0 : \|x(t)\| \geq \frac{\varepsilon}{2}\}$.) Therefore

$$\frac{\varepsilon}{2} \leq \|x(t'') - x(t')\|. \tag{3.80}$$

Using (3.76) and (3.68) as in the proof of Theorem 3.53, we can show that

$$\int_{t_0}^{t''} D^+_{(3.1)} V(t, x(t)) dt \geq V(t'', x(t'')) - V(t_0 + 0, \varphi_0(0)). \tag{3.81}$$

From (3.81), (3.80), (3.78), (3.76) and condition (3) of Theorem 3.54, we obtain

$$V(t'', x(t'')) \leq V(t_0 + 0, \varphi_0(0)) + \int_{t_0}^{t''} D^+_{(3.1)} V(t, x(t)) dt$$

$$= V(t_0 + 0, \varphi_0(0)) + \int_{t_0}^{t'} D^+_{(3.1)} V(t, x(t)) dt + \int_{t'}^{t''} D^+_{(3.1)} V(t, x(t)) dt$$

$$\leq V(t_0 + 0, \varphi_0(0)) + \int_{t_0}^{t'} |p_1(t)| dt - \delta_1 \int_{t'}^{t''} \| f(t, x_t) \| dt$$

$$+ \int_{t'}^{t''} |p_2(t)| dt$$

$$\leq V(t_0 + 0, \varphi_0(0)) + \int_{t_0}^{\infty} [|p_1(t)| + |p_2(t)|] dt - \delta_1 \| x(t'') - x(t') \|$$

$$< \frac{\delta_1 \varepsilon}{4} + \frac{\delta_1 \varepsilon}{4} - \frac{\delta_1 \varepsilon}{2},$$

which contradicts (3.74).   □

Theorem 3.55. *Let the conditions of Theorem 3.53 hold, and let a function $b \in K$ exist such that*

$$V(t, x) \leq b(\|x\|), \quad (t, x) \in [t_0, \infty) \times \Omega. \tag{3.82}$$

Then the set $x = 0$ is an uniformly eventually stable set of system (3.1).

Proof. Let $\varepsilon > 0$ be given. Choose $\delta = \delta(\varepsilon) < b^{-1}\left(\frac{1}{2} a(\varepsilon)\right)$, $0 < \delta < \varepsilon$ and $\Gamma = \Gamma(\varepsilon) > 0$ so that $|q(t, x)| \leq \Gamma$ for $(t, x) \in \overline{B}_\delta$. Let the number $T = T(\varepsilon) > 0$ be chosen so that

$$\int_t^\infty |p(s)| ds < \frac{b(\delta)}{\Gamma}, \quad t \geq T. \tag{3.83}$$

Let $t_0 \geq T$, $\varphi_0 \in PC[[-r, 0], \Omega]$: $\|\varphi_0\|_r < \delta$ and let $x(t) = x(t; t_0, \varphi_0)$ be the solution of problem (3.1), (3.2). From (3.67)–(3.69), (3.82) and (3.83), we have

$$a(\|x(t; t_0, \varphi_0)\|) \leq V(t_0 + 0, \varphi_0(0)) + \int_{t_0}^t D^+_{(3.1)} V(s, x(s)) ds$$

$$\leq b(\|\varphi_0(0)\|) + \Gamma \int_{t_0}^t |p(s)| ds < 2b(\delta) < a(\varepsilon)$$

for $t \geq t_0$. Therefore, $\|x(t; t_0, \varphi_0))\| < \varepsilon$ for $t \geq t_0$.   □

Since (3.10) is a special case of (3.1), the following theorems follow directly from Theorems 3.53–3.55.

Theorem 3.56. *Assume that:*

(1) *Conditions H1.1, H1.2, H1.3, H1.9, H1.11, H1.12, H1.13, H2.2 and H2.3 hold.*

(2) *There exists a function $V \in V_0$ such that H2.5 and (3.67) hold,*

$$V(t + 0, x + I_k(x)) \leq V(t, x), \quad x \in \Omega, \; t = t_k, \; k = 1, 2, \tag{3.84}$$

and the inequality

$$D^+_{(3.10)} V(t, x(t)) \leq p(t) q(t, x(t)), \; t \neq t_k, \; k = 1, 2, \ldots$$

is valid for $t \in [t_0, \infty)$, $x \in \Omega_1$, $p : [t_0, \infty) \to \mathbb{R}$, $q : [t_0, \infty) \times \Omega \to \mathbb{R}$.

(3) *Conditions (3) and (4) of Theorem 3.53 hold.*

Then the set $x = 0$ is an eventually stable set of system (3.10).

Theorem 3.57. *Assume that:*

(1) *Condition (1) of Theorem 3.56 holds.*

(2) *There exists a function $V \in V_0$ such that H2.5, (3.74) and (3.84) hold,*

$$\|x + I_k(x)\| \leq \|x\|, \; x \in \Omega, \; t = t_k, \; k = 1, 2, \ldots,$$

and the inequality

$$D^+_{(3.10)} V(t, x(t)) \leq |p_1(t)|, \; t \neq t_k, \; k = 1, 2, \ldots$$

is valid for $t \in [t_0, \infty)$, $x \in \Omega_1$, $p_1 : [t_0, \infty) \to \mathbb{R}$.

(3) *For each ε_1, ε_2, $0 < \varepsilon_1 < \varepsilon_2$ there exist $\delta_1 = \delta_1(\varepsilon_1, \varepsilon_2) > 0$ and $T_1 = T_1(\varepsilon_1, \varepsilon_2) > 0$ such that*

$$D^+_{(3.10)} V(t, x(t)) \leq -\delta_1 \|f(t, x_t)\| + |p_2(t)|, \; t \neq t_k, \; k = 1, 2, \ldots$$

for $t \geq T_1$, $x \in \Omega_1 \cap (\overline{S}_{\varepsilon_2} \setminus S_{\varepsilon_1})$, $p_2 : [t_0, \infty) \to \mathbb{R}$.

(4) $\displaystyle \int_{t_0}^{\infty} |p_i(t)| dt < \infty, \; i = 1, 2.$

Then the set $x = 0$ is an eventually stable set of system (3.10).

Theorem 3.58. *Let the conditions of Theorem 3.56 hold, and let a function $b \in K$ exist such that*

$$V(t, x) \leq b(\|x\|), \; (t, x) \in [t_0, \infty) \times \Omega.$$

Then the set $x = 0$ is an uniformly eventually stable set of system (3.10).

Example 3.59. Consider the equation:

$$\begin{cases} \dot{x}(t) = p(t)x(t - r(t)), \ t \neq t_k \\ \Delta x(t_k) = c_k, \ t_k > 0, \ k = 1, 2, \dots, \end{cases} \tag{3.85}$$

where $t \geq 0$; $x \in \mathbb{R}_+$; $0 < r(t) \leq r$; $p \in C[\mathbb{R}_+, \mathbb{R}]$; $c_k < 0$ and $|c_k + x| < |x|$ for $k = 1, 2, \dots$, $0 < t_1 < t_2 < \cdots$ and $\lim_{k \to \infty} t_k = \infty$.

The set $x = 0$ is not stable in the sense of Lyapunov, because it is not an equilibrium for the equation (3.85).

Let $\alpha > 0$. Consider the function $V(t, x) = |x|$. Then the set

$$\Omega_1 = \{x \in PC[\mathbb{R}_+, S_\alpha] : |x(s)| \leq |x(t)|, \ t - r \leq s \leq t\}.$$

For $t \geq 0$, $t \neq t_k$ and $x \in \Omega_1$, we have

$$D^+_{(3.85)} V(t, x(t)) = \text{sgn}(x(t))[p(t)x(t - r(t))]$$
$$= |p(t)| \, |x(t - r(t))| \leq |p(t)| \, |x(t)|.$$

Also, for $t = t_k$, $k = 1, 2, \dots$, we obtain

$$V(t + 0, x(t) + c_k) = |c_k + x(t)| < V(t, x(t)).$$

If $\int_0^\infty |p(t)| dt < \infty$, then all conditions of Theorem 3.56 are satisfied, and the set $x = 0$ is an eventually stable set with respect to (3.85).

Eventual boundedness

Consider the system of impulsive functional differential equations with variable impulsive perturbations (3.18). Let $\varphi_0 \in PC[[-r, 0], \mathbb{R}^n]$. Denote by $x(t) = x(t; t_0, \varphi_0)$ the solution of system (3.18) satisfying the initial conditions (3.19).

Definition 3.60. The solutions of (3.18) are said to be:

(a) *eventually equi-bounded*, if

$$(\forall \alpha > 0)(\exists T = T(\alpha) > 0)(\forall t_0 \geq T)(\exists \beta = \beta(t_0, \alpha) > 0)$$

$$(\forall \varphi_0 \in PC[[-r, 0], \mathbb{R}^n] : \|\varphi_0\|_r < \alpha)(\forall t \geq t_0) : \|x(t; t_0, \varphi_0)\| < \beta;$$

(b) *uniformly eventually bounded*, if the number β in (a) is independent of $t_0 \in \mathbb{R}$.

The proofs of the next theorems are similar to the proofs of Theorems 3.53 and 3.55. Piecewise continuous Lyapunov functions $V : [t_0, \infty) \times \mathbb{R}^n \to \mathbb{R}_+$, $V \in V_0$ are used.

Theorem 3.61. *Let the conditions of Theorem* 3.53 *hold for* $\Omega \equiv \mathbb{R}^n$, *and* $a(u) \to \infty$
as $u \to \infty$.
   Then the solutions of system (3.18) *are eventually equi-bounded.*

Theorem 3.62. *Let the conditions of Theorem* 3.61 *hold, and let a function* $b \in K$
exist such that
$$V(t, x) \le b(\|x\|), \ (t, x) \in [t_0, \infty) \times \mathbb{R}^n.$$

   Then the solutions of system (3.18) *are uniformly eventually bounded.*

   The next results for the impulsive system of functional differential equations (3.24)
follow directly from Theorem 3.61 and Theorem 3.62.

Theorem 3.63. *Let the conditions of Theorem* 3.56 *hold for* $\Omega \equiv \mathbb{R}^n$, *and* $a(u) \to \infty$
as $u \to \infty$.
   Then the solutions of system (3.24) *are eventually equi-bounded.*

Theorem 3.64. *Let the conditions of Theorem* 3.63 *hold, and let a function* $b \in K$
exist such that
$$V(t, x) \le b(\|x\|), \ (t, x) \in [t_0, \infty) \times \mathbb{R}^n.$$

   Then the solutions of system (3.24) *are uniformly eventually bounded.*

Remark 3.65. As in the continuous case [223], similar definitions apply for eventual
stability and boundedness with respect to sets of a sufficiently general type.

3.5   Practical stability

In the study of Lyapunov stability, an interesting set of problems deals with bringing
sets close to a certain state, rather than the state $x = 0$. The desired state of a system
may be unstable in sense of Lyapunov and yet a solution of the system may oscillate
sufficiently near this state that its performance is acceptable. Such considerations led to
the notion of *practical stability*. The main results in this prospect are due to Martynyuk
and his collaborators [131, 132, 155]. See, also [89, 133] and the references cited
therein.
   In this section, we shall use piecewise continuous vector Lyapunov functions to
study practical stability of impulsive functional differential systems.
   Let $t_0 \in \mathbb{R}$, $r > 0$, Ω be a bounded domain in $\mathbb{R}^n$ containing the origin, and
$\|x\| = (\sum_{i=1}^{n} x_i^2)^{\frac{1}{2}}$ be the norm of the element $x \in \mathbb{R}^n$. Consider the system of impulsive functional differential equations (3.10). Let $\varphi_0 \in PC[[-r, 0], \Omega]$. Denote by
$x(t; t_0, \varphi_0)$ the solution of system (3.10), satisfying the initial conditions (3.2).

   We shall give the following definitions of practical stability of system (3.10).

Definition 3.66. The system (3.10) is said to be:

(a) *practically stable* with respect to (λ, A), if given (λ, A) with $0 < \lambda < A$, we have that $\|\varphi_0\|_r < \lambda$ implies $\|x(t; t_0, \varphi_0)\| < A$, $t \geq t_0$ for some $t_0 \in \mathbb{R}$;

(b) *uniformly practically stable* with respect to (λ, A), if (a) holds for every $t_0 \in \mathbb{R}$;

(c) *practically asymptotically stable* with respect to (λ, A), if (a) holds and $\lim_{t \to \infty} \|x(t; t_0, \varphi_0)\| = 0$;

(d) *practically unstable* with respect to (λ, A), if (a) does not hold.

Together with system (3.10) we shall consider the system (1.32).

In the successive investigations, we shall use piecewise continuous auxiliary vector functions $V : [t_0, \infty) \times \Omega \to \mathbb{R}^m_+$, $V = \mathrm{col}(V_1, \ldots, V_m)$ such that $V_j \in V_0$, $j = 1, 2, \ldots, m$ and the comparison principle.

Theorem 3.67. *Assume that:*

(1) *The conditions of Theorem 1.23 and H2.2, H2.3 hold.*

(2) $0 < \lambda < A$ *is given and* $S_A \subset \Omega$.

(3) $F(t, 0) = 0$ *for* $t \in [t_0, \infty)$.

(4) $J_k(0) = 0$, $k = 1, 2, \ldots$.

(5) *There exist functions* $a, b \in K$ *such that*

$$a(\|x\|) \leq L_0(t, x) \leq b(\|x\|), \quad (t, x) \in [t_0, \infty) \times \Omega, \tag{3.86}$$

where $L_0(t, x) = \sum_{i=1}^m V_i(t, x)$.

(6) $b(\lambda) < a(A)$.

Then, the practical stability properties of the system (1.32) *with respect to* $(b(\lambda), a(A))$ *imply the corresponding practical stability properties of the system* (3.10) *with respect to* (λ, A).

Proof. We shall first prove practical stability of (3.10). Suppose that (1.32) is practically stable with respect to $(b(\lambda), a(A))$. Then, we have that

$$\sum_{i=1}^m u_{i0} < b(\lambda) \quad \text{implies} \quad \sum_{i=1}^m u_i^+(t; t_0, u_0) < a(A), \quad t \geq t_0 \tag{3.87}$$

for some given $t_0 \in \mathbb{R}$, where $u_0 = (u_{10}, \ldots, u_{m0})^T$ and the maximal solution $u^+(t; t_0, u_0)$ of (1.32) is defined in the interval $[t_0, \infty)$.

Setting $u_0 = V(t_0 + 0, \varphi_0(0))$, we get by Theorem 1.23,

$$V(t, x(t; t_0, \varphi_0)) \leq u^+(t; t_0, V(t_0 + 0, \varphi_0(0))) \text{ for } t \geq t_0. \tag{3.88}$$

Let

$$\|\varphi_0\|_r < \lambda. \tag{3.89}$$

Then, from (3.86) and (3.89), it follows

$$L_0(t_0 + 0, \varphi_0(0)) \le b(\|\varphi_0(0)\|) \le b(\|\varphi_0\|_r) < b(\lambda)$$

which due to (3.87) implies

$$\sum_{i=1}^{m} u_i^+(t; t_0, V(t_0 + 0, \varphi_0(0))) < a(A), \quad t \ge t_0. \tag{3.90}$$

Consequently, from (3.86), (3.88) and (3.90), we obtain

$$a(\|x(t; t_0, \varphi_0)\|) \le L_0(t, x(t; t_0, \varphi_0))$$

$$\le \sum_{i=1}^{m} u_i^+(t; t_0, V(t_0, \varphi_0(0))) < a(A), \quad t \ge t_0.$$

Hence, $\|x(t; t_0, \varphi_0)\| < A$, $t \ge t_0$ for the given $t_0 \in \mathbb{R}$, which proves the practical stability of (3.10).

Suppose now, that (1.32) is uniformly practically stable with respect to $(b(\lambda), a(A))$. Therefore, we have that

$$\sum_{i=1}^{m} u_{i0} < b(\lambda) \quad \text{implies} \quad \sum_{i=1}^{m} u_i^+(t; t_0, u_0) < a(A), \quad t \ge t_0 \tag{3.91}$$

for every $t_0 \in \mathbb{R}$.

We claim that $\|\varphi_0\|_r < \lambda$ implies $\|x(t; t_0, \varphi_0)\| < A$, $t \ge t_0$ for every $t_0 \in \mathbb{R}$. If the claim is not true, there exists $t_0 \in \mathbb{R}$, a corresponding solution $x(t; t_0, \varphi_0)$ of (3.10) with $\|\varphi_0\|_r < \lambda$, and $t^* > t_0$ such that,

$$\|x(t^*; t_0, \varphi_0)\| \ge A, \quad \|x(t; t_0, \varphi_0)\| < A, \quad t_0 \le t \le t_k,$$

where $t^* \in (t_k, t_{k+1}]$ for some k. Then, due to H1.9 and condition (5) of Theorem 1.23, we can find $t^0 \in (t_k, t^*]$ such that

$$\|x(t^0; t_0, \varphi_0)\| \ge A \quad \text{and} \quad x(t^0; t_0, \varphi_0) \in \Omega. \tag{3.92}$$

Hence, setting $u_0 = V(t_0, \varphi_0(t^0 - t_k))$, since all the conditions of Theorem 1.23 are satisfied, we get

$$V(t, x(t; t_0, \varphi_0)) \le u^+(t; t_0, V(t_0, \varphi_0(t^0 - t_k))) \quad \text{for } t_0 \le t \le t^0. \tag{3.93}$$

From (3.92), (3.86), (3.93) and (3.91), it follows that

$$a(A) \le a(\|x(t^0; t_0, \varphi_0)\|) \le L_0(t^0, x(t; t_0, \varphi_0))$$

$$\le \sum_{i=1}^{m} u_i^+ (t^0; t_0, V(t_0, \varphi_0(t^0 - t_k))) < a(A).$$

The contradiction obtained proves that (3.10) is uniformly practically stable.   □

Remark 3.68. In Theorem 3.67, we have used the function $L_0(t, x) = \sum_{i=1}^{m} V_i(t, x)$ as a measure and, consequently we need to modify the definition of practical stability of (1.32) as follows: for example, (1.32) is practically stable with respect to $(b(\lambda), a(A))$ if (3.87) is satisfied for some given $t_0 \in \mathbb{R}$. We could use other convenient measures such as

$$L_0(t, x) = \max_{1 \le i \le m} V_i(t, x),$$

$$L_0(t, x) = \sum_{i=1}^{m} d_i V_i(t, x),$$

where $d \in \mathbb{R}_+^m$, or

$$L_0(t, x) = Q(V(t, x)),$$

where $Q : \mathbb{R}_+^m \to \mathbb{R}_+$ and $Q(u)$ is non-decreasing in u, and appropriate modifications of practical stability definitions are employed for the system (3.86).

The following example will demonstrate Theorem 3.67.

Example 3.69. Consider the system

$$
\begin{cases}
\dot{x}(t) = n(t)y(t) + m(t)x(t)[x^2(t - h) + y^2(t - h)], \ t \ne t_k, \ t \ge 0 \\
\dot{y}(t) = -n(t)x(t) + m(t)y(t)[x^2(t - h) + y^2(t - h)], \ t \ne t_k, \ t \ge 0 \quad (3.94) \\
\Delta x(t_k) = c_k x(t_k), \ \Delta y(t_k) = d_k y(t_k), \ k = 1, 2, \ldots,
\end{cases}
$$

where $x, y \in \mathbb{R}$; $h > 0$; the functions $n(t)$ and $m(t)$ are continuous in $\mathbb{R}_+$; $-1 < c_k \le 0, -1 < d_k \le 0, k = 1, 2, \ldots;\ 0 < t_1 < t_2 < \cdots$, $\lim_{k \to \infty} t_k = \infty$.
Let

$$
\begin{cases}
x(s) = \varphi_1(s), \ s \in [-h, 0] \\
y(s) = \varphi_2(s), \ s \in [-h, 0],
\end{cases}
$$

where the functions φ_1 and φ_2 are continuous in $[-h, 0]$.
Choose

$$V(t, x, y) = x^2 + y^2 = \rho^2.$$

Then

$$\Omega_1 = \{\text{col}(x(t), y(t)) \in PC[\mathbb{R}_+, \mathbb{R}^2] : \rho^2(s) \le \rho^2(t), \ t - h \le s \le t\}, \quad (3.95)$$

and for $t \geq 0$, $t \neq t_k$, $(x, y) \in \Omega_1$, we have

$$D^+_{(3.94)} V(t, x(t), y(t)) = 2m(t)x^2(t)\rho^2(t - h) + 2m(t)y^2(t)\rho^2(t - h)$$
$$\leq 2m(t)V^2(t, x(t), y(t)).$$

Also,

$$V(t_k + 0, x(t_k) + c_k x(t_k), y(t_k) + d_k y(t_k)) = (1 + c_k)^2 x^2(t_k) + (1 + d_k)^2 y^2(t_k)$$
$$\leq V(t_k, x(t_k), y(t_k)), \quad k = 1, 2, \ldots .$$

Consider the comparison system

$$\begin{cases} \dot{u}(t) = 2m(t)u^2(t), \ t \neq t_k, \ t \geq 0 \\ u(0) = u_0 \\ u(t_k + 0) = u(t_k), \ k = 1, 2, \ldots, \end{cases} \tag{3.96}$$

where $u \in \mathbb{R}_+$ and $u_0 = \varphi_1^2(0) + \varphi_2^2(0) = \rho^2(0)$.
The general solution of the system (3.96) is given by

$$u(t) = \left[u_0^{-1} - 2\int_0^t m(s)\,ds\right]^{-1}. \tag{3.97}$$

It is clear that the trivial solution of (3.96) is stable if $m(t) \leq 0$, $t \geq 0$. If $m(t) > 0$, $t \geq 0$, then the trivial solution of (3.96) is stable when the integral

$$\int_0^t m(s)\,ds \tag{3.98}$$

is bounded and unstable when (3.98) is unbounded.

Let $A = 2\lambda$. We can take $a(u) = b(u) = u^2$. Suppose that $\int_0^t m(t)\,dt = \beta > 0$. It therefore follows, from (3.97), that the system (3.96) is practically stable if $\beta \leq \frac{3}{8\lambda^2}$ and practically unstable if $\beta > \frac{3}{8\lambda^2}$.

Hence, we get, by Theorem 3.67, that the system (3.94) is practically stable if $\beta \leq \frac{3}{8\lambda^2}$ and practically unstable if $\beta > \frac{3}{8\lambda^2}$.

In Example 3.69, we have used the single Lyapunov function $V(t, x)$. In this case the function $L_0(t, x) = V(t, x)$. To demonstrate the advantage of employing several Lyapunov functions, let us consider the following example.

Example 3.70. Consider the system

$$\begin{cases} \dot{x}(t) = e^{-t}x(t - h(t)) + y(t - h(t))\sin t - (x^3 + xy^2)\sin^2 t, \ t \neq t_k \\ \dot{y}(t) = x(t - h(t))\sin t + e^{-t}y(t - h(t)) - (x^2y + y^3)\sin^2 t, \ t \neq t_k \\ \Delta x(t) = a_k x(t) + b_k y(t), \ t = t_k, \ k = 1, 2, \ldots \\ \Delta y(t) = b_k x(t) + a_k y(t), \ t = t_k, \ k = 1, 2, \ldots, \end{cases} \tag{3.99}$$

where $t \geq 0$; $0 < h(t) < h$; $a_k = \frac{1}{2}(\sqrt{1 + c_k} + \sqrt{1 + d_k} - 2)$, $b_k = \frac{1}{2}(\sqrt{1 + c_k} - \sqrt{1 + d_k})$, $-1 < c_k \leq 0$, $-1 < d_k \leq 0$, $k = 1, 2, \ldots$; $0 < t_1 < t_2 < \cdots$ and $\lim_{k \to \infty} t_k = \infty$.

Suppose that we choose a single Lyapunov function $V(t, x, y) = x^2 + y^2$. Then the set Ω_1 is given by (3.95). Hence, using the inequality $2|ab| \leq a^2 + b^2$ and observing that $(x^2 + y^2)^2 \sin^2 t \geq 0$, we get

$$D^+_{(3.99)} V(t, x(t), y(t)) = 2x(t)\dot{x}(t) + 2y(t)\dot{y}(t)$$
$$\leq 2[|e^{-t}| + |\sin t|]V(t, x(t), y(t)),$$

for $t \geq 0$, $t \neq t_k$ and $(x, y) \in \Omega_1$.

Also,

$$V(t_k + 0, x(t_k) + a_k x(t_k) + b_k y(t_k), y(t_k) + b_k x(t_k) + a_k y(t_k))$$
$$= [(1 + a_k)x(t_k) + b_k y(t_k)]^2 + [(1 + a_k)y(t_k) + b_k x(t_k)]^2$$
$$\leq V(t_k, x(t_k), y(t_k)) + 2|c_k - d_k|V(t_k, x(t_k), y(t_k)), \quad k = 1, 2, \ldots.$$

It is clear that

$$\begin{cases} \dot{u}(t) = 2[|e^{-t}| + |\sin t|]u(t), \ t \neq t_k, \ t \geq 0 \\ \Delta u(t_k) = 2|c_k - d_k|u(t_k), \ k = 1, 2, \ldots, \end{cases}$$

where $u \in \mathbb{R}_+$, is not practically stable and, consequently we cannot deduce any information about the practical stability of the system (3.99) from Theorem 3.67, even though the system (3.99) is practically stable [132].

Now, let us take the function $V = (V_1, V_2)$, where the functions V_1 and V_2 are defined by $V_1(t, x, y) = \frac{1}{2}(x + y)^2$, $V_2(t, x, y) = \frac{1}{2}(x - y)^2$ so that $L_0(t, x, y) = x^2 + y^2$. This means that we can take $a(u) = b(u) = u^2$. Then

$$\Omega_1 = \left\{(x, y) \in PC[\mathbb{R}_+, \mathbb{R}^2_+] : V(s, x(s), y(s)) \leq V(t, x(t), y(t)), t - h \leq s \leq t\right\}.$$

Moreover, for $t \geq 0$, the inequalities

$$D^+_{(3.99)} V(t, x(t), y(t)) \leq F(t, V(t, x(t), y(t))), (x, y) \in \Omega_1, t \neq t_k, k - 1, 2, \ldots,$$

$$V(t_k + 0, x(t_k) + \Delta x(t_k), y(t_k) + \Delta y(t_k)) \leq \psi_k(V(t_k, x(t_k), y(t_k))), k = 1, 2, \ldots,$$

are satisfied with $F = (F_1, F_2)$, where

$$F_1(t, u_1, u_2) = 2(e^{-t} + \sin t)u_1,$$

$$F_2(t, u_1, u_2) = 2(e^{-t} - \sin t)u_2,$$

and $\psi_k(u) = u + C_k u$, $k = 1, 2, \ldots$, $C_k = \begin{pmatrix} c_k & 0 \\ 0 & d_k \end{pmatrix}$.

It is obvious that the functions F and ψ_k satisfy the conditions of Theorem 1.23 and the comparison system

$$
\begin{cases}
\dot{u}_1(t) = 2(e^{-t} + \sin t)u_1(t), \ t \neq t_k \\
\dot{u}_2(t) = 2(e^{-t} - \sin t)u_2(t), \ t \neq t_k \\
\Delta u_1(t_k) = c_k u_1(t_k), \ \Delta u_2(t_k) = d_k u_2(t_k), \ k = 1, 2, \ldots
\end{cases}
$$

is practically stable for any $0 < \lambda < A$, which satisfy, for example, $\exp(e^{-t_0} + 2) < (\frac{A}{\lambda})^2$, $t_0 \in \mathbb{R}_+$ [132]. Hence, Theorem 3.67 implies that the system (3.99) is also practically stable.

We have assumed, in Theorem 3.67, stronger requirements on L_0 only to unify all the practical results in one theorem. This puts burden on the comparison system (3.86). However, to obtain only non-uniform practical stability criteria, we could weaken certain assumptions of Theorem 3.67 as in the next result.

Theorem 3.71. *Assume that the conditions of Theorem 3.67 hold with the following changes in conditions* (5) *and* (6):

(5*) *There exist functions* $a \in K$ *and* $b(t, \cdot) \in K$, *such that*

$$
a(\|x\|) \le L_0(t, x) \le b(t, \|x\|), \ (t, x) \in [t_0, \infty) \times \Omega.
$$

(6*) $b(t_0, \lambda) < a(A)$ *for some* $t_0 \in \mathbb{R}$.

Then, the uniform or non-uniform practical stability properties of the system (1.32) *with respect to* $(b(t_0, \lambda), a(A))$ *imply the corresponding non-uniform practical stability properties of the system* (3.10) *with respect to* (λ, A).

We shall next consider a result which gives practical asymptotic stability of (3.10). We will use two Lyapunov like functions.

Theorem 3.72. *Assume that:*

(1) *Conditions H2.2, H2.3 hold.*

(2) $0 < \lambda < A$ *is given and* $S_A \subset \Omega$.

(3) *The functions* $V, W \in V_0$ *and* $a, c \in K$, $b(t, .) \in K$ *are such that*

$$
a(\|x\|) \le L_0(t, x) \le b(t, \|x\|), \ (t, x) \in [t_0, \infty) \times \Omega, \tag{3.100}
$$

$$
c(\|x\|)e \le W(t, x), \ (t, x) \in [t_0, \infty) \times \Omega, \tag{3.101}
$$

where $e \in \mathbb{R}_+^m$, $e = (1, 1, \ldots, 1)$,

$$
V(t_k + 0, x(t_k) + I_k(x(t_k))) \le V(t_k, x(t_k)), \ k = 1, 2, \ldots, \tag{3.102}
$$

$$
W(t_k + 0, x(t_k) + I_k(x(t_k))) \le W(t_k, x(t_k)), \ k = 1, 2, \ldots, \tag{3.103}
$$

and the inequality

$$D^+_{(3.10)}V(t, x(t)) \leq -d(L_1(t, x(t)))e, \quad t \neq t_k, \quad k = 1, 2, \ldots, \qquad (3.104)$$

is valid for $t \geq t_0$, $x \in \Omega_1$, $L_1(t, x) = \sum_{i=1}^{m} W_i(t, x)$, $d \in K$.

(4) *The function $D^+_{(3.10)}W(t, x(t))$ is bounded in G.*

(5) $b(t_0, \lambda) < a(A)$ *for some $t_0 \in \mathbb{R}$.*

Then the system (3.10) *is practically asymptotically stable with respect to* (λ, A).

Proof. By Theorem 3.67 with $F(t, u) \equiv -d(u)e$ and $\psi_k(u) \equiv u$, $t \geq t_0$, $k = 1, 2, \ldots$, it follows because of conditions for the function $W \in V_0$ that the system (3.10) is practically stable. Hence, it is enough to prove that every solution $x(t) = x(t; t_0, \varphi_0)$ with $\|\varphi_0\|_r < \lambda$ satisfies $\lim_{t \to \infty} \|x(t; t_0, \varphi_0)\| = 0$.

Suppose that this is not true. Then there exist $\varphi_0 \in PC[[-r, 0], \Omega]$: $\|\varphi_0\|_r < \lambda$, $\beta > 0$, $\gamma > 0$ and a sequence $\{\xi_k\}_{k=1}^{\infty} \in [t_0, \infty)$ such that for $k = 1, 2, \ldots$ the following inequalities are valid:

$$\xi_k - \xi_{k-1} \geq \beta, \quad \|x(\xi_k; t_0, \varphi_0)\| \geq \gamma.$$

From the last inequality and (3.101), we get

$$W(\xi_k, x(\xi_k; t_0, \varphi_0)) \geq c(\gamma)e, \quad k = 1, 2, \ldots . \qquad (3.105)$$

From condition (4) of Theorem 3.72, it follows that there exists a constant $M \in \mathbb{R}_+$ such that

$$\sup\{D^+_{(3.10)}W(t, x(t)) : \quad t \in G\} \leq Me. \qquad (3.106)$$

By (3.103), (3.105) and (3.106), we obtain

$$W(t, x(t; t_0, \varphi_0)) \geq W(\xi_k, x(\xi_k; t_0, \varphi_0)) + \int_{\xi_k}^{t} D^+_{(3.10)}W(s, x(s; t_0, \varphi_0)) \, ds$$

$$= W(\xi_k, x(\xi_k; t_0, \varphi_0)) - \int_{t}^{\xi_k} D^+_{(3.10)}W(s, x(s; t_0, \varphi_0)) \, ds$$

$$\geq c(\gamma)e - Me(\xi_k - t) \geq c(\gamma)e - Me\varepsilon > \frac{c(\gamma)e}{2}$$

for $t \in [\xi_k - \varepsilon, \xi_k]$, where $0 < \varepsilon < \min\{\beta, \frac{c(\gamma)}{2M}\}$.

From the estimate obtained, making use of (3.104) and (3.102), we conclude that for $\xi_R \in \{\xi_k\}_{k=1}^{\infty}$, we have

$$0 \leq V(\xi_R, x(\xi_R; t_0, \varphi_0))$$

$$\leq V(t_0 + 0, \varphi_0(0)) + \int_{t_0}^{\xi_R} D_{(3.10)}^{+} V(s, x(s; t_0, \varphi_0))\, ds$$

$$\leq V(t_0 + 0, \varphi_0(0)) + \sum_{k=1}^{R} \int_{\xi_k - \varepsilon}^{\xi_k} D_{(3.10)}^{+} V(s, x(s; t_0, \varphi_0))\, ds$$

$$\leq V(t_0 + 0, \varphi_0(0)) - \sum_{k=1}^{R} \int_{\xi_k - \varepsilon}^{\xi_k} d(L_1(s, x(s; t_0, \varphi_0)))\, ds$$

$$\leq V(t_0 + 0, \varphi_0(0)) - Rd\left(\frac{mc(\gamma)}{2}\right)\varepsilon e,$$

which contradicts (3.100) for large R.

Thus, $\lim_{t \to \infty} \|x(t; t_0, \varphi_0)\| = 0$.   □

Corollary 3.73. *In Theorem* 3.72, *the following choices of* $W(t, x)$ *are admissible to yield the same conclusion:*

(i) $W(t, x) = \|x\| e$ *provided that* f *is bounded on* $(t_0, \infty) \times S_A$;

(ii) $W(t, x) = V(t, x)$.

3.6   Lipschitz stability

In the present section, Lipschitz stability of the zero solution of a system of impulsive functional differential equations will be considered. For nonlinear systems of differential equations without impulses, this notion was introduced by Dannan and Elaydi (1986) [80].

Consider the system of impulsive functional differential equations (3.10) for $\Omega \equiv S_\rho$, $\rho = \text{const} > 0$. Let $\varphi_0 \in PC[[-r, 0], S_\rho]$. Denote by $x(t; t_0, \varphi_0)$ the solution of system (3.10), satisfying the initial conditions (3.2).

Definition 3.74. The zero solution $x(t) \equiv 0$ of system (3.10) is said to be *uniformly Lipschitz stable*, if

$$(\exists M > 0)(\exists \delta > 0)(\forall \varphi_0 \in PC[[-r, 0], S_\rho] : \|\varphi_0\|_r < \delta)$$

$$(\forall t \geq t_0) : \|x(t; t_0, \varphi_0)\| \leq M\|\varphi_0\|_r.$$

Together with system (3.10), we shall consider the comparison equation

$$\begin{cases} \dot{u}(t) = g(t, u), \ t \geq t_0, \ t \neq t_k \\ \Delta u(t_k) = B_k(u(t_k)), \ t_k > t_0, \ k = 1, 2, \ldots, \end{cases} \tag{3.107}$$

where $g : [t_0, \infty) \times \mathbb{R}_+ \to \mathbb{R}$, $B_k : \mathbb{R}_+ \to \mathbb{R}$, $k = 1, 2, \ldots$.

Let $u_0 \in \mathbb{R}_+$. Denote by $u^+(t; t_0, u_0)$ the maximal solution of system (3.107), satisfying the initial condition $u^+(t_0) = u_0$.

Definition 3.75. The zero solution of equation (3.107) is said to be:

(a) *uniformly Lipschitz stable*, if

$$(\exists M > 0)(\exists \delta > 0)(\forall u_0 \in \mathbb{R}_+ : u_0 < \delta)$$

$$(\forall t \geq t_0) : u^+(t; t_0, u_0) \leq M u_0;$$

(b) *uniformly globally Lipschitz stable*, if

$$(\exists M > 0)(\forall u_0 \in \mathbb{R}_+)(\forall t \geq t_0) :$$

$$u^+(t; t_0, u_0) \leq M u_0.$$

Introduce the following conditions:

H3.14. $g(t, 0) = 0$, $t \in [t_0, \infty)$.

H3.15. $B_k \subset C[\mathbb{R}_+, \mathbb{R}]$, $B_k(0) = 0$ and $\psi_k(u) = u + B_k(u)$, $\psi_k : [0, \rho_0) \to [0, \rho)$, $k = 1, 2, \ldots$, are non-decreasing in u, $\rho_0 = \text{const} > 0$.

H3.16. For any $x \in S_\rho$ and any $k = 1, 2, \ldots$, the following inequalities are valid: $\|x + I_k(x)\| \leq \psi_k(\|x\|)$.

H3.17. The zero solution of equation (3.107) is uniformly Lipschitz stable.

In the proofs of the main results, we shall use the piecewise continuous auxiliary scalar functions $V : [t_0, \infty) \times S_\rho \to \mathbb{R}_+$, such that $V \in V_0$, and the comparison principle.

The main results are obtained by means of Theorem 1.24 for $\Omega \equiv S_\rho$, and by the following lemma.

Lemma 3.76 ([31]). *Assume that:*

(1) *The functions u, m satisfy $u, m \in PC[[t_0, \infty), \mathbb{R}_+]$.*

(2) $c_0 = \text{const} > 0$, $\beta_k = \text{const} \geq 0$, $k = 1, 2, \ldots$.

(3) *The function $p \in C[\mathbb{R}_+, \mathbb{R}_+]$ is non-decreasing in $\mathbb{R}_+$ and positive in $(0, \infty)$.*

(4) *For $t \geq t_0$, the following inequality is valid:*

$$u(t) \leq c_0 + \int_{t_0}^{t} m(s) p(u(s)) \, ds + \sum_{t_0 < t_k < t} \beta_k u(t_k).$$

(5) $\psi_k = \displaystyle\int_{c_k}^{u} \frac{ds}{p(s)}, \quad k = 1, 2, \ldots,$ *where* $c_k = (1 + \beta_k) \psi_{k-1}^{-1} \left(\displaystyle\int_{t_{k-1}}^{t_k} m(s) \, ds \right).$

(6) $\psi_0 = \displaystyle\int_{u_0}^{u} \frac{ds}{p(s)}, \quad u \geq u_0 > 0.$

Then

$$u(t) \leq \psi_k^{-1} \left(\int_{t_k}^{t} m(s) \, ds \right), \quad t_{k-1} < t \leq t_k, \quad k = 1, 2, \ldots.$$

Theorem 3.77. *Assume that:*

(1) *Conditions (1), (2) and (4) of Theorem 1.24 and H2.2, H2.3, H3.14–H3.17 hold.*

(2) *For $t \geq t_0$, $x \in \Omega_1$ and for sufficiently small $\sigma > 0$ the inequality*

$$\|x(t) + \sigma f(t, x_t)\| \leq \|x(t)\| + \sigma g(t, \|x(t)\|) + \varepsilon(\sigma), \quad t \neq t_k, \quad k = 1, 2, \ldots$$

is valid, where $\frac{\varepsilon(\sigma)}{\sigma} \to 0$ as $\sigma \to 0^+$.

Then the zero solution of system (3.10) is uniformly Lipschitz stable.

Proof. Let $\rho^* = \min(\rho, \rho_0)$. From condition H3.17, it follows that there exist constants $M > 0$ and $\delta > 0$ ($M\delta < \rho^*$) such that for $0 \leq u_0 < \delta$ and $t \geq t_0$ we have

$$u^+(t; t_0, u_0) \leq M u_0. \tag{3.108}$$

We shall prove that $\|x(t; t_0, \varphi_0)\| \leq M \|\varphi_0\|_r$ for $\|\varphi_0\|_r < \delta$ and $t \geq t_0$.

Suppose that this is not true. Then, there exist a solution $x(t) = x(t; t_0, \varphi_0)$ of system (3.10) for which $\|\varphi_0\|_r < \delta$ and $t^* \in (t_k, t_{k+1}]$ for some positive integer k such that $\|x(t^*)\| > M \|\varphi_0\|_r$ and $\|x(t)\| \leq M \|\varphi_0\|_r$ for $t_0 \leq t \leq t_k$.

From condition H3.16, it follows that

$$\|x(t_k + 0)\| = \|x(t_k) + I_k(x(t_k))\| \leq \psi_k(\|x(t_k)\|)$$
$$\leq \psi_k(M \|\varphi_0\|_r) \leq \psi_k(M\delta) \leq \psi_k(\rho^*) < \rho.$$

From the above estimate, it follows that there exists $t^0, t_k < t^0 \leq t^*$, such that

$$M \|\varphi_0\|_r < \|x(t^0)\| < \rho \quad \text{and} \quad \|x(t)\| < \rho, \quad t_0 \leq t \leq t^0. \tag{3.109}$$

Introduce the notations $V(t, x(t)) = \|x(t)\|$ and $u_0 = \|\varphi_0\|_r$. Since condition (2) of Theorem 3.77 is satisfied, then for $x \in \Omega_1, t \in [t_0, t^0], t \neq t_j, j = 1, 2, \ldots, k$, the following inequalities are valid:

$$D^+_{(3.10)}V(t, x(t)) = \lim_{\sigma \to 0+} \sup \frac{1}{\sigma}[\|x(t+\sigma)\| - \|x(t)\|]$$

$$\leq \lim_{\sigma \to 0+} \sup \frac{1}{\sigma}[\|x(t+\sigma)\| + \sigma g(t, \|x(t)\|) + \varepsilon(\sigma) - \|x(t) + \sigma f(t, x_t)\|]$$

$$\leq g(t, \|x(t)\|) + \lim_{\sigma \to 0+} \frac{\varepsilon(\sigma)}{\sigma} + \lim_{\sigma \to 0+} \|\frac{1}{\sigma}[x(t+\sigma) - x(t)] - f(t, x_t)\|$$

$$= g(t, \|x(t)\|) \quad = \quad g(t, V(t, x(t))).$$

From condition H3.16 for $j = 1, 2, \ldots, k$, we derive the inequalities

$$V(t_j + 0, x(t_j + 0)) = \|x(t_j + 0)\| = \|x(t_j) + I_j(x(t_j))\|$$
$$\leq \psi_j(\|x(t_j)\|) = \psi_j(V(t_j, x(t_j))).$$

Since

$$V(t_0 + 0, \varphi_0(0)) = \|\varphi_0(0)\| \leq \|\varphi_0\|_r = u_0,$$

then, from Theorem 1.24, there follows the estimate

$$\|x(t)\| = V(t, x(t)) \leq u^+(t; t_0, u_0), \quad t_0 \leq t \leq t^0. \tag{3.110}$$

From (3.108), (3.109) and (3.110), we are led to the inequalities

$$M\|\varphi_0\|_r < \|x(t^0)\| = V(t^0, x(t^0)) \leq u^+(t^0; t_0, u_0)$$
$$\leq Mu_0 = M\|\varphi_0\|_r.$$

The contradiction obtained shows that

$$\|x(t; t_0, \varphi_0)\| \leq M\|\varphi_0\|_r$$

for $\|\varphi_0\|_r < \delta$ and $t \geq t_0$. □

Theorem 3.78. *Assume that:*

(1) *Condition (1) of Theorem 3.77 holds.*

(2) *For $t \geq t_0$ and $x \in \Omega_1$, the inequality*

$$[x(t), f(t, x_t)]_+ \leq g(t, \|x(t)\|), \quad t \neq t_k, \quad k = 1, 2, \ldots$$

is valid, where $[x, y]_+ = \lim_{\sigma \to 0+} \sup \frac{1}{\sigma}[\|x + \sigma y\| - \|x\|], \quad x, y \in \mathbb{R}^n$.

Then the zero solution of system (3.10) is uniformly Lipschitz stable.

The proof of Theorem 3.78 is analogous to the proof of Theorem 3.77.

Consider the system of impulsive functional differential equations (3.24). Let $\varphi_0 \in PC[[-r, 0], \mathbb{R}^n]$. Denote by $x(t; t_0, \varphi_0)$ the solution of system (3.24), satisfying the initial conditions (3.19).

Definition 3.79. The zero solution $x(t) \equiv 0$ of system (3.24) is said to be uniformly globally Lipschitz stable, if

$$(\exists M > 0)(\forall \varphi_0 \in PC[[-r, 0], \mathbb{R}^n])(\forall t \geq t_0):$$

$$\|x(t; t_0, \varphi_0)\| \leq M \|\varphi_0\|_r.$$

Results on the uniform global Lipschitz stability of the zero solution of system (3.24) can be obtained, if in Theorem 3.77 and in Theorem 3.78, the set S_ρ is replaced by $\mathbb{R}^n$ and the condition H3.17 is replaced by the condition:

H3.17*. The zero solution of equation (3.107) is uniformly globally Lipschitz stable.

Theorem 3.80. *Assume that:*

(1) *Conditions* (1), (2), *and* (4) *of Theorem* 1.24 *and* H2.2, H2.3, H3.14, H3.15, H3.16, H3.17* *for* $x \in \mathbb{R}^n$ *hold.*

(2) *For* $t \geq t_0$, $x \in \Omega_1$ *and for sufficiently small* $\sigma > 0$, *the inequality*

$$\|x(t) + \sigma f(t, x_t)\| \leq \|x(t)\| + \sigma g(t, \|x(t)\|) + \varepsilon(\sigma), \ t \neq t_k, \ k = 1, 2, \ldots$$

is valid, where $\frac{\varepsilon(\sigma)}{\sigma} \to 0$ *as* $\sigma \to 0^+$.

Then the zero solution of system (3.24) *is uniformly globally Lipschitz stable.*

Theorem 3.81. *Suppose that condition* (1) *of Theorem* 3.80, *and condition* (2) *of Theorem* 3.78 *hold.*
  Then the zero solution of system (3.24) *is uniformly globally Lipschitz stable.*

Introduce the following conditions:

H3.18. The function $p \in C[\mathbb{R}_+, \mathbb{R}_+]$ is non-decreasing in $\mathbb{R}_+$, positive in $(0, \infty)$ and submultiplicative, i.e.

$$p(\lambda u) \leq p(\lambda) p(u) \quad \text{for} \quad \lambda > 0, u > 0.$$

H3.19. $p(\lambda u) \geq \mu(\lambda) p(u)$ for $\lambda > 0, u > 0$, where $\mu(\lambda) > 0$ for $\lambda > 0$.

H3.20. There exists a function $m \in C[(t_0, \infty), \mathbb{R}_+]$ such that the inequality

$$\|f(t, x_t)\| \leq m(t) p(\|x\|)$$

is valid for $(t, x_t) \in [t_0, \infty) \times PC[[-r, 0], \mathbb{R}^n]$.

H3.21. For $x \in \mathbb{R}^n$ and any $k = 1, 2, \ldots$ the inequalities

$$\|I_k(x)\| \le \beta_k \|x\|$$

are valid, where $\beta_k = \text{const} > 0$.

Theorem 3.82. *Assume that:*

(1) *Conditions H1.12, H1.13, H2.2, H2.3, H2.6–H2.8, H2.18, H3.18–H3.21 hold.*

(2) $\psi_k = \int_{c_k}^u \frac{ds}{p(s)}$, *where* $c_k = (1 + \beta_k)\psi_{k-1}^{-1}\left(\int_{t_{k-1}}^{t_k} m(s)\, ds\right)$, $k = 1, 2, \ldots$ *and*
$\psi_0 = \int_c^u \frac{ds}{p(s)}$, $u \ge c > 0$.

(3) $\psi_k(\infty) = \infty$, $k = 0, 1, 2, \ldots$.

(4) *For each* $k = 0, 1, 2, \ldots$, $t \in (t_k, t_{k+1}]$ *and* $\varphi_0 \in PC[[-r, 0], \mathbb{R}^n]$ *the following inequalities are valid:*

$$\psi_k^{-1}\left(\frac{p(\|\varphi_0\|_r)}{\|\varphi_0\|_r}\int_{t_k}^t m(s)\, ds\right) \le M, \quad 0 < M = \text{const}.$$

Then the zero solution of system (3.24) *is uniformly globally Lipschitz stable.*

Proof. For $t_k < t \le t_{k+1}, k = 0, 1, 2, \ldots$ the function $x(t) = x(t; t_0, \varphi_0)$ satisfies the integral equation

$$x(t) = x(t_k) + I_k(x(t_k)) + \int_{t_k}^t f(s, x_s)\, ds.$$

From this, we obtain inductively

$$x(t) = x(t_0 + 0) + \sum_{t_0 < t_k < t} I_k(x(t_k)) + \int_{t_0}^t f(s, x_s)\, ds, \quad t > t_0.$$

From conditions H3.18–H3.21 and the above equality, we get to the inequalities

$$\|x(t; t_0, \varphi_0)\| \le \|\varphi_0(0)\| + \sum_{t_0 < t_k < t} \|I_k(x(t_k))\| + \int_{t_0}^t \|f(s, x_s)\|\, ds$$

$$\le \|\varphi_0\|_r + \sum_{t_0 < t_k < t} \beta_k \|x(t_k)\| + \int_{t_0}^t m(s)p(\|x(s)\|)\, ds,$$

from which we obtain the estimates

$$\frac{\|x(t; t_0, \varphi_0)\|}{\|\varphi_0\|_r}$$

$$\le 1 + \sum_{t_0 < t_k < t} \beta_k \frac{\|x(t_k; t_0, \varphi_0)\|}{\|\varphi_0\|_r} + \int_{t_0}^t \frac{m(s)}{\|\varphi_0\|_r} p\left(\|\varphi_0\|_r \frac{\|x(s; t_0, \varphi_0)\|}{\|\varphi_0\|_r}\right) ds$$

$$\le 1 + \sum_{t_0 < t_k < t} \beta_k \frac{\|x(t_k; t_0, \varphi_0)\|}{\|\varphi_0\|_r} + \int_{t_0}^t \frac{p(\|\varphi_0\|_r)}{\|\varphi_0\|_r} m(s) p\left(\frac{\|x(s; t_0, \varphi_0)\|}{\|\varphi_0\|_r}\right) ds,$$

for $t \ge t_0$.

To the last inequality, we apply Lemma 3.76 and we are led to the inequality

$$\|x(t; t_0, \varphi_0)\| \leq \|\varphi_0\|_r \psi_k^{-1}\left(\frac{p(\|\varphi_0\|_r)}{\|\varphi_0\|_r} \int_{t_0}^t m(s)\, ds\right), \quad t \in (t_k, t_{k+1}], \ k = 0, 1, 2, \dots.$$

From the last inequality and condition (4) of Theorem 3.82, it follows that $\|x(t; t_0, \varphi_0)\| \leq M \|\varphi_0\|_r$ for $\varphi_0 \in PC[[-r, 0], \mathbb{R}^n]$ and $t \geq t_0$. ☐

Example 3.83. Consider the linear impulsive system of functional differential equations:

$$\begin{cases} \dot{x}(t) = Ax(t) + Bx(t - r(t)), \ t \neq t_k, \ t \geq 0 \\ \Delta x(t_k) = C_k x(t_k), \ t_k > t_0, \quad k = 1, 2, \dots, \end{cases} \tag{3.111}$$

where $x \in \mathbb{R}^n$; $0 < r(t) \leq r$; A, B and C_k, $k = 1, 2, \dots$ are constant matrices of type $(n \times n)$; $0 < t_1 < t_2 < \cdots$ and $\lim_{k \to \infty} t_k = \infty$.

Consider the Lyapunov function $V(t, x) = \|x\|$. Then, the set Ω_1 is

$$\Omega_1 = \{x \in PC[(0, \infty), \mathbb{R}^n] : \|x(s)\| \leq \|x(t)\|, \ t - r \leq s \leq t\}.$$

For $x \in \Omega_1$ and for $t \geq 0$, $t \neq t_k$, $k = 1, 2, \dots$, we have

$$[x(t), Ax(t) + Bx(t - r(t))]_+ \leq \mu(A + B)\|x(t)\|,$$

where $\mu(A + B)$ is Lozinskii's "logarithmic norm" of the matrix $A + B$ defined by the equality

$$\mu(A + B) = \lim_{\sigma \to 0^+} \sup \frac{1}{\sigma}[\|E + \sigma(A + B)\|_1 - 1],$$

$\|G\|_1 = \sup_{\|x\| \leq 1} |Gx|$ is the norm of the $(n \times n)$-matrix G and E is unit $(n \times n)$-matrix.

If $\mu(A + B) \leq 0$ and there exist constants $d_k > 0$, $k = 1, 2, \dots$ such that

$$\|E + C_k\|_1 \leq d_k, \quad \prod_{k=1}^{\infty} d_k < \infty,$$

then the zero solution of system

$$\begin{cases} \dot{u}(t) = \mu(A + B)u(t), \ t \geq 0, \ t \neq t_k \\ \Delta u(t_k) = (d_k - 1)u(t_k), \ t_k > 0, \ k = 1, 2, \dots \end{cases}$$

is uniformly globally Lipschitz stable. According to Theorem 3.81, the zero solution of system (3.111) is uniformly globally Lipschitz stable.

Example 3.84. Gopalsamy [91] proposed a model of a single-species population exhibiting the so-called *Allee effect* in which the per-capita growth rate is a quadratic function of the density and subject to delays. In particular, he studied the equation

$$\dot{N}(t) = N(t)[a + bN(t - r) - cN^2(t - r)], \quad t \geq 0, \quad (3.112)$$

where $a, c \in (0, \infty)$; $b \in \mathbb{R}$; and $r \in [0, \infty)$.

We consider the following model:

$$\begin{cases} \dot{N}(t) = N(t)[a(t) + b(t)N(t - r(t)) - c(t)N^2(t - r(t))], \quad t \neq t_k, \ t \geq 0 \\ \Delta N(t_k) = N(t_k + 0) - N(t_k) = I_k(N(t_k)), \ k = 1, 2, \ldots, \end{cases}$$

$$(3.113)$$

where $0 < r(t) \leq r; a, b, c$ are continuous functions, a and c are positive functions; $0 < t_1 < t_2 < \cdots$, $\lim_{k \to \infty} t_k = \infty$ and $I_k : \mathbb{R}_+ \to \mathbb{R}$ are functions which characterize the magnitude of the impulse effect at the moments t_k.

Let $\phi : [-r, 0] \to \mathbb{R}_+$ be a continuous function. The initial conditions for (3.113) are assumed to be as follows:

$$N(s) = \phi(s) \geq 0 \text{ for } -r \leq s < 0, \ N(0) > 0.$$

Define the function $V(t, N) = |N|$. Then, the set

$$\Omega_1 = \{N \in PC[(0, \infty), \mathbb{R}_+] : |N(s)| \leq |N(t)|, \ t - r \leq s \leq t\}.$$

Let the following conditions hold:

(1) There exist functions $p \in C[\mathbb{R}_+, \mathbb{R}]$ and $q \in K$ such that

$$[N(t), N(t)(a(t) + b(t)N(t - r(t)) - c(t)N^2(t - r(t)))]_+ \leq p(t)q(N(t)),$$

for $t \geq 0$, $N \in \Omega_1$ and $|N(t)| < \rho$, $\rho = \text{const} > 0$.

(2) There exist functions $B_k \in K$ and $\psi_k(N) = N + B_k(N)$, $\psi_k : [0, \rho_0) \to [0, \rho)$, $k = 1, 2, \ldots$ such that

$$|N + I_k(N)| \leq \psi_k(|N|), \quad k = 1, 2, \ldots$$

for $|N| < \rho$.

(3) For any $h \in (0, \rho_0)$, the inequality

$$\int_{t_k}^{t_{k+1}} p(s) \, ds + \int_h^{\psi_k(h)} \frac{ds}{q(s)} \leq 0, \quad k = 1, 2, \ldots$$

is valid.

Then, the zero solution of the system

$$\begin{cases} \dot{N}(t) = p(t)q(N(t)), \ t \geq 0, \ t \neq t_k \\ N(t_k + 0) = \psi_k(N(t_k)), \ t_k > t_0, \ k = 1, 2, \ldots \end{cases}$$

is uniformly Lipschitz stable. According to Theorem 3.78, the zero solution of system (3.113) is uniformly Lipschitz stable.

3.7   Stability in terms of two measures

In this section, we shall consider a more general case and develop the stability theory of the impulsive functional differential equations in terms of two measures. The priorities of this approach are useful and well known in the investigations on the stability and boundedness of the solutions of differential equations, as well as in the generalizations obtained by this method [131–133].

Consider the system of impulsive functional differential equations (3.10). Let $\varphi_0 \in PC[[-r, 0], \Omega]$. Denote by $x(t; t_0, \varphi_0)$ the solution of system (3.10), satisfying the initial conditions (3.2).

In the further considerations, we shall use piecewise continuous auxiliary functions $V : [t_0 - r, \infty) \times \Omega \to \mathbb{R}_+$ which belong to the class V_0 and the comparison principle.

We shall use the following notations:

$$CK = \{a \in C[[t_0 - r, \infty) \times \mathbb{R}_+, \mathbb{R}_+] : a(t, \cdot) \in K \text{ for any fixed } t \in [t_0 - r, \infty) \};$$

$$\Gamma = \{h \in V_0 : \inf_{(t,x)} h(t, x) = 0\}.$$

Definition 3.85. Let $h, h^0 \in \Gamma$ and define, for $\varphi \in PC[[-r, 0], \mathbb{R}^n]$,

$$
\begin{cases}
h_0(t, \varphi) = \sup_{-r \leq s \leq 0} h^0(t + s, \varphi(s)) \\
\bar{h}(t, \varphi) = \sup_{-r \leq s \leq 0} h(t + s, \varphi(s)).
\end{cases}
\tag{3.114}
$$

Then:

(a) h_0 is *finer* than $\bar{h}$, if there exist a number $\delta > 0$ and a function $\phi \in K$ such that $h_0(t, \varphi) < \delta$ implies $\bar{h}(t, \varphi) \leq \phi(h_0(t, \varphi))$;

(b) h_0 is *weakly finer* than $\bar{h}$, if there exist a number $\delta > 0$ and a function $\phi \in CK$ such that $h_0(t, \varphi) < \delta$ implies $\bar{h}(t, \varphi) \leq \phi(t, h_0(t, \varphi))$.

Definition 3.86. Let $h, h^0 \in \Gamma$ and $V \in V_0$. The function V is said to be:

(a) *h-positively definite*, if there exist a number $\delta > 0$ and a function $a \in K$ such that $h(t, x) < \delta$ implies $V(t, x) \geq a(h(t, x))$;

(b) *h_0-decrescent*, if there exist a number $\delta > 0$ and a function $b \in K$ such that $h_0(t, \varphi) < \delta$ implies $V(t + 0, x) \leq b(h_0(t, \varphi))$;

(c) *h_0-weakly decrescent*, if there exist a number $\delta > 0$ and a function $b \in CK$ such that $h_0(t, \varphi) < \delta$ implies $V(t + 0, x) \leq b(t, h_0(t, \varphi))$.

We shall use the following definitions of stability of the system (3.24) in terms of two different measures, that generalize various classical notions of stability.

Definition 3.87. Let $h, h^0 \in \Gamma$ and h_0 be defined by (3.114). The system (3.10) is said to be:

(a) (h_0, h)-*stable*, if

$$(\forall t_0 \in \mathbb{R})(\forall \varepsilon > 0)(\exists \delta = \delta(t_0, \varepsilon) > 0)$$

$$(\forall \varphi \in PC[[-r, 0], \Omega] : h_0(t_0, \varphi) < \delta)(\forall t \geq t_0) : \quad h(t, x(t; t_0, \varphi)) < \varepsilon;$$

(b) (h_0, h)-*uniformly stable*, if the number δ in (a) is independent of t_0;

(c) (h_0, h)-*equi-attractive*, if

$$(\forall t_0 \in \mathbb{R})(\exists \delta = \delta(t_0) > 0)(\forall \varepsilon > 0)(\exists T = T(t_0, \varepsilon) > 0)$$

$$(\forall \varphi \in PC[[-r, 0], \Omega] : \quad h_0(t_0, \varphi) < \delta)(\forall t \geq t_0 + T) : \quad h(t, x(t; t_0, \varphi)) < \varepsilon;$$

(d) (h_0, h)-*uniformly attractive*, if the numbers δ and T in (c) are independent of t_0;

(e) (h_0, h)-*equi-asymptotically stable*, if it is (h_0, h)-stable and (h_0, h)-equi-attractive;

(f) (h_0, h)-*uniformly asymptotically stable*, if it is (h_0, h)-uniformly stable and (h_0, h)-uniformly attractive;

(g) (h_0, h)-*unstable*, if (a) fails to hold.

For a concrete choice of the measures h_0 and h, Definition 3.87 is reduced to the following particular cases:

(1) Lyapunov stability of the zero solution of (3.10), if

$$h_0(t_0, \varphi) = \|\varphi\|_r = \sup_{t \in [t_0 - r, t_0]} \|\varphi(t - t_0)\|, \quad h(t, x) = \|x\|.$$

(2) stability by part of the variables of the zero solution of (3.10), if

$$h_0(t_0, \varphi) = \|\varphi\|_r, \quad h(t, x) = \|x\|_k = \sqrt{x_1^2 + \cdots + x_k^2},$$

$$x = (x_1, \ldots, x_n), \quad 1 \leq k \leq n.$$

(3) Lyapunov stability of the non-null solution $x_0(t) = x_0(t; t_0, \phi_0)$ of (3.10), if
$h_0(t_0, \varphi) = \|\varphi - \phi_0\|_r, \quad h(t, x) = \|x - x_0(t)\|.$

(4) stability of conditionally invariant set B with respect to the set A, where $A \subset B \subset \mathbb{R}^n$, if

$$h_0(t_0, \varphi) = \sup_{t \in [t_0 - r, t_0]} d(\phi(t - t_0), A), \quad h(t, x) = d(x, B),$$

d being the distance function.

(5) eventual stability of (3.10), if $h(t, x) = \|x\|$ and $h_0(t_0, \varphi) = \|\varphi\|_r + \alpha(t - t_0)$, $\alpha \in K$ and $\lim_{t \to \infty} \alpha(t) = 0$.

Together with the system (3.10), we consider the scalar impulsive differential equation (3.107). Let $u_0 \in \mathbb{R}_+$. Denote by $u^+(t; t_0, u_0)$ the maximal solution of (3.107), satisfying the initial condition $u^+(t_0) = u_0$.

Assume $\rho > 0$, $h, h^0 \in \Gamma$, h_0 is defined by (3.114) and let

$$S(h, \rho) = \{(t, x) \in [t_0 - r, \infty) \times \mathbb{R}^n : h(t, x) < \rho\};$$

$$S(h_0, \rho) = \{(t, \varphi) \in [t_0, \infty) \times PC[[-r, 0], \mathbb{R}^n] : h_0(t, \varphi) < \rho\}.$$

Introduce the following conditions:

H3.22. $B_k \in C[\mathbb{R}_+, \mathbb{R}]$, $B_k(0) = 0$ and $\psi_k(u) = u + B_k(u)$ are non-decreasing with respect to u, $k = 1, 2, \ldots$.

H2.23. There exists ρ_0, $0 < \rho_0 < \rho$, such that for $x \in \Omega$, $h(t_k, x) < \rho_0$ implies $h(t_k + 0, x + I_k(x)) < \rho$, $k = 1, 2, \ldots$.

Theorem 3.88. *Assume that:*

(1) *Conditions (1), (2) and (4) of Theorem 1.24 and H2.2, H2.3, H3.14, H3.22, H3.23 hold.*

(2) *$h, h^0 \in \Gamma$ and h_0 is finer than $\bar{h}$, where $h_0, \bar{h}$ are defined by (3.114).*

(3) *For $\rho > 0$, there exists $V \in V_0$ such that H2.5 holds, $V : S(h, \rho) \cap S(h_0, \rho) \to \mathbb{R}_+$, V is h-positively definite and h_0-decrescent,*

$$V(t + 0, x + I_k(x)) \le \psi_k(V(t, x)), \quad x \in \Omega, \ t = t_k, \ t > t_0,$$

and the inequality

$$D^+_{(3.10)} V(t, x(t)) \le g(t, V(t, x(t))), \quad t \ne t_k$$

is valid for each $t \ge t_0$, $x \in \Omega_1$.

Then the stability properties of the trivial solution of the equation (3.107) imply the corresponding (h_0, h)-stability properties of system (3.10).

Proof. Let us first prove (h_0, h)-stability.

Since V is h-positively definite on $S(h, \rho) \cap S(h_0, \rho)$, then there exists a function $b \in K$ such that

$$V(t, x) \ge b(h(t, x)) \quad \text{as} \quad h(t, x) < \rho. \tag{3.115}$$

Let $0 < \varepsilon < \rho_0$, $t_0 \in \mathbb{R}$ be given and suppose that the trivial solution of equation (3.107) is stable. Then for given $b = b(\varepsilon) > 0$, there exists $\delta_0 = \delta_0(t_0, \varepsilon) > 0$ such that

$$u^+(t; t_0, u_0) < b(\varepsilon) \quad \text{as} \quad 0 \le u_0 < \delta_0, \ t \ge t_0. \tag{3.116}$$

We choose now $u_0 = V(t_0 + 0, \varphi(0))$. Since V is h_0-decrescent, there exist a number $\delta_1 > 0$ and a function $a \in K$ such that, for $h_0(t, \varphi)) < \delta_1$,

$$V(t + 0, x) \le a(h_0(t, \varphi)). \tag{3.117}$$

On the other hand h_0 is finer than $\bar{h}$ and there exist a number $\delta_2 > 0$ and a function $\phi \in K$ such that $h_0(t_0, \varphi) < \delta_2$ implies

$$\bar{h}(t_0, \varphi) \le \phi(h_0(t_0, \varphi)), \tag{3.118}$$

where $\delta_2 > 0$ is such that $\phi(\delta_2) < \rho$. Hence, by (3.114), we have

$$\begin{cases} h(t_0 + 0, \varphi(0)) \le \bar{h}(t_0, \varphi) \le \phi(h_0(t_0, \varphi)) < \phi(\delta_2) < \rho \\ h^0(t_0 + 0, \varphi(0)) \le h_0(t_0, \varphi) < \delta_2. \end{cases} \tag{3.119}$$

Setting $\delta_3 = \min(\delta_1, \delta_2)$. It follows, from (3.115), (3.119) and (3.117), that $h_0(t_0, \varphi) < \delta_3$ implies

$$b(h(t_0 + 0, \varphi(0))) \le V(t_0 + 0, \varphi(0)) \le a(h_0(t_0, \varphi)). \tag{3.120}$$

Choose $\delta = \delta(t_0, \varepsilon) > 0$ such that $0 < \delta < \delta_3$, $a(\delta) < \delta_0$ and let $x(t) = x(t; t_0, \varphi)$ to be such solution of system (3.10) that $h_0(t_0, \varphi) < \delta$. Then (3.120) shows that $h(t_0 + 0, \varphi(0)) < \varepsilon$, since $\delta_0 < b(\varepsilon)$.

We claim that

$$h(t, x(t)) < \varepsilon \quad \text{as} \quad t \ge t_0.$$

If it is not true, then there would exists a $t^* > t_0$ such that $t_k < t^* \le t_{k+1}$ for some fixed integer k and

$$h(t^*, x(t^*)) \ge \varepsilon \quad \text{and} \quad h(t, x(t)) < \varepsilon, \quad t_0 \le t \le t_k.$$

Since $0 < \varepsilon < \rho_0$, condition H3.23 shows that

$$h(t_k + 0, x(t_k + 0)) = h(t_k + 0, x(t_k) + I_k(x(t_k))) < \rho.$$

Therefore, there exists t^0, $t_k < t^0 \le t^*$ such that

$$\varepsilon \le h(t^0, x(t^0)) < \rho \quad \text{and} \quad h(t, x(t)) < \rho, \quad t_0 \le t \le t^0. \tag{3.121}$$

Applying now Theorem 1.24 for the interval $[t_0, t^0]$ and $u_0 = V(t_0 + 0, \varphi(0))$, we obtain

$$V(t, x(t; t_0, \varphi)) \le u^+(t; t_0, V(t_0 + 0, \varphi(0))), \quad t_0 \le t \le t^0. \tag{3.122}$$

So the implications (3.121), (3.115), (3.122) and (3.116) lead to

$$b(\varepsilon) \le b(h(t^0, x(t^0))) \le V(t^0, x(t^0))$$
$$\le u^+(t^0; t_0, V(t_0 + 0, \varphi(0))) < b(\varepsilon).$$

The contradiction, we have already obtained, shows that $h(t, x(t)) < \varepsilon$ for each $t \geq t_0$. Therefore, the system (3.10) is (h_0, h)-stable.

If we suppose that the trivial solution of (3.107) is uniformly stable, then it is clear that the number δ can be chosen independently of t_0 and thus we get the (h_0, h)-uniform stability of the system (3.10).

Let us suppose next that the trivial solution of (3.107) is equi-asymptotically stable, which implies that the system (3.10) is (h_0, h)-stable. So, for each $t_0 \in \mathbb{R}$ there exists a number $\delta_{01} = \delta_{01}(t_0, \rho) > 0$ such that if $h_0(t_0, \varphi) < \delta_{01}$ then $h(t, x(t; t_0, \varphi)) < \rho$ as $t \geq t_0$.

Let $0 < \varepsilon < \rho_0$ and $t_0 \in \mathbb{R}$. The equi-asymptotical stability of the null solution of the equation (3.107) implies that there exist $\delta_{02} = \delta_{02}(t_0) > 0$ and $T = T(t_0, \varepsilon) > 0$ such that for $0 \leq u_0 < \delta_{02}$ and $t \geq t_0 + T$ the next inequality holds:

$$u^+(t; t_0, u_0) < b(\varepsilon). \tag{3.123}$$

Choosing $u_0 = V(t_0 + 0, \varphi(0))$ as before, we find $\delta_{03} = \delta_{03}(t_0)$, $0 < \delta_{03} \leq \delta_{02}$ such that

$$a(\delta_{03}) < \delta_{02}. \tag{3.124}$$

It follows, from (3.117) and (3.124), that if $h_0(t_0, \varphi) < \delta_{03}$ then

$$V(t_0 + 0, \varphi(0)) < a(h_0(t_0, \varphi)) \leq a(\delta_{03}) < \delta_{02}.$$

In the case, by means of (3.123), we would have

$$u^+(t; t_0, V(t_0 + 0, \varphi(0))) < b(\varepsilon), \quad t \geq t_0 + T. \tag{3.125}$$

Assume $\delta_0 = \min(\delta_{01}, \delta_{02}, \delta_{03})$ and let $h_0(t_0, \varphi) < \delta_0$. Theorem 1.24 shows that if $x(t) = x(t; t_0, \varphi)$ is an arbitrary solution of the system (3.10), then the estimate (3.122) holds for all $t \geq t_0 + T$. Therefore, we obtain from (3.115), (3.122) and (3.125) that the inequalities

$$b(h(t, x(t))) \leq V(t, x(t)) \leq u^+(t; t_0, V(t_0 + 0, \varphi(0))) < b(\varepsilon)$$

hold for each $t \geq t_0 + T$. Hence, $h(t, x(t)) < \varepsilon$ as $t \geq t_0 + T$ which shows that the system (3.10) is (h_0, h)-equi-attractive.

In case we suppose that the trivial solution of (3.107) is uniformly asymptotically stable, we get that (3.10) is also (h_0, h)-uniformly asymptotically stable, since δ_0 and T will be independent of t_0. $\qquad \square$

Remark 3.89. It is well known that, in the stability theory of functional differential equations, the condition $D^+_{(3.10)} V(t, x(t)) \leq g(t, V(t, x(t)))$ allows the derivative of the Lyapunov function to be positive which may not even guarantee the stability of a functional differential system (see [98, 99]). However, as we can see from Theorem 3.88, impulses have played an important role in stabilizing a functional differential system [146, 213].

We have assumed in Theorem 3.88 stronger requirements on V, h, h_0 only to unify all the stability criteria in one theorem. This obviously puts burden on the comparison equation (3.107). However, to obtain only non-uniform stability criteria, we could weaken certain assumption of Theorem 3.88, as in the next result.

Theorem 3.90. *Assume that:*

(1) *Condition (1) of Theorem 3.88 holds.*

(2) $h, h^0 \in \Gamma$ *and* h_0 *is weakly finer than* $\bar{h}$, *where* $h_0, \bar{h}$ *are defined by (3.114).*

(3) *For* $\rho > 0$, *there exists* $V \in V_0$ *such that* H2.5 *holds*, $V : S(h, \rho) \cap S(h_0, \rho) \rightarrow \mathbb{R}_+$, V *is* h-*positively definite and* h_0-*weakly decrescent*,

$$V(t + 0, x + I_k(x)) \leq \psi_k(V(t, x)), \quad x \in \Omega, \, t = t_k, \, t > t_0,$$

and the inequality

$$D^+_{(3.10)} V(t, x(t)) \leq g(t, V(t, x(t))), \quad t \neq t_k$$

is valid for each $t \geq t_0$, $x \in \Omega_1$.

Then the uniform and non-uniform stability properties of the trivial solution of the equation (3.107) imply the corresponding non-uniform (h_0, h)-*stability properties of system (3.10).*

The proof of Theorem 3.90 is analogous to the proof of Theorem 3.88. However, Definition 3.85 (b) is used instead of Definition 3.85 (a), and Definition 3.86 (c) is used instead of Definition 3.86 (b).

Corollary 3.91. *Assume that:*

(1) *Conditions* H1.1, H1.3, H1.9, H1.11, H1.12, H1.13, H2.2, H2.3 *hold.*

(2) *Condition (2) of Theorem 3.88 holds.*

(3) *For* $\rho > 0$, *there exists* $V \in V_0$ *such that* H2.5 *holds*, $V : S(h, \rho) \cap S(h_0, \rho) \rightarrow \mathbb{R}_+$, V *is* h-*positively definite and* h_0-*decrescent*,

$$V(t + 0, x + I_k(x)) \leq V(t, x)), \quad x \in \Omega, \, t = t_k,$$

and the inequality

$$D^+_{(3.10)} V(t, x(t)) \leq 0, \quad t \neq t_k, \, k = 1, 2, \ldots$$

is valid for each $t \geq t_0$, $x \in \Omega_1$.

Then the system (3.10) is (h_0, h)-*uniformly stable.*

The proof of Corollary 3.91 could be done in the same way as in Theorem 3.88, using Corollary 1.25 this time.

Example 3.92. Consider the impulsive functional differential equation

$$\begin{cases} \dot{x}(t) = a(t)x^3(t) + b(t)x(t)x^2(t - r(t)), \ t \ge 0, \ t \ne t_k \\ \Delta x(t_k) = I_k(x(t_k)), \ k = 1, 2, \ldots, \end{cases} \tag{3.126}$$

where $x \in PC[\mathbb{R}_+, \mathbb{R}_+]$; $0 < r(t) \le r$; $a(t)$ and $b(t)$ are continuous in $\mathbb{R}_+$, $b(t) \ge 0$, $a(t) + b(t) \le -a < 0$; $I_k(x)$, $k = 1, 2, \ldots$ are continuous in $\mathbb{R}$ and such that $x + I_k(x) > 0$ and $|x + I_k(x)| \le |x|$ for $x > 0$; $0 < t_1 < t_2 < \cdots$ and $\lim_{k \to \infty} t_k = \infty$.

Let $x(t) = \varphi_1(t)$, $t \in [-r, 0]$. Let $h_0(t, \varphi_1) = \sup_{s \in [-r,0]} |\varphi_1(s)|$ and $h(t, x) = |x|$. We consider the function

$$V(t, x) = \begin{cases} \alpha e^{-\frac{1}{x^2}}, & \text{for } x > 0 \\ 0, & \text{for } x = 0. \end{cases}$$

The set Ω_1 is defined by

$$\Omega_1 = \left\{ x \in PC[\mathbb{R}_+, \mathbb{R}_+] : \ x^2(s) \le x^2(t), \ t - r < s \le t \right\}.$$

If $t \ge 0$ and $x \in \Omega_1$, we have

$$D^+_{(3.126)} V(t, x(t)) = \alpha e^{-\frac{1}{x^2(t)}} \cdot \frac{2}{x^3(t)} [a(t)x^3(t) + b(t)x(t)x^2(t - r(t))]$$

$$\le -2a V(t, x(t)), \quad t \ne t_k, \ k = 1, 2, \ldots.$$

Moreover,

$$V(t_k + 0, x(t_k) + I_k(x(t_k))) = \alpha e^{-\frac{1}{(x(t_k) + I_k(x(t_k)))^2}}$$

$$\le V(t_k, x(t_k)), \ k = 1, 2, \ldots.$$

Since the trivial solution of the equation

$$\begin{cases} \dot{u}(t) = -2au(t), \ t \ge 0, \ t \ne t_k \\ \Delta u(t_k) = 0, \end{cases}$$

is equi-asymptotically stable ([129]), then Theorem 3.90 with $g(t, u) = -2au$ and $B_k(u) = 0$, $k = 1, 2, \ldots$ shows that equation (3.126) is (h_0, h)-equi-asymptotically stable.

Example 3.93. Consider the impulsive functional differential equation

$$\begin{cases} \dot{x}(t) = -ax(t) + bx(t - r(t)) - e(t)g(x(t)), \ t \ge 0, \ t \ne t_k \\ \Delta x(t_k) = -\alpha_k x(t_k), \ k = 1, 2, \ldots, \end{cases} \tag{3.127}$$

where $x \in \mathbb{R}_+$; $a, b > 0$; $0 < r(t) \leq r$; $e(t) \geq 0$ is a continuous function; $g(0) = 0$ and $xg(x) > 0$ if $x > 0$; $0 \leq \alpha_k \leq 2$, $k = 1, 2, \ldots$; $0 < t_1 < t_2 < \cdots$ and $\lim_{k \to \infty} t_k = \infty$.

Let $x(t) = \varphi_2(t)$, $t \in [-r, 0]$. Let $h_0(t, \phi_2) = \sup_{s \in [-r,0]} |\varphi_2(s)|$ and $h(t, x) = |x|$. We consider the function $V(t, x) = x^2$. The set Ω_1 is defined by

$$\Omega_1 = \left\{ x \in PC[\mathbb{R}_+, \mathbb{R}_+] : x^2(s) \leq x^2(t), \ t - r < s \leq t \right\}.$$

If $t \geq 0$, we have

$$D^+_{(3.127)} V(t, x(t)) = -2ax^2(t) + 2bx(t)x(t - r(t)) - 2e(t)x(t)g(x(t))$$
$$\leq 2V(t, x(t))[-a + b], \ x \in \Omega_1, \ t \neq t_k, \ k = 1, 2, \ldots.$$

Moreover,

$$V(t_k + 0, x(t_k) - \alpha_k x(t_k)) = (1 - \alpha_k)^2 V(t_k, x(t_k))$$
$$\leq V(t_k, x(t_k)), \ k = 1, 2, \ldots.$$

Assume the inequality $a \geq b$ holds. Then Corollary 3.91 shows that the equation (3.127) is (h_0, h)-uniformly stable.

Let the inequality $b \leq a - \varepsilon$ hold for some positive ε. Applying Theorem 3.88, we obtain that (3.127) is (h_0, h)-uniformly asymptotically stable.

3.8   Boundedness in terms of two measures

Consider the system of impulsive functional differential equations (3.24). Let $\varphi \in PC[[-r, 0], \mathbb{R}^n]$. Denote by $x(t; t_0, \varphi)$ the solution of system (3.24), satisfying the initial conditions (3.19).

Definition 3.94. Let h, $h^0 \in \Gamma$ and h_0 be defined by (3.114). The system (3.24) is said to be:

(a) (h_0, h)-*uniformly bounded*, if

$$(\forall \alpha > 0)(\exists \beta = \beta(\alpha) > 0)(\forall t_0 \in \mathbb{R})$$

$$(\forall \varphi \in PC[[-r, 0], \mathbb{R}^n] : h_0(t_0, \varphi) < \alpha)(\forall t \geq t_0) : h(t, x(t; t_0, \varphi)) < \beta;$$

(b) (h_0, h)-*quasi-uniformly ultimately bounded*, if

$$(\exists B > 0)(\forall \alpha > 0)(\exists T = T(\alpha) > 0)(\forall t_0 \in \mathbb{R})$$

$$(\forall \varphi \in PC[[-r, 0], \mathbb{R}^n] : h_0(t_0, \varphi) < \alpha)(\forall t \geq t_0 + T) : h(t, x(t; t_0, \varphi)) < B;$$

(c) (h_0, h)-*uniformly ultimately bounded*, if (a) and (b) hold together.

Let $\rho > 0$. We shall use also the following classes of functions:

$$S^c(h^0, \rho) = \{(t, x) \in [t_0 - r), \infty) \times \mathbb{R}^n : h^0(t, x) \geq \rho\};$$
$$S^c(h_0, \rho) = \{(t, \varphi) \in [t_0, \infty) \times PC[[-r, 0], \mathbb{R}^n] : h_0(t, \varphi) \geq \rho\}.$$

Introduce the following condition:

H3.24. There exists ρ_0, $\rho_0 \geq \rho > 0$ such that $h^0(t_k, x) \geq \rho_0$ implies $h^0(t_k + 0, x + I_k(x)) \geq \rho$, $k = 1, 2, \ldots$.

Theorem 3.95. *Assume that:*

(1) *Conditions H1.1, H1.3, H1.9, H1.12, H1.13, H2.12, H3.22 and H3.24 hold.*

(2) *$h, h^0 \in \Gamma$ and $\bar{h}(t, \varphi) \leq \phi(h_0(t, \varphi))$ for some $\phi \in K$ where $h_0, \bar{h}$ are defined by (3.114).*

(3) *For $\rho > 0$, there exists $V \in V_0$ such that*

$$V(t, x) \geq a(h(t, x)) \quad \text{for} \quad (t, x) \in S^c(h^0, \rho), \tag{3.128}$$

$$V(t + 0, x) \leq b(h_0(t, \varphi)) \quad \text{for} \quad (t, \varphi) \in S^c(h_0, \rho), \tag{3.129}$$

where $a, b \in K$ and $a(u) \to \infty$ as $u \to \infty$,

$$V(t + 0, x(t) + I_k(x(t))) \leq V(t, x(t)), \ x \in \mathbb{R}^n, \ t = t_k, \ t > t_0, \tag{3.130}$$

and the inequality

$$D^+_{(3.24)}V(t, x(t)) \leq 0, \ (t, x) \in S^c(h^0, \rho), t \neq t_k \tag{3.131}$$

is valid for each $t \geq t_0$, $x \in \Omega_1$.

Then the system (3.24) is (h_0, h)-uniformly bounded.

Proof. Let $\alpha > \rho_0$ be given. Choose $\beta = \beta(\alpha) > 0$ so that

$$\beta > \max\{\rho_0, \ \phi(\alpha), \ a^{-1}(b(\alpha))\}.$$

Let $t_0 \in \mathbb{R}$ and $\varphi \in PC[[-r, 0], \mathbb{R}^n]$. Consider the solution $x(t) = x(t; t_0, \varphi)$ of (3.24) with $h_0(t_0, \varphi) < \alpha$. By the condition (2) of Theorem 3.95, we have

$$h(t_0 + 0, \varphi(0)) \leq \bar{h}(t_0, \varphi) \leq \phi(h_0(t_0, \varphi)) < \phi(\alpha) < \beta.$$

We claim that

$$h(t, x(t)) < \beta, \quad t > t_0.$$

If it is not true, then there exists some solution $x(t) = x(t; t_0, \varphi)$ of (3.24) with $h_0(t_0, \varphi) < \alpha$ and a $t^* > t_0$ such that $t_k < t^* \leq t_{k+1}$ for some fixed integer k and

$$h(t^*, x(t^*)) \geq \beta \quad \text{and} \quad h(t, x(t)) < \beta, \quad t_0 \leq t \leq t_k.$$

Applying now Corollary 1.25 for the interval $[t_0, t_k]$, we obtain

$$V(t, x(t; t_0, \varphi)) \leq V(t_0 + 0, \varphi(0)), \quad t_0 \leq t \leq t_k. \tag{3.132}$$

Since $h^0(t_k, x(t_k)) \geq \rho_0$, condition H3.24 shows that

$$h^0(t_k + 0, x(t_k + 0)) = h^0(t_k + 0, x(t_k) + I_k(x(t_k))) \geq \rho,$$

i.e. $(t_k + 0, x(t_k + 0)) \in S^c(h^0, \rho)$.
   So the implications (3.128), (3.130), (3.132) and (3.129) lead to

$$a(h(t_k + 0, x(t_k + 0))) \leq V(t_k + 0, x(t_k + 0)) = V(t_k + 0, x(t_k) + I_k(x(t_k)))$$
$$\leq V(t_k, x(t_k)) \leq V(t_0 + 0, \varphi(0))$$
$$\leq b(h_0(t_0, \varphi)) < b(\alpha) < a(\beta).$$

Therefore

$$h(t_k + 0, x(t_k + 0)) < \beta.$$

Thus, there exist $t_1^*, t_2^*, t_k \leq t_1^* < t_2^* \leq t^*$ such that

$$h^0(t_1^*, x(t_1^*)) = \alpha, \; h_0(t_1^*, x_{t_1^*}) = \alpha,$$

$$h(t_2^*, x(t_2^*)) = \beta, \; \bar{h}(t_2^*, x_{t_2^*}) = \beta$$

and

$$(t, x(t)) \in \overline{S}^c(h^0, \alpha) \cap \overline{S}(h, \beta),$$
$$(t, x_t) \in \overline{S}^c(h_0, \alpha) \cap \overline{S}(\bar{h}, \beta), \; t \in [t_1^*, t_2^*]. \tag{3.133}$$

By (3.129) we have

$$V(t_1^* + 0, x(t_1^* + 0)) = V(t_1^*, x(t_1^*)) \leq b(h_0(t_1^*, x_{t_1^*})) = b(\alpha) < a(\beta).$$

   We want to show that

$$V(t, x(t)) < a(\beta), \; t \in [t_1^*, t_2^*]. \tag{3.134}$$

Suppose that this is not true and let

$$\xi = \inf\{t_2^* \geq t > t_1^* : V(t, x(t)) \geq a(\beta)\}.$$

Since $V(t, x(t))$ is continuous at $\xi \in (t_1^*, t_2^*]$, we see that

$$V(\xi + 0, x(\xi + 0)) \geq a(\beta)$$

holds which implies that

$$D_{(3.24)}^+ V(\xi, x(\xi)) > 0,$$

which contradicts to (3.131). Hence, (3.134) holds. On the other hand, using (3.133) and (3.128), we have

$$V(t_2^*, x(t_2^*)) \geq a(h(t_2^*, x_{t_2^*})) = a(\beta),$$

which contradicts (3.134). Thus

$$h(t, x(t)) < \beta, \quad t \geq t_0$$

for any solution $x(t) = x(t; t_0, \varphi)$ of (3.24) with $h_0(t_0, \varphi) < \alpha$ and the system (3.24) is (h_0, h)-uniformly bounded.   □

Corollary 3.96. *If in Theorem 3.95 condition (3.131) is valid for $x \in \Omega_P$, then the conclusion of Theorem 3.95 is still valid.*

Theorem 3.97. *Assume that:*

(1) *Conditions (1) and (2) of Theorem 3.95 hold.*

(2) *For $\rho > 0$, there exists $V \in V_0$ such that (3.130) holds,*

$$a(\bar{h}(t, \varphi)) \leq V(t + 0, x) \leq b(h_0(t, \varphi)) \quad for \quad (t, \varphi) \in S^c(h_0, \rho), \qquad (3.135)$$

where $a, b \in K$ and $a(u) \to \infty$ as $u \to \infty$, and the inequality

$$D^+_{(3.24)} V(t, x(t)) \leq 0, \quad (t, x) \in S^c(h^0, \rho), t \neq t_k$$

is valid for each $t \geq t_0$, $x \in \Omega_P$.

Then the system (3.24) is (h_0, h)-uniformly bounded.

The proof of Theorem 3.97 is analogous to the proof of Theorem 3.95 and we shall omit it.

Theorem 3.98. *Assume that:*

(1) *Conditions (1) and (2) of Theorem 3.95 hold.*

(2) *For $\rho > 0$, there exists $V \in V_0$ such that (3.130) and (3.135) hold, and the inequality*

$$D^+_{(3.24)} V(t, x(t)) \leq -c(h_0(t, x_t)), \quad (t, x) \in S^c(h^0, \rho), t \neq t_k \qquad (3.136)$$

is valid for each $t \geq t_0$, $x \in \Omega_P$, $c \in K$.

Then the system (3.24) is (h_0, h)-uniformly ultimately bounded.

Proof. The system (3.24) is (h_0, h)-uniformly bounded by means of Theorem 3.97. Then, there exists a positive number B such that for each $t_0 \in \mathbb{R}$

$$h_0(t_0, \varphi) < \delta_0 \text{ implies } h(t, x(t; t_0, \varphi)) < B, \quad t \ge t_0.$$

Now, we consider the solution $x(t) = x(t; t_0, \varphi)$ of (3.24) with $h_0(t_0, \varphi) < \alpha$, where α is arbitrary number and $\delta_0 > \alpha > \rho_0$. Then there exists a positive number $\beta = \beta(\alpha) > \max\{\rho_0, \phi(\rho), a^{-1}(b(\alpha))\}$ and $\beta < B$ such that

$$h(t, x(t)) < \beta, \quad t \ge t_0.$$

Now, let the function $P : \mathbb{R}_+ \to \mathbb{R}_+$ be continuous and non-decreasing on $\mathbb{R}_+$, and $P(u) > u$ as $u > 0$. We set

$$\eta = \inf\{P(u) - u : \; a(\phi(\rho)) \le u \le a(\beta)\}.$$

Then

$$P(u) > u + \eta \text{ as } a(\phi(\rho)) \le u \le a(\beta), \tag{3.137}$$

and we choose the integer v such that

$$a(\phi(\rho)) + v\eta > a(\beta). \tag{3.138}$$

If $V(t + 0, x(t + 0)) \ge a(\phi(\rho_0))$ for some $t \ge t_0$ then

$$V(t, x(t)) \ge V(t + 0, x(t + 0)) \ge a(\phi(\rho_0)) \ge a(\phi(\rho)),$$

$$b(h_0(t, x_t)) \ge V(t + 0, x(t + 0)) \ge a(\phi(\rho_0)) \ge a(\phi(\rho))$$

and therefore

$$h_0(t, x_t) \ge b^{-1}(a(\phi(\rho))) = \delta_1.$$

Hence,

$$c(h_0(t, x_t)) \ge c(\delta_1) = \delta_2. \tag{3.139}$$

Let us denote

$$\xi_k = t_0 + k\frac{\eta}{\delta_2}, \quad k = 0, 1, 2, \dots, v.$$

We want to prove

$$V(t, x(t)) < a(\phi(\rho)) + (v - k)\eta, \quad t \ge \xi_k \tag{3.140}$$

for all $k = 0, 1, 2, \dots, v$.

Indeed, using Corollary 1.25, (3.135) and (3.138), we obtain

$$V(t, x(t; t_0, \varphi)) \le V(t_0 + 0, \varphi(0)) \le b(h_0(t_0, \varphi))$$

$$< b(\alpha) < a(\beta) < a(\phi(\rho)) + v\eta, \quad t > t_0 = \xi_0$$

that means the validity of (3.140) for $k = 0$.

Assume (3.140) to be fulfilled for some integer k, $0 < k < \nu$, i.e.

$$V(s, x(s)) < a(\phi(\rho)) + (\nu - k)\eta, \quad s \geq \xi_k. \tag{3.141}$$

We suppose now that

$$V(t, x(t)) \geq a(\phi(\rho)) + (\nu - k - 1)\eta, \quad \xi_k \leq t \leq \xi_{k+1}.$$

Then
$$a(\phi(\rho)) \leq V(t, x(t)) \leq V(t_0 + 0, \varphi(0)) \leq b(h_0(t_0, \varphi))$$
$$< b(\alpha) < a(\beta), \quad \xi_k \leq t \leq \xi_{k+1}$$

and (3.137) and (3.141) imply

$$P(V(t, x(t))) > V(t, x(t)) + \eta \geq a(\phi(\rho)) + (\nu - k)\eta$$
$$> V(s, x(s)), \quad \xi_k \leq s \leq t \leq \xi_{k+1}.$$

Therefore $x(\cdot) \in \Omega_P$ as $\xi_k \leq s \leq t \leq \xi_{k+1}$. Then conditions of Theorem 3.98 and (3.139) yield

$$V(\xi_{k+1}, x(\xi_{k+1})) \leq V(\xi_k + 0, x(\xi_k + 0)) - \int_{\xi_k}^{\xi_{k+1}} c(h_0(s, x_s)) \, ds$$
$$< a(\phi(\rho)) + (\nu - k)\eta - \delta_2[\xi_{k+1} - \xi_k] = a(\phi(\rho)) + (\nu - k - 1)\eta$$
$$< V(\xi_k, x(\xi_k)),$$

which contradicts to the fact that $x(\cdot) \in \Omega_P$ as $\xi_k \leq s \leq t \leq \xi_{k+1}$. Therefore, there exists t^*, $\xi_k \leq t^* \leq \xi_{k+1}$ such that

$$V(t^*, x(t^*)) < a(\phi(\rho)) + (\nu - k - 1)\eta$$

and (3.130) implies

$$V(t^* + 0, x(t^* + 0)) < a(\phi(\rho)) + (\nu - k - 1)\eta.$$

We will prove

$$V(t, x(t)) < a(\phi(\rho)) + (\nu - k - 1)\eta, \quad t \geq t^*.$$

Supposing the opposite, we set

$$\mu = \inf\{t \geq t^* : V(t, x(t)) \geq a(\phi(\rho)) + (\nu - k - 1)\eta\}.$$

It follows, from (3.130) and (3.136), that $\mu \neq t_k, k = 1, 2, \ldots$, whence $V(\mu, x(\mu)) = a(\phi(\rho)) + (\nu - k - 1)\eta$. Then, for sufficiently close to zero $\sigma > 0$, we have

$$V(\mu + \sigma, x(\mu + \sigma)) \geq a(\phi(\rho)) + (\nu - k - 1)\eta,$$

whence

$$D^+_{(3.24)} V(\mu, x(\mu)) \geq 0.$$

On the other hand, we can prove as above that $x(\cdot) \in \Omega_P$ as $t^* \leq s \leq t \leq \mu$ and therefore

$$D^+_{(3.24)} V(\mu, x(\mu)) \leq -\delta_2 < 0.$$

The contradiction we have already obtained yields

$$V(t, x(t)) < a(\phi(\rho)) + (\nu - k - 1)\eta, \quad t \geq \xi_{k+1}.$$

It follows that (3.140) holds for all $k = 0, 1, 2, \ldots, \nu$.

Let $T = T(\alpha) = \nu \frac{\eta}{\delta_2}$. Then (3.140) implies

$$V(t, x(t)) < a(\phi(\rho)) \quad \text{as } t \geq t_0 + T. \tag{3.142}$$

Finally, the conditions of Theorem 3.98 and (3.142) lead us to

$$a(h(t, x(t))) \leq a(\bar{h}(t, x_t)) \leq V(t + 0, x(t + 0))$$
$$\leq V(t, x(t)) < a(\phi(\rho)) < a(\beta) < a(B) \quad \text{as } t \geq t_0 + T.$$

Therefore,

$$h_0(t_0, \varphi) < \alpha \quad \text{implies} \quad h(t, x(t)) < B \quad \text{as } t \geq t_0 + T$$

and (3.24) is a (h_0, h)-uniformly ultimately bounded system.   □

Example 3.99. Consider the impulsive functional differential equation with an infinite delay

$$\begin{cases} \dot{x}(t) = a(t, x(t)) + b(t, x(t - r)) + \displaystyle\int_{-\infty}^{0} h(t, s, x(t + s))ds, \quad t \neq t_k \\[2mm] \Delta x(t_k) = I_k(x(t_k)), \quad k = 1, 2, \ldots, \end{cases}$$

$$\tag{3.143}$$

where $x \in \mathbb{R}$; $t \geq 0$; $r > 0$; $a, b \in C[\mathbb{R}_1 \times \mathbb{R}, \mathbb{R}]$, $a(t, 0) - 0$, $|b(t, x)| \leq \beta(t)|x|$, $\beta \in C[\mathbb{R}_+, \mathbb{R}_+]$; $h \in C[\mathbb{R}_+ \times (-\infty, 0] \times \mathbb{R}, \mathbb{R}]$, $|h(t, s, v)| \leq m(s)|v|$, $m \in C[(-\infty, 0], \mathbb{R}_+]$; $|x + I_k(x)| \leq |x|$, $k = 1, 2, \ldots$; $-\infty < t_1 < t_2 < \cdots$ and $\lim_{k \to \infty} t_k = \infty$.

Let $h_0(t, x) = |x|_\infty = \sup_{s \in (-\infty, t]} |x(s)|$ and $h(t, x) = |x|$. We consider the function $V(t, x) = \frac{1}{2}x^2$. Let $\rho = 1$. Then $S^c_\rho = \{x \in \mathbb{R} : |x| \geq 1\}$.

Suppose that there exist constants $\mu > 1$ and $L > 0$ such that

$$2\mu \int_{-\infty}^{0} m(s)ds \leq 2L \leq -\frac{a(t, x)}{x} - \mu\beta(t), \quad t \geq 0, \; x \neq 0, \tag{3.144}$$

then the equation (3.143) is (h_0, h)-uniformly bounded. In fact, we can choose $a(u) = b(u) = u^2$. Set $P(u) = \mu^2 u^2$. Then the set Ω_P is defined by

$$\Omega_P = \{x \in PC[\mathbb{R}_+, \mathbb{R}] : x^2(s) \le \mu^2 x^2(t), \ -\infty < s \le t\}.$$

From (3.144), we have

$$D^+_{(3.143)} V(t, x(t))$$

$$\le x(t) a(t, x(t)) + \beta(t)|x(t)| \, |x(t - r)| + |x(t)| \int_{-\infty}^t m(v - t)|x(v)| dv$$

$$\le |x(t)|^2 \left[\frac{a(t, x(t))}{x(t)} + \mu \left(\beta(t) + \int_{-\infty}^0 m(s) ds \right) \right]$$

$$\le -L|x(t)|^2, \ x \in \Omega_P, \ |x| \ge 1, \ t \ge 0, \ t \ne t_k, \ k = 1, 2, \dots .$$

Moreover,

$$V(t_k + 0, x(t_k) + I_k(x(t_k))) = \frac{1}{2}(x(t_k) + I_k(x(t_k)))^2 \le V(t_k, x(t_k)), \ k = 1, 2, \dots .$$

Applying Corollary 3.96, we obtain that (3.143) is (h_0, h)-uniformly bounded.

Notes and comments

The idea of stability of sets was initiated by Yoshizawa in [223]. Theorems 3.2–3.7 are new. Close to them are the results of Stamova in [196]. Similar results for impulsive differential-difference equations are given by Bainov and Stamova in [44]. Theorems 3.12–3.18 are new. Similar results are given by Stamova in [193]. Theorems 3.21, 3.22 and 3.23 are taken from Stamova and Stamov [210]. Similar results for impulsive differential-difference equations are given by Bainov and Stamova in [49] and for linear impulsive differential-difference equations by Bainov, Stamova and Vatsala in [53].

The results in Section 3.2 are new. Similar results are given by Bainov and Stamova in [51] and by Stamova and Stamov in [208].

The parametric stability notion was introduced by Siljak in collaboration with Ikeda and Ohta in [186]. The results on the parametric stability for impulsive functional differential equations, listed in Section 3.3, are taken from Stamova [203].

The results in Section 3.4 are new.

The systematic study of practical stability was made in [132]. The results in Section 3.5 on the practical stability of systems under consideration are due to Stamova [201]. Similar results for impulsive differential-difference equations are given by Bainov and Stamova in [37] and by Bainov, Dishliev and Stamova in [26].

The notion of Lipschitz stability was introduced by Dannan and Elaydi [80]. The contents in Section 3.6 are from Bainov and Stamova [50]. Similar results are given for impulsive differential-difference equations in [34] and for linear impulsive differential-difference equations in [36].

The stability and boundedness notions in terms of two measures, listed in Section 3.7 and in Section 3.8, are generalizations of all previously considered notions of stability and boundedness. See Lakshmikantham and Liu [133]. The results in Section 3.7 are taken from Stamova [200]. The results in Section 3.8 are from [202]. For related results see Lakshmikantham, Leela and Martynyuk [131] and Stamova and Eftekhar [205].

Chapter 4

Applications

In the present chapter, we shall consider some applications to real world problems to illustrate the theory developed in the previous chapters.

Section 4.1 will deal with models of population dynamics. Uniform stability and uniform asymptotic stability of the equilibria will be discussed for impulsive Lotka–Volterra models with finite and infinite delays. We shall show that by means of appropriate impulsive perturbations we can control the system's population dynamics.

In Section 4.2, we shall consider impulsive neural networks with delays. The problems of global asymptotic and global exponential stability will be studied. We shall establish several stability criteria by employing Lyapunov functions and Razumikhin technique. These results can easily be used to design and verify globally stable networks.

In Section 4.3, we shall present models from economics. We shall again demonstrate the utility of the Lyapunov direct method. We shall show, also, that the role of impulses in changing the behavior of solutions of impulsive differential equations is very important.

4.1 Population models

Impulsive n-species Lotka–Volterra models with finite delays

The dynamical behavior of Lotka–Volterra models have been investigated by many authors. See, for example, [2–8, 65, 78, 91, 105, 109–112, 127, 135, 138–141, 144, 145, 165, 166, 176, 198, 211, 215, 218–220, 222, 224, 226, 230] and the references cited therein.

The classical n-species Lotka–Volterra model can be expressed as follows:

$$\dot{x}_i(t) = x_i(t)\left[b_i(t) - \sum_{j=1}^{n} a_{ij}(t)x_j(t)\right], \quad i = 1,\dots,n, \tag{4.1}$$

where $t \geq 0$; $x_i(t)$ represents the density of species i at the moment t; $b_i(t)$ is the reproduction rate function; and $a_{ij}(t)$ are functions which describe the effect of the jth population upon the ith population, which is positive if it enhances, and negative if it inhibits the growth.

The Lotka–Volterra type systems (4.1) are very important in the models of multi-species population dynamics. During the past few decades, a lot of work has been done on the problem of stability-complexity relationship in ecosystem's model, especially in the case of predator-prey type interactions described by the systems of type (4.1). These kinds of systems are of great interest not only for population dynamics or in chemical kinetics, but they are important in ecological modeling and all fields of science, from plasma physics to neural nets.

It is well known that the time delay is quite common for a natural population. There are considerable works on the study of the asymptotic stability of Lotka–Volterra type systems with time delays that have been developed in [4, 85, 105, 135, 138, 165, 166, 211, 214, 215, 218, 219, 222, 230]. In addition to these, the books of Gopalsamy [91] and Kuang [127] are good sources for these topics of Lotka–Volterra type systems with time delays,

$$\dot{x}_i(t) = x_i(t)\left[b_i(t) - a_{ii}(t)x_i(t) - \sum_{\substack{j=1 \\ j \neq i}}^n a_{ij}(t)x_j(t - \tau_{ij}(t))\right], \qquad (4.2)$$

where $i, j = 1, \ldots, n$; $a_{ij}, \tau_{ij} \in C[\mathbb{R}_+, \mathbb{R}_+]$; $b_i \in C[\mathbb{R}_+, \mathbb{R}]$; $0 \leq \tau_{ij} \leq \tau, \tau =$ const.

If at certain moments of time the evolution of the process is subject to sudden changes, then the population numbers vary by jumps. Therefore, it is important to study the behavior of the solutions of Lotka–Volterra systems with impulsive perturbations.

In this part of Section 4.1, we shall investigate the following n-species Lotka–Volterra type impulsive system with several deviating arguments:

$$\begin{cases} \dot{x}_i(t) = x_i(t)\left[b_i(t) - a_{ii}(t)x_i(t) - \displaystyle\sum_{\substack{j=1 \\ j \neq i}}^n a_{ij}(t)x_j(t - \tau_{ij}(t))\right], \ t \neq t_k \\ x_i(t_k^+) = x_i(t_k) + I_{ik}(x_i(t_k)), \ i = 1, \ldots, n, \ k = 1, 2, \ldots, \end{cases} \qquad (4.3)$$

where $n \geq 2$; $t \geq 0$; $I_{ik} : \mathbb{R}_+ \to \mathbb{R}$, $i = 1, \ldots, n, k = 1, 2, \ldots$; $0 < t_1 < t_2 < \cdots < t_k < \cdots$ are fixed impulsive points and $\lim_{k \to \infty} t_k = \infty$. In mathematical ecology, the system (4.3) denotes a model of the dynamics of an n-species system in which each individual competes with all others of the system for a common resource and the intra-species and inter-species competition involves deviating arguments τ_{ij} such that $0 \leq \tau_{ij}(t) \leq \tau$, where τ is a constant. The numbers $x_i(t_k)$ and $x_i(t_k^+)$ are, respectively, the population densities of species i before and after impulse perturbation at the moment t_k; and I_{ik} are functions which characterize the magnitude of the impulse effect on the species i at the moments t_k.

Let $\|x\| = |x_1| + \cdots + |x_n|$ denote the norm of $x \in \mathbb{R}^n$. Let $J \subset \mathbb{R}$ be an interval. Define the following class of functions:

$$CB[J, \mathbb{R}] = \{\sigma \in C[J, \mathbb{R}] : \sigma(t) \text{is bounded on } J\}.$$

Let $\varphi \in CB[[-\tau, 0], \mathbb{R}^n]$, $\varphi = \mathrm{col}(\varphi_1, \varphi_2, \ldots, \varphi_n)$. We denote by $x(t) = x(t; 0, \varphi) = \mathrm{col}(x_1(t; 0, \varphi), x_2(t; 0, \varphi), \ldots, x_n(t; 0, \varphi))$ the solution of system (4.3), satisfying the initial conditions

$$\begin{cases} x_i(s; 0, \varphi) = \varphi_i(s), \ s \in [-\tau, 0] \\ x_i(0^+; 0, \varphi) = \varphi_i(0), \ i = 1, \ldots, n, \end{cases} \tag{4.4}$$

and by $J^+(0, \varphi)$ the maximal interval of type $[0, \beta)$ in which the solution $x(t; 0, \varphi)$ is defined.

Let $\|\varphi\|_\tau = \max_{s \in [-\tau, 0]} \|\varphi(s)\|$ be the norm of the function $\varphi \in CB[[-\tau, 0], \mathbb{R}^n]$.

Introduce the following conditions:

H4.1. $b_i \in C[\mathbb{R}_+, \mathbb{R}]$, $i = 1, 2, \ldots, n$.

H4.2. $a_{ij}, \tau_{ij} \in C[\mathbb{R}_+, \mathbb{R}_+]$, $i, j = 1, 2, \ldots, n$.

H4.3. $0 < t_1 < t_2 < \cdots$ and $\lim_{k \to \infty} t_k = \infty$.

H4.4. $I_{ik} \in C[\mathbb{R}_+, \mathbb{R}]$, $i = 1, 2, \ldots, n, k = 1, 2, \ldots$.

H4.5. $x_i + I_{ik}(x_i) \geq 0$ for $x_i \in \mathbb{R}_+, i = 1, 2, \ldots, n, k = 1, 2, \ldots$.

Given a continuous function $g(t)$ which is defined on J, $J \subseteq \mathbb{R}$, we set

$$g^L = \inf_{t \in J} g(t), \quad g^M = \sup_{t \in J} g(t).$$

In our subsequent analysis, we shall use piecewise continuous functions $V : [0, \infty) \times \mathbb{R}_+^n \to \mathbb{R}_+$ which belong to the class V_0.

For $V \in V_0$ and for any $(t, x) \in [t_{k-1}, t_k) \times \mathbb{R}_+^n$, the right-hand derivative of the function $V \in V_0$ with respect to system (4.3) is defined by

$$D_{(4.3)}^+ V(t, x(t)) = \lim_{h \to 0^+} \sup \frac{1}{h} \Big[V(t + h, x(t + h)) - V(t, x(t)) \Big].$$

For a function $V \in V_0$ and for some $t \geq 0$ we shall use, also, the class

$$\Omega_1 = \{x \in PC[[0, \infty), \mathbb{R}_+^n] : V(s, x(s)) \leq V(t, x(t)), \ t - \tau \leq s \leq t\}.$$

In the proofs of the main theorems, we shall use the following lemmas.

Lemma 4.1. *Let the conditions H4.1–H4.4 hold. Then* $J^+(0, \varphi) = [0, \infty)$.

Proof. Lemma 4.1 follows from Theorem 1.17.

Indeed, since the conditions H4.1 and H4.2 hold then from the existence theorem for the corresponding system without impulses [91, 127, 135], it follows that the solution $x(t) = x(t; 0, \varphi)$ of problem (4.3), (4.4) is defined on $[0, t_1] \cup (t_k, t_{k+1}], k = 1, 2, \ldots.$ From conditions H4.3 and H4.4, we conclude that it is continuable for $t \geq 0$.   □

Lemma 4.2. *Assume that:*

(1) *Conditions H4.1–H4.5 hold.*

(2) $x(t) = x(t; 0, \varphi) = \mathrm{col}(x_1(t; 0, \varphi), x_2(t; 0, \varphi), \ldots, x_n(t; 0, \varphi))$ *is a solution of* (4.3), (4.4) *such that*

$$x_i(s) = \varphi_i(s) \geq 0, \quad \sup \varphi_i(s) < \infty, \quad \varphi_i(0) > 0,$$

$1 \leq i \leq n.$

Then $x_i(t) > 0, \ 1 \leq i \leq n, \ t \in [0, \infty).$

Proof. Since $\varphi_i(0) > 0$, the condition H4.5 holds, and the solution of (4.3) is defined by

$$x_i(t) = \varphi_i(0) \exp \left\{ \int_0^t \left[b_i(s) - a_{ii}(s)x_i(s) - \sum_{\substack{j=1 \\ j \neq i}}^n a_{ij}(s)x_j(s - \tau_{ij}(s)) \right] ds \right\},$$

$t \in [0, t_1],$

$$x_i(t) = x_i(t_k^+) \exp \left\{ \int_{t_k}^t \left[b_i(s) - a_{ii}(s)x_i(s) - \sum_{\substack{j=1 \\ j \neq i}}^n a_{ij}(s)x_j(s - \tau_{ij}(s)) \right] ds \right\},$$

$t \in (t_k, t_{k+1}],$

$$x_i(t_k^+) = x_i(t_k) + I_{ik}(x_i(t_k)), \quad i = 1, 2, \ldots, n, \quad k = 1, 2, \ldots,$$

then the solution of (4.3) is positive for $t \in [0, \infty).$ □

Lemma 4.3. *Assume that:*

(1) *The conditions of Lemma 4.2 hold.*

(2) *The function* $U_i(t) \geq 0$ *is the maximal solution of the logistic system*

$$\begin{cases} \dot{U}_i(t) = U_i(t) \left[|b_i^M| - a_{ii}^L U_i(t) \right], \ t \neq t_k \\ U_i(t_k^+) = U_i(t_k) + I_{ik}^M, \end{cases}$$

where $I_{ik}^M = \max\{I_{ik}(U_i(t_k))\}$ *for* $1 \leq i \leq n$ *and* $k = 1, 2, \ldots.$

(3) *The function $V_i(t) \geq 0$ is the minimal solution of the system*

$$
\begin{cases}
\dot{V}_i(t) = V_i(t)\left[b_i^L - a_{ii}^M V_i(t) - \sum_{\substack{j=1 \\ j \neq i}}^{n} a_{ij}^M \sup_{t-\tau \leq s \leq t} U_j(s) \right], & t \neq t_k \\
V_i(t_k^+) = V_i(t_k) + I_{ik}^L,
\end{cases}
$$

where $I_{ik}^L = \min\{I_{ik}(V_i(t_k))\}$ for $1 \leq i \leq n$ and $k = 1, 2, \ldots$.

(4) $0 \leq V_i(0^+) \leq \varphi_i(0) \leq U_i(0^+)$, $1 \leq i \leq n$.

Then

$$
V_i(t) \leq x_i(t) \leq U_i(t), \quad 1 \leq i \leq n, \quad t \in [0, \infty). \tag{4.5}
$$

Proof. Since all conditions of Lemma 4.2 are satisfied, the domain $\{\text{col}(x_1, x_2, \ldots, x_n) : x_i > 0, i = 1, 2, \ldots, n\}$ is positive invariant with respect to system (4.3).

From (4.3) for $i = 1, 2, \ldots, n$, we have

$$
\begin{cases}
\dot{x}_i(t) \leq x_i(t)\left[|b_i^M| - a_{ii}^L x_i(t) \right], & t \neq t_k \\
x_i(t_k^+) \leq x_i(t_k) + I_{ik}^M, & k = 1, 2, \ldots,
\end{cases}
$$

and

$$
\begin{cases}
\dot{x}_i(t) \geq x_i(t)\left[b_i^L - a_{ii}^M x_i(t) - \sum_{\substack{j=1 \\ j \neq i}}^{n} a_{ij}^M \sup_{t-\tau \leq s \leq t} x_j(s) \right], & t \neq t_k \\
x_i(t_k^+) \geq x_i(t_k) + I_{ik}^L, & k = 1, 2, \ldots.
\end{cases}
$$

Then from the differential inequalities for the piecewise continuous functions $V_i(t)$, $U_i(t)$ and $x_i(t)$ [130], we obtain that (4.5) is valid for $t \in [0, \infty)$ and $1 \leq i \leq n$. $\square$

Lemma 4.4. *Let the conditions of Lemma 4.2 hold and*

$$
b_i^L \geq \sum_{\substack{j=1 \\ j \neq i}}^{n} \frac{a_{ij}^M b_j^M}{a_{ii}^L}, \quad i, j = 1, 2, \ldots, n.
$$

Then for all $t \in [0, t_1] \cup (t_k, t_{k+1}]$, $k = 1, 2, \ldots$ and $1 \leq i \leq n$ the following inequalities are valid:

$$
\alpha_i \leq x_i(t) \leq \beta_i, \tag{4.6}
$$

where

$$
\alpha_i = \frac{b_i^L - \sum_{\substack{j=1 \\ j \neq i}}^{n} \frac{a_{ij}^M b_j^M}{a_{ii}^L}}{a_{ii}^M}, \quad \beta_i = \frac{|b_i^M|}{a_{ii}^L}.
$$

If, in addition, the functions I_{ik} are such that

$$\alpha_i \leq x_i + I_{ik}(x_i) \leq \beta_i$$

for $x_i \in \mathbb{R}_+$, $i = 1, 2, \ldots, n$, $k = 1, 2, \ldots$, then the inequalities (4.6) are valid for all $t \in [0, \infty)$ and $1 \leq i \leq n$.

Proof. From Lemma 4.3, we have that (4.5) are valid for $t \in [0, \infty)$ and $1 \leq i \leq n$.
  We shall prove that there exist positive constants α_i and β_i such that

$$\alpha_i \leq V_i(t) \leq U_i(t) \leq \beta_i \tag{4.7}$$

for all $t \in [0, t_1] \cup (t_k, t_{k+1}]$, $k = 1, 2, \ldots$ and $1 \leq i \leq n$.
  First, we shall prove that

$$U_i(t) \leq \beta_i \tag{4.8}$$

for all $t \in [0, t_1] \cup (t_k, t_{k+1}]$, $k = 1, 2, \ldots$ and $1 \leq i \leq n$.
  If $t \subset [0, \infty)$, $t \neq t_k$ and for some i, $i = 1, 2, \ldots, n$, $U_i(t) > \beta_i$, then for $t \in [0, t_1] \cup (t_k, t_{k+1}]$, $k = 1, 2, \ldots$, we will have

$$\dot{U}_i(t) < U_i(t) \left[|b_i^M| - a_{ii}^L U_i(t) \right] < 0.$$

This proves that (4.8) holds for all $t \in [0, t_1] \cup (t_k, t_{k\,|\,1}]$, $k = 1, 2, \ldots$ and $i = 1, 2, \ldots, n$, as long as $U_i(t)$ is defined.
  The inequality $\alpha_i \leq V_i(t)$ is proved by analogous way.
  Hence, the incqualities (4.7) are valid for all $t \in [0, t_1] \cup (t_k, t_{k+1}]$, $k = 1, 2, \ldots$ and $1 \leq i \leq n$.
  If, in addition, the functions I_{ik} are such that $\alpha_i \leq x_i(t_k) + I_{ik}(x_i(t_k)) \leq \beta_i$ for $x_i \in \mathbb{R}_+$, $i = 1, 2, \ldots, n$, $k = 1, 2, \ldots$, then inequalities (4.7) are valid for all $i = 1, 2, \ldots, n$ and $t \in [0, \infty)$. $\square$

Corollary 4.5. *Let the conditions of Lemma 4.4 hold, and the functions I_{ik} are such that*

$$\alpha_i \leq x_i + I_{ik}(x_i) \leq \beta_i \quad \text{for} \quad x_i \in \mathbb{R}_+, \quad i = 1, 2, \ldots, n, \quad k = 1, 2, \ldots \ .$$

Then:

(1) *The system (4.3) is uniformly ultimately bounded.*

(2) *There exist positive constants m and $M < \infty$ such that*

$$m \leq x_i(t) \leq M, \quad t \in [0, \infty). \tag{4.9}$$

Let $\phi \in CB[[-\tau, 0], \mathbb{R}^n]$, $\phi = \mathrm{col}(\phi_1, \phi_2, \ldots, \phi_n)$ and $x^*(t) = x^*(t; 0, \phi) = \mathrm{col}(x_1^*(t; 0, \phi), x_2^*(t; 0, \phi), \ldots, x_n^*(t; 0, \phi))$ be a solution of system (4.3), satisfying the initial conditions

$$\begin{cases} x_i^*(s; 0, \phi) = \phi_i(s), \ s \in [-\tau, 0] \\ x_i^*(0^+; 0, \phi) = \phi_i(0), \quad i = 1, 2, \ldots, n. \end{cases}$$

In the next, we shall suppose that

$$\varphi_i(s) \geq 0, \quad \sup \varphi_i(s) < \infty, \quad \varphi_i(0) > 0,$$

$$\phi_i(s) \geq 0, \quad \sup \phi_i(s) < \infty, \quad \phi_i(0) > 0, \quad i = 1, 2, \ldots, n.$$

Theorem 4.6. *Assume that:*

(1) *The conditions of Lemma 4.4 hold.*

(2) $m \leq x_i + I_{ik}(x_i) \leq M$ *for* $m \leq x_i \leq M$, $i = 1, 2, \ldots, n$, $k = 1, 2, \ldots$.

(3) *The following inequalities are valid*

$$m \min_{1 \leq i \leq n} a_{ii}(t) \geq M \max_{1 \leq i \leq n} \left(\sum_{\substack{j=1 \\ j \neq i}}^{n} a_{ji}(t) \right), \quad t \neq t_k, \quad k = 1, 2, \ldots.$$

Then the solution $x^*(t)$ *of system (4.3) is uniformly stable.*

Proof. Define a Lyapunov function

$$V(t, x(t)) = \sum_{i=1}^{n} V^i(t, x(t)) = \sum_{i=1}^{n} \left| \ln \frac{x_i(t)}{x_i^*(t)} \right|. \tag{4.10}$$

By the Mean Value Theorem and by (4.9), it follows that for any closed interval contained in $[0, t_1] \cup (t_k, t_{k+1}]$, $k = 1, 2, \ldots$ and for all $i = 1, 2, \ldots$

$$\frac{1}{M} |x_i(t) - x_i^*(t)| \leq |\ln x_i(t) - \ln x_i^*(t)| \leq \frac{1}{m} |x_i(t) - x_i^*(t)|. \tag{4.11}$$

From the inequalities (4.11), we obtain

$$V(0^+, x(0^+)) = \sum_{i=1}^{n} |\ln x_i(0^+) - \ln x_i^*(0^+)|$$

$$\leq \frac{1}{m} \sum_{i=1}^{n} |\varphi_i(0) - \phi_i(0)| \leq \frac{1}{m} \|\varphi - \phi\|_\tau. \tag{4.12}$$

For $t > 0$ and $t = t_k$, $k = 1, 2, \ldots$, we have

$$V(t_k^+, x(t_k^+)) = \sum_{i=1}^{n} \left| \ln \frac{x_i(t_k^+)}{x_i^*(t_k^+)} \right|$$

$$= \sum_{i=1}^{n} \left| \ln \frac{x_i(t_k) + I_{ik}(x_i(t_k))}{x_i^*(t_k) + I_{ik}(x_i^*(t_k))} \right| \leq \sum_{i=1}^{n} \left| \ln \frac{M}{m} \right| = \sum_{i=1}^{n} \left| \ln \frac{m}{M} \right|$$

$$\leq \sum_{i=1}^{n} \left| \ln \frac{x_i(t_k)}{x_i^*(t_k)} \right| = V(t_k, x(t_k)). \tag{4.13}$$

Consider the upper right-hand derivative $D_{(4.3)}^{+} V(t, x(t))$ of the function $V(t, x(t))$ with respect to system (4.3). For $t \geq 0$ and $t \neq t_k$, $k = 1, 2, \ldots$, we derive the estimate

$$D_{(4.3)}^{+} V(t, x(t)) = \sum_{i=1}^{n} \left(\frac{\dot{x}_i(t)}{x_i(t)} - \frac{\dot{x}_i^*(t)}{x_i^*(t)} \right) \operatorname{sgn} \left(x_i(t) - x_i^*(t) \right)$$

$$\leq \sum_{i=1}^{n} \left[-a_{ii}(t) |x_i(t) - x_i^*(t)| + \sum_{\substack{j=1 \\ j \neq i}}^{n} a_{ij}(t) |x_j(t - \tau_{ij}(t)) - x_j^*(t - \tau_{ij}(t))| \right]$$

$$\leq - \min_{1 \leq i \leq n} a_{ii}(t) \sum_{i=1}^{n} |x_i(t) - x_i^*(t)|$$

$$+ \max_{1 \leq i \leq n} \left(\sum_{\substack{j=1 \\ j \neq i}}^{n} a_{ji}(t) \right) \sum_{i=1}^{n} \sup_{s \in [t-\tau, t]} |x_i(s) - x_i^*(s)|.$$

From (4.11) for $x \in \Omega_1$, $t \geq 0$, $t \neq t_k$, $k = 1, 2, \ldots$, we have

$$\frac{1}{M} \sum_{i=1}^{n} |x_i(s) - x_i^*(s)| \leq V(s, x(s))$$

$$\leq V(t, x(t)) < \frac{1}{m} \sum_{i=1}^{n} |x_i(t) - x_i^*(t)|, \quad s \in [t - \tau, t],$$

and hence

$$\sum_{i=1}^{n} |x_i(s) - x_i^*(s)| \leq \frac{M}{m} \sum_{i=1}^{n} |x_i(t) - x_i^*(t)|.$$

Then

$$D_{(4.3)}^{+} V(t, x(t)) \leq 0, \tag{4.14}$$

$t \geq 0$ and $t \neq t_k$, $k = 1, 2, \ldots$.

Since all conditions of Theorem 2.8 are true, then the solution $x^*(t)$ of system (4.3) is uniformly stable.

Indeed, given $0 < \varepsilon < M$, choose $\delta = \frac{\varepsilon m}{2M}$. Then, from (4.11), (4.14) and (4.13), for $\|\varphi - \phi\|_\tau \leq \delta$ we obtain

$$\|x(t) - x^*(t)\| \leq MV(t, x(t)) \leq MV(0^+, x(0^+))$$

$$\leq \frac{M}{m}\|\varphi - \phi\|_\tau \leq \varepsilon,$$

$t \geq 0$. This shows that the solution $x^*(t)$ of system (4.3) is uniformly stable.   □

Theorem 4.7. *In addition to the assumptions of Theorem 4.6, suppose there exists a nonnegative constant μ such that*

$$m \min_{1 \leq i \leq n} a_{ii}(t) \geq \mu + M \max_{1 \leq i \leq n} \left(\sum_{\substack{j=1 \\ j \neq i}}^{n} a_{ji}(t) \right), \quad t \neq t_k, \quad k = 1, 2, \dots . \quad (4.15)$$

Then the solution $x^*(t)$ of system (4.3) is uniformly asymptotically stable.

Proof. We consider again the Lyapunov function (4.10). From (4.11) and (4.15), we obtain

$$D^+_{(4.3)}V(t, x(t)) \leq -\frac{\mu}{m} \sum_{i=1}^{n} |x_i(t) - x_i^*(t)|,$$

$t \geq 0$ and $t \neq t_k, k = 1, 2, \dots .$

Since all conditions of Theorem 2.9 are satisfied, the solution $x^*(t)$ of system (4.3) is uniformly asymptotically stable.   □

The results obtained can be applied in the investigation of the stability of any solution which is of interest.

One of the solutions which is an object of investigations for the systems of type (4.2) is the positive *periodic solution*. To consider periodic environmental factors, it is reasonable to study the Lotka–Volterra systems with periodic coefficients. The assumption of periodicity of the parameters b_i, a_{ij}, τ_{ij} is a way of incorporating of the environment periodicity (e.g. seasonal effects of weather condition, food supplies, temperature, etc). A very basic and important ecological problem associated with the study of multispecies population interaction in a periodic environment is the existence and asymptotic stability of periodic solutions. Such questions also arise in many other situations. The problem of existence of periodic solutions of population growth models without impulsive perturbations has been investigated by many authors [64, 65, 85, 211, 222, 230]. The main results are based on the coincidence degree theory [88]. Efficient sufficient conditions which guarantee the existence of periodic solutions for impulsive Lotka–Volterra systems are given in [135]. In [109] similar conditions are proved for neutral impulsive Lotka–Volterra systems.

The existence and stability of *equilibrium* states of some special cases of (4.3) without impulses has been studied extensively in the literature. In this case we do not need of the assumptions of the parameters periodicity. Many authors [2, 91, 127, 139, 140, 157, 211, 230] considered the following two-species competition Lotka–Volterra system with constant delays

$$\begin{cases} \dot{x}(t) = x(t)\,[r_1 - a_{11}x(t) - a_{12}y(t - \tau_{12})] \\ \dot{y}(t) = y(t)\,[r_2 - a_{21}x(t - \tau_{21}) - a_{22}y(t)]\,, \end{cases}$$

where $x(t)$ and $y(t)$ represent the population densities of two species at the moment t; constants $r_1 > 0$ and $r_2 > 0$ are the intrinsic growth rates; constants $a_{11} > 0$ and $a_{22} > 0$ are coefficients of intra-species competitions; $a_{12} > 0$ and $a_{21} > 0$ are inter-specific coefficients; and $\tau_{12} > 0$ and $\tau_{21} > 0$ are constant delays.

Example 4.8. For the system

$$\begin{cases} \dot{x}(t) = x(t)\,[7 - 12x(t) - y(t - \tau_{12})] \\ \dot{y}(t) = y(t)\,[8 - 2x(t - \tau_{21}) - 7y(t)]\,, \end{cases} \tag{4.16}$$

with parameters $r_1 = 7$, $r_2 = 8$, $a_{11} = 12$, $a_{22} = 7$, $a_{12} = 1$ and $a_{21} = 2$ one can show that the point $(x^*, y^*) = (\frac{1}{2}, 1)$ is an equilibrium which is uniformly asymptotically stable [85, 91].

Now, we consider the impulsive Lotka–Volterra system

$$\begin{cases} \dot{x}(t) = x(t)\,[7 - 12x(t) - y(t - \tau_{12})]\,, \ t \neq t_k \\ \dot{y}(t) = y(t)\,[8 - 2x(t - \tau_{21}) - 7y(t)]\,, \ t \neq t_k \\ \Delta x(t_k) = -\dfrac{3}{5}\left(x(t_k) - \dfrac{1}{2}\right),\ k = 1, 2, \ldots \\ \Delta y(t_k) = -\dfrac{4}{5}\left(y(t_k) - 1\right),\ k = 1, 2, \ldots, \end{cases} \tag{4.17}$$

where $0 < t_1 < t_2 < \cdots$ and $\lim_{k \to \infty} t_k = \infty$.

For the system (4.17), the point $(x^*, y^*) = (\frac{1}{2}, 1)$ is an equilibrium and all conditions of Theorem 4.7 are satisfied. In fact, for $\mu = \frac{3}{2}$, $m = \frac{1}{2}$ and $M = 1$, we have

$$\frac{1}{2} \le x(t_k) + I_{1k}(x(t_k)) = \frac{4x(t_k) + 3}{10} \le 1,$$

$$\frac{1}{2} \le y(t_k) + I_{2k}(y(t_k)) = \frac{y(t_k) + 4}{5} \le 1$$

for $\frac{1}{2} \le x(t_k) \le 1$, $\frac{1}{2} \le y(t_k) \le 1$, $k = 1, 2, \ldots$.

Therefore, the equilibrium $(x^*, y^*) = (\frac{1}{2}, 1)$ is uniformly asymptotically stable.

If, in the system (4.17), we consider the impulsive perturbations of the form:

$$\begin{cases} \Delta x(t_k) = -2\left(x(t_k) - \dfrac{1}{2}\right), \ k = 1, 2, \ldots \\ \Delta y(t_k) = -\dfrac{4}{5}\left(y(t_k) - 1\right), \ k = 1, 2, \ldots, \end{cases}$$

then the point $(x^*, y^*) = (\frac{1}{2}, 1)$ is again an equilibrium, but there is nothing we can say about its uniform asymptotic stability, because for $\frac{1}{2} \le x(t_k) \le 1$, we have $0 \le x(t_k) + I_{1k}(x(t_k)) \le \frac{1}{2}, k = 1, 2, \ldots$.

The example shows that by means of appropriate impulsive perturbations we can control the system's population dynamics. We can see that impulses are used to keep the stability properties of the system. On the other hand, a well-behaved system may lose its (asymptotic) stability due to uncontrolled impulsive inputs. Theorem 4.7 provides a set of sufficient conditions under which the asymptotic stability properties of a Lotka–Volterra system can be preserved under impulsive perturbations.

Example 4.9. The system

$$\begin{cases} \dot{x}(t) = x(t)\left[4 - 12x(t) - y(t - \tau_{12})\right] \\ \dot{y}(t) = y(t)\left[\dfrac{61}{4} - x(t - \tau_{21}) - 15y(t)\right], \end{cases} \tag{4.18}$$

with parameters $r_1 = 4$, $r_2 = \frac{61}{4}$, $a_{11} = 12$, $a_{22} = 15$, $a_{12} = 1$ and $a_{21} = 1$ has a uniformly asymptotically stable [85, 91] equilibrium point $(x^*, y^*) = (\frac{1}{3}, 0)$ which implies the second species will driven to extinction.

However, for the impulsive Lotka–Volterra system

$$\begin{cases} \dot{x}(t) = x(t)\left[4 - 12x(t) - y(t - \tau_{12})\right], \ t \ne t_k \\ \dot{y}(t) = y(t)\left[\dfrac{61}{4} - x(t - \tau_{21}) - 15y(t)\right], \ t \ne t_k \\ \Delta x(t_k) = -\dfrac{1}{2}\left(x(t_k) - \dfrac{1}{4}\right), \ k = 1, 2, \ldots \\ \Delta y(t_k) = -\dfrac{1}{3}\left(y(t_k) - 1\right), \ k = 1, 2, \ldots, \end{cases}$$

where $0 < t_1 < t_2 < \cdots$ and $\lim_{k \to \infty} t_k = \infty$, the point $(x^*, y^*) = (\frac{1}{4}, 1)$ is an equilibrium which is uniformly asymptotically stable. In fact, all conditions Theorem 4.7 are satisfied for $\mu = 2$, $m = \frac{1}{4}$ and $M = 1$ and

$$\frac{1}{4} \le x(t_k) + I_{1k}(x(t_k)) = \frac{4x(t_k) + 1}{8} \le 1,$$

$$\frac{1}{4} \le y(t_k) + I_{2k}(y(t_k)) = \frac{2y(t_k) + 1}{3} \le 1$$

for $\frac{1}{4} \le x(t_k) \le 1, \frac{1}{4} \le y(t_k) \le 1, k = 1, 2, \ldots$.

This shows that the impulsive perturbations can prevent the population from going extinct.

Impulsive n-species Lotka–Volterra cooperation models with finite delays

In this part of Section 4.1, we shall study asymptotic behavior of some n-species Lotka–Volterra cooperation systems with finite delays and impulsive perturbations at fixed moments of time.

Let $0 < t_1 < t_2 < \cdots$ and $\lim_{k \to \infty} t_k = \infty$. Consider the system:

$$
\begin{cases}
\dot{x}_i(t) = x_i(t)\left[r_i(t) - \dfrac{x_i(t - \tau_{ii}(t))}{a_i(t) + \sum_{\substack{j=1 \\ j \neq i}}^{n} b_j(t)x_j(t - \tau_{ij}(t))} - c_i(t)x_i(t) \right], \quad t \neq t_k \\
x_i(t_k^+) = x_i(t_k) + I_{ik}(x_i(t_k)), \quad i = 1, \dots, n, \ k = 1, 2, \dots,
\end{cases}
$$

$$(4.19)$$

where $t \geq 0$; $x_i(t)$ denotes the density of species i at the moment t; $r_i(t)$, $a_i(t)$, $b_i(t)$, $c_i(t)$ $(i = 1, 2, \dots, n)$ are the system parameters; $0 \leq \tau_{ij} \leq \tau$, $\tau = \text{const}$, $i, j = 1, 2, \dots, n$.

Let $\varphi \in CB[[-\tau, 0], \mathbb{R}^n]$, $\varphi = \text{col}(\varphi_1, \varphi_2, \dots, \varphi_n)$. We denote by $x(t) = x(t; 0, \varphi) = \text{col}(x_1(t; 0, \varphi), x_2(t; 0, \varphi), \dots, x_n(t; 0, \varphi))$ the solution of system (4.19), satisfying the initial conditions

$$
\begin{cases}
x_l(s; 0, \varphi) = \varphi_i(s), \quad s \in [-\tau, 0] \\
x_i(0^+; 0, \varphi) = \varphi_i(0), \quad i = 1, \dots, n.
\end{cases}
$$

$$(4.20)$$

Introduce the following condition:

H4.6. The functions $r_i(t)$, $a_i(t)$, $b_i(t)$ and $c_i(t)$ are continuous, positive and bounded on $\mathbb{R}_+$.

Lemma 4.10. *Assume that:*

(1) *Conditions H4.3–H4.6 hold.*

(2) $x(t) = x(t; 0, \varphi) = \text{col}(x_1(t; 0, \varphi), x_2(t; 0, \varphi), \dots, x_n(t; 0, \varphi))$ *is a solution of (4.19), (4.20) such that*

$$
x_i(s) = \varphi_i(s) \geq 0, \quad \sup \varphi_i(s) < \infty, \quad \varphi_i(0) > 0,
$$

$1 \leq i \leq n$.

Then $x_i(t) > 0$, $1 \leq i \leq n$, $t \in [0, \infty)$.

Proof. The proof of Lemma 4.10 is analogous to the proof of Lemma 4.2.   □

Theorem 4.11. *Assume that:*

(1) *The conditions of Lemma 4.10 hold.*

(2) *The functions I_{ik} are such that*

$$-x_i \le I_{ik}(x_i) \le 0 \ \text{ for } \ x_i \in \mathbb{R}_+, \ i = 1, 2, \ldots, n, \ k = 1, 2, \ldots .$$

Then the system (4.19) is uniformly ultimately bounded.

Proof. From the condition H4.6 and from the corresponding theorem for the continuous case ([214, 219, 222]), it follows that for all $t \in [0, t_1] \cup (t_k, t_{k+1}], k = 1, 2, \ldots$ and $1 \le i \le n$ there exist positive constants m_i^* and M_i^* such that the following inequalities are valid:

$$m_i^* \le x_i(t) \le M_i^*.$$

Using Lemma 4.10 and condition (2) of Theorem 4.11, we obtain

$$0 < x_i(t_k + 0) = x_i(t_k) + I_{ik}(x_i(t_k)) \le x_i(t_k) \le M_i^*.$$

Therefore, there exist positive constants m_i and M_i such that

$$m_i \le x_i(t) \le M_i,$$

$i = 1, 2, \ldots, n, \ t \in [0, \infty)$.   □

Corollary 4.12. *Let the conditions of Theorem 4.11 hold. Then there exist positive constants m and $M < \infty$ such that the inequalities (4.9) are valid.*

Let $\phi \in CB[[-\tau, 0], \mathbb{R}^n], \ \phi = \text{col}(\phi_1, \phi_2, \ldots, \phi_n)$ and $x^*(t) = x^*(t; 0, \phi) = \text{col}(x_1^*(t; 0, \phi), x_2^*(t; 0, \phi), \ldots, x_n^*(t; 0, \phi))$ be a solution of system (4.19), satisfying the initial conditions

$$\begin{cases} x_i^*(s; 0, \phi) = \phi_i(s), \ s \in [-\tau, 0] \\ x_i^*(0^+; 0, \phi) = \phi_i(0), \ i = 1, 2, \ldots, n. \end{cases}$$

In the following, we shall suppose that

$$\varphi_i(s) \ge 0, \quad \sup \varphi_i(s) < \infty, \quad \varphi_i(0) > 0,$$

$$\phi_i(s) \ge 0, \quad \sup \phi_i(s) < \infty, \quad \phi_i(0) > 0, \ i = 1, 2, \ldots, n.$$

Theorem 4.13. *Assume that:*

(1) *The conditions of Theorem 4.11 hold.*

(2) $m \le x_i + I_{ik}(x_i) \le M \ \text{ for } \ m \le x_i \le M, \ i = 1, 2, \ldots, n, \ k = 1, 2, \ldots .$

(3) *There exists a nonnegative constant μ such that*

$$m \min_{1 \le i \le n} c_i^L \ge \mu + M^2 \sum_{i=1}^{n} \max_{j \ne i} \frac{b_j^M}{\left(a_i^L + m \sum_{\substack{s=1 \\ s \ne i}}^{n} b_s^L\right)^2} > 0.$$

Then the solution $x^*(t)$ of system (4.19) is uniformly asymptotically stable.

Proof. Consider the Lyapunov function

$$V(t, x(t)) = \sum_{i=1}^{n} \left| \ln \frac{x_i(t)}{x_i^*(t)} \right|.$$

For $t = t_k, k = 1, 2, \ldots, (4.13)$ is valid.
For $t \ge 0$ and $t \ne t_k, k = 1, 2, \ldots,$ we have

$$D_{(4.19)}^{+} V(t, x(t)) = \sum_{i=1}^{n} \left(\frac{\dot{x}_i(t)}{x_i(t)} - \frac{\dot{x}_i^*(t)}{x_i^*(t)} \right) \operatorname{sgn}\left(x_i(t) - x_i^*(t)\right)$$

$$\le \sum_{i=1}^{n} \left\{ -c_i(t) |x_i(t) - x_i^*(t)| \right.$$

$$- \frac{1}{a_i(t) + \sum_{\substack{j=1 \\ j \ne i}}^{n} b_j(t) x_j^*(t - \tau_{ij}(t))} |x_j(t - \tau_{ij}(t)) - x_j^*(t - \tau_{ij}(t))|$$

$$+ \sum_{\substack{j=1 \\ j \ne i}}^{n} \frac{b_j(t) x_i(t - \tau_{ii}(t)) |x_j(t - \tau_{ij}(t)) - x_j^*(t - \tau_{ij}(t))|}{\left(a_i(t) + \sum_{\substack{s=1 \\ s \ne i}}^{n} b_s(t) x_s(t - \tau_{is}(t))\right) \left(a_i(t) + \sum_{\substack{s=1 \\ s \ne i}}^{n} b_s(t) x_s^*(t - \tau_{is}(t))\right)} \right\}$$

$$\le \sum_{i=1}^{n} \left\{ -c_i^L |x_i(t) - x_i^*(t)| \right.$$

$$+ \sum_{\substack{j=1 \\ j \ne i}}^{n} \frac{M b_j^M}{\left(a_i^L + m \sum_{\substack{s=1 \\ s \ne i}}^{n} b_s^L\right)^2} |x_j(t - \tau_{ij}(t)) - x_j^*(t - \tau_{ij}(t))| \right\}$$

$$\le - \min_{1 \le i \le n} c_i^L \sum_{i=1}^{n} |x_i(t) - x_i^*(t)|$$

$$+ \left(\sum_{i=1}^{n} \max_{j \ne i} \frac{M b_j^M}{\left(a_i^L + m \sum_{\substack{s=1 \\ s \ne i}}^{n} b_s^L\right)^2} \right) \sum_{i=1}^{n} \sup_{s \in [t - \tau, t]} |x_i(s) - x_i^*(s)|.$$

From (4.11) for any solution $x(t)$ of (4.3) such that

$$V(s, x(s)) \le V(t, x(t)), \quad t - \tau \le s \le t, t \ne t_k, k = 1, 2, \ldots,$$

we have

$$\sum_{i=1}^{n} |x_i(s) - x_i^*(s)| \le \frac{M}{m} \sum_{i=1}^{n} |x_i(t) - x_i^*(t)|.$$

Then

$$D_{(4.19)}^{+} V(t, x(t)) \le -\frac{\mu}{m} \sum_{i=1}^{n} |x_i(t) - x_i^*(t)| \le -\mu V(t, x(t)),$$

$t \ge 0$ and $t \ne t_k, k = 1, 2, \ldots$.

From the last estimate and (4.13), we get

$$V(t, x(t)) \le V(0^+, x(0^+))e^{-\mu t}, \quad t \in [0, \infty).$$

So,

$$\|x(t) - x^*(t)\| = \sum_{i=1}^{n} |x_i(t) - x_i^*(t)| \le M V(t, x(t))$$

$$\le M V(0^+, x(0^+))e^{-\mu t} \le \frac{M}{m} \|\varphi - \phi\|_\tau e^{-\mu t}, \quad t \in [0, \infty),$$

and this completes the proof of the theorem.   □

Example 4.14. The system

$$\begin{cases} \dot{x}(t) = x(t) \left[\dfrac{907}{224} - \dfrac{x(t - \tau_{11})}{1 + 4y(t - \tau_{12})} - 16x(t) \right] \\[4mm] \dot{y}(t) = y(t) \left[15 - \dfrac{y(t - \tau_{22})}{1 + 2x(t - \tau_{21})} - 14y(t) \right], \end{cases} \tag{4.21}$$

with parameters $r_1 = \frac{907}{224}, r_2 = 15, a_1 = a_2 = 1, b_1 = 2, b_2 = 4, c_1 = 16$ and $c_2 = 14$ has a uniformly asymptotically stable [214] equilibrium point $(x^*, y^*) = (0, 1)$ which implies the first species will go extinct.

However, for the impulsive Lotka–Volterra system

$$\begin{cases} \dot{x}(t) = x(t) \left[\dfrac{907}{224} - \dfrac{x(t - \tau_{11})}{1 + 4y(t - \tau_{12})} - 16x(t) \right], \; t \ne t_k \\[4mm] \dot{y}(t) = y(t) \left[15 - \dfrac{y(t - \tau_{22})}{1 + 2x(t - \tau_{21})} - 14y(t) \right], \; t \ne t_k \\[4mm] \Delta x(t_k) = -\dfrac{1}{4}\left(x(t_k) - \dfrac{1}{4} \right), \; k = 1, 2, \ldots \\[4mm] \Delta y(t_k) = -\dfrac{11}{15}\left(y(t_k) - \dfrac{45}{44} \right), \; k = 1, 2, \ldots, \end{cases}$$

where $0 < t_1 < t_2 < \cdots$ and $\lim_{k \to \infty} t_k = \infty$, the point $(x^*, y^*) = (\frac{1}{4}, \frac{45}{44})$ is an equilibrium which is uniformly asymptotically stable. In fact, all conditions of Theorem 4.13 are satisfied for $\mu = 1.524$, $m = \frac{1}{4}$ and $M = \frac{45}{44}$ and

$$\frac{1}{4} \le x(t_k) + I_{1k}(x(t_k)) = \frac{12x(t_k) + 1}{16} \le \frac{45}{44},$$

$$\frac{1}{4} \le y(t_k) + I_{2k}(y(t_k)) = \frac{4y(t_k)}{15} + \frac{3}{4} \le \frac{45}{44}$$

for $\frac{1}{4} \le x(t_k) \le \frac{45}{44}, \frac{1}{4} \le y(t_k) \le \frac{45}{44}, k = 1, 2, \dots$.

This example again shows that the impulsive perturbations can prevent the population from going extinct. In short, by impulsive controls of the population numbers of the first and the second species at fixed moments, such as stocking and harvesting, we can control the system's population dynamics.

Impulsive n-species Lotka–Volterra models with infinite delays

Gopalsamy [91] studied the existence of periodic solutions of the equation

$$\dot{x}_i(t) = x_i(t) \left[b_i(t) - a_{ii}(t) x_i(t) - \sum_{\substack{j=1 \\ j \ne i}}^{n} \int_{-\infty}^{t} k_i(t, s) a_{ij}(t) x_j(s) \, ds \right], \quad t \in \mathbb{R}, \quad (4.22)$$

$i = 1, 2, \dots, n$, when the delay kernel $k_i(t, s) = k_l(t - s)$ is of convolution type.

Ahmad and Rao [4] investigated the existence of asymptotically periodic solutions of a nonatonomous competitive Lotka–Volterra system of integro-differential equations with infinite delay

$$\dot{x}_i(t) = x_i(t) \left[b_i(t) - f_i(t, x_i(t)) - \sum_{\substack{j=1 \\ j \ne i}}^{n} \int_{-\infty}^{t} k_i(t, s) h_{ij}(t, x_j(s)) \, ds \right], \quad (4.23)$$

$i = 1, 2, \dots, n, t \in \mathbb{R}$. The paper [4] improves the results of Gopalsamy and some of the earlier results on this topic of interest.

In this part of Section 4.1, we shall consider equation (4.23) with impulsive perturbations of the population density at fixed moments of time. Impulses can be considered as a control. Sufficient conditions for uniform stability and uniform asymptotic stability of solutions will be investigated.

Let $\|x\| = \sum_{i=1}^{n} |x_i|$ define the norm of $x \in \mathbb{R}^n$, $t_0 \in \mathbb{R}$ and $t_0 < t_1 < t_2 < \cdots$, $\lim_{k \to \infty} t_k = \infty$. Consider the impulsive nonautonomous competitive Lotka–Volterra system of integro-differential equations with infinite delay

$$
\begin{cases}
\dot{x}_i(t) = x_i(t)\left[b_i(t) - f_i(t, x_i(t)) - \sum\limits_{\substack{j=1 \\ j \neq i}}^{n} \int_{-\infty}^{t} k_i(t,s) h_{ij}(t, x_j(s))\, ds \right], \ t \neq t_k, \\[4mm]
x_i(t_k^+) = x_i(t_k) + g_{ik} x_i(t_k) + c_i, \ k = 1, 2, \dots,
\end{cases}
$$

$$(4.24)$$

where $i = 1, \dots, n$, $n \geq 2$, and $t \in [t_0, \infty)$.

We assume that b_i, f_i, k_i, and h_{ij} are nonnegative continuous functions, g_{ik} are real and c_i are nonnegative constants.

Let $\varphi \in CB[(-\infty, 0], \mathbb{R}^n]$, $\varphi = \mathrm{col}(\varphi_1, \varphi_2, \dots, \varphi_n)$. We denote by $x(t) = x(t; t_0, \varphi) = \mathrm{col}(x_1(t; t_0, \varphi), x_2(t; t_0, \varphi), \dots, x_n(t; t_0, \varphi))$ the solution of system (4.24), satisfying the initial conditions

$$
\begin{cases}
x_i(t; t_0, \varphi) = \varphi_i(t - t_0), \ t \in (-\infty, t_0] \\[2mm]
x_i(t_0^+; t_0, \varphi) = \varphi_i(0), \ i = 1, \dots, n,
\end{cases}
$$

$$(4.25)$$

and by $J^+ = J^+(t_0, \varphi)$ the maximal interval of type $[t_0, \beta)$ in which the solution $x(t; t_0, \varphi)$ is defined.

Let $\|\varphi\|_\infty = \max_{t \in (-\infty, t_0]} \|\varphi(t - t_0)\|$ be the norm of the function $\varphi \in CB[(-\infty, 0]$, $\mathbb{R}^n]$.

Introduce the following conditions:

H4.7. The delay kernel $k_i : \mathbb{R}^2 \to \mathbb{R}_+$ is continuous, and there exist positive numbers μ_i such that

$$
\int_{-\infty}^{t} k_i(t,s)\, ds \leq \mu_i < \infty
$$

for all $t \geq t_0$, $t \neq t_k$, $k = 1, 2, \dots$ and $i = 1, 2, \dots, n$.

H4.8. $f_i(t, x_i) > 0$ for $x_i > 0$, $f_i(t, 0) = 0$, and there exist positive continuous functions $a_{ii}(t)$ such that

$$
|f_i(t, x_i) - f_i(t, y_i)| \geq a_{ii}(t)|x_i - y_i|
$$

for all x_i, $y_i \in \mathbb{R}$, $t \geq t_0$, $t \neq t_k$, $k = 1, 2, \dots$, and $(x_i - y_i)[f_i(t, x_i) - f_i(t, y_i)] > 0$ for $x_i \neq y_i$, $i = 1, 2, \dots, n$.

H4.9. $h_{ij}(t, x_i) > 0$ for $x_i > 0$, $h_{ij}(t, 0) = 0$, and there exist positive continuous functions $a_{ij}(t)$ such that

$$
|h_{ij}(t, x_i) - h_{ij}(t, y_i)| \leq a_{ij}(t)|x_i - y_i|
$$

for all x_i, $y_i \in \mathbb{R}$, and $a_{ij}(t)$ is non-increasing for $t \geq t_0$, $t \neq t_k$, $k = 1, 2, \dots$ and $i, j = 1, 2, \dots, n$, $i \neq j$.

H4.10. $c^M < \infty$, $c^L > 0$, where $c^M = \max\{c_i\}$ and $c^L = \min\{c_i\}$ for $1 \le i \le n$.

H4.11. $t_0 < t_1 < t_2 < \cdots$ and $\lim_{k \to \infty} t_k = \infty$.

In the proofs of the main theorems we shall use the following lemmas.

Lemma 4.15. *Let the conditions H4.7–H4.11 hold, and*

$$\int_{-\infty}^{t} k_i(t, s) h_{ij}(t, x_j(s)) \, ds$$

be continuous for all $t \ge t_0$, $i, j = 1, 2, \ldots, n$.
  Then $J^+(t_0, \varphi) = [t_0, \infty)$.

Proof. If conditions H4.7, H4.8 and H4.9 hold and $\int_{-\infty}^{t} k_i(t, s) h_{ij}(t, x_j(s)) \, ds$ is continuous for all $t \ge t_0$, then it follows [4, 91] that the solution $x(t) = x(t; t_0, \varphi)$ of problem (4.24), (4.25) is defined on $[t_0, t_1] \cup (t_k, t_{k+1}]$, $k = 1, 2, \ldots$. From conditions H4.10 and H4.11, we conclude that $J^+(t_0, \varphi) = [t_0, \infty)$. $\square$

Lemma 4.16. *Assume that:*

(1) *The conditions of Lemma 4.15 hold.*

(2) $x(t) = x(t; t_0, \varphi) = \mathrm{col}(x_1(t; t_0, \varphi), x_2(t; t_0, \varphi), \ldots, x_n(t; t_0, \varphi))$ *is a solution of* (4.24), (4.25) *such that*

$$x_i(t) = \varphi_i(t - t_0) \ge 0, \quad \sup \varphi_i(s) < \infty, \quad \varphi_i(0) > 0, \qquad (4.26)$$

$1 \le i \le n$.

(3) *For each $1 \le i \le n$ and $k = 1, 2, \ldots$*

$$1 + g_{ik} > 0.$$

Then

$$x_i(t) > 0, \quad 1 \le i \le n, \quad t \ge t_0.$$

Proof. By integrating (4.24) in the interval $[t_0, t_1]$, we have

$$x_i(t) = x_i(t_0^+) \exp \left(\int_{t_0}^{t} F_i(s) \, ds \right), \quad t \in [t_0, t_1],$$

where

$$F_i(t) = b_i(t) - f_i(t, x_i(t)) - \sum_{\substack{j=1 \\ j \ne i}}^{n} \int_{-\infty}^{t} k_i(t, s) h_{ij}(t, x_j(s)) \, ds, \quad 1 \le i \le n.$$

Since, in the interval $[t_0, t_1]$ we have no points of discontinuity of $x_i(t)$, from (4.26) it is obvious that $x_i(t) > 0$ for $t \in (t_0, t_1]$. Then $x(t_1) > 0$.

We have from (4.24) that

$$x_i(t_1^+) = x_i(t_1) + g_{i1}x_i(t_1) + c_i, \quad 1 \le i \le n.$$

From condition (3) of Lemma 4.16 and H4.10, it follows that

$$x_i(t_1^+) = (1 + g_{i1})x_i(t_1) + c_i > 0, \quad 1 \le i \le n.$$

We now integrate (4.24) in the interval $(t_1, t_2]$ and we have

$$x_i(t) = x_i(t_1^+) \exp\left(\int_{t_1}^t F_i(s)\,ds\right), \quad t \in (t_1, t_2].$$

From the above relation it follows that $x_i(t) > 0$ for $t \in (t_1, t_2]$.
   By similar arguments, we can obtain that

$$x_i(t) = x_i(t_k^+) \exp\left(\int_{t_k}^t F_i(s)\,ds\right), \quad t \in (t_k, t_{k+1}].$$

for $1 \le i \le n, k = 1, 2, \ldots$, so $x_i(t) > 0$ for $t \ge t_0$. $\square$

Lemma 4.17. *Assume that:*

(1) *The conditions of Lemma* 4.16 *hold.*

(2) *For all* $i = 1, 2, \ldots, n$ *there exist functions* $P_i, Q_i \in PC^1[[t_0, \infty), \mathbb{R}]$ *such that* $P_i(t_0^+) \le \varphi_i(0) \le Q_i(t_0^+).$

Then

$$P_i(t) \le x_i(t) \le Q_i(t) \tag{4.27}$$

for all $t \ge t_0$ *and* $i = 1, 2, \ldots, n.$

Proof. First we shall proof that

$$x_i(t) \le Q_i(t) \tag{4.28}$$

for all $t \ge t_0$ and $i = 1, 2, \ldots, n$, where $Q_i(t)$ is the maximal solution of the initial value problem

$$\begin{cases} \dot{q}_i(t) = q_i(t)\,[b_i(t) - f_i(t, q_i(t))], \; t \ne t_k \\ q_i(t_0^+) = q_{i0} > 0 \\ q_i(t_k^+) = q_i(t_k) + g^M q_i(t_k) + c^M, \; k = 1, 2, \ldots, \end{cases} \tag{4.29}$$

where $g^M = \max\{g_{ik}\}$ for $1 \le i \le n$ and $k = 1, 2, \ldots.$

The maximal solution $Q_i(t) = Q_i(t; t_0, q_0)$, $q_0 = \mathrm{col}(q_{10}, q_{20}, \ldots, q_{n0})$ of (4.29) is defined by the equality

$$
Q_i(t; t_0, q_0) =
\begin{cases}
q_i^0(t; t_0, Q_i^0 + 0), \ t_0 < t \leq t_1 \\
q_i^1(t; t_1, Q_i^1 + 0), \ t_1 < t \leq t_2 \\
\vdots \\
q_i^k(t; t_k, Q_i^k + 0), \ t_k < t \leq t_{k+1} \\
\vdots
\end{cases}
$$

where $q_i^k(t; t_k, Q_i^k + 0)$ is the maximal solution of the equation without impulses $\dot{q}_i(t) = q_i(t)[b_i(t) - f_i(t, q_i(t))]$ in the interval $(t_k, t_{k+1}]$, $k = 0, 1, 2, \ldots$, for which $Q_i^k + 0 = (1 + g^M)q_i^{k-1}(t_k; t_{k-1}, Q_i^{k-1} + 0) + c^M$, $k = 1, 2, \ldots, 1 \leq i \leq n$ and $Q_i^0 + 0 = q_{i0}$.

By the condition (1) of Lemma 4.17, it follows that

$$
\dot{x}_i(t) \leq x_i(t)[b_i(t) - f_i(t, x_i(t))], \quad t \neq t_k, \tag{4.30}
$$

$1 \leq i \leq n$ and $k = 1, 2, \ldots$.

Let $t \in [t_0, t_1]$. If $0 < \varphi_i(0) \leq Q_i(t_0^+)$, $i = 1, 2, \ldots, n$, then elementary differential inequality [130] yields that

$$
x_i(t) \leq Q_i(t)
$$

for all $t \in [t_0, t_1]$, i.e. the inequality (4.28) is valid for $t \in [t_0, t_1]$.

Suppose that (4.28) is satisfied for $t \in (t_{k-1}, t_k]$, $k > 1$.

Then, using hypothesis H4.10 and the fact that (4.28) is satisfied for $t = t_k$, we obtain

$$
x_i(t_k^+) = x_i(t_k) + g_{ik} x_i(t_k) + c_i \leq x_i(t_k) + g^M x_i(t_k) + c^M
$$

$$
\leq Q_i(t_k) + g^M Q_i(t_k) + c^M = (1 + g^M)q_i^{k-1}(t_k; t_{k-1}, Q_i^{k-1} + 0) + c^M
$$

$$
= Q_i^k + 0.
$$

We again apply the comparison result (4.30) in the interval $(t_k, t_{k+1}]$ and obtain

$$
x_i(t; t_0, \varphi) \leq q_i^k(t; t_k, Q_i^k + 0) = Q_i(t; t_0, q_0),
$$

i.e. the inequality (4.28) is valid for $(t_k, t_{k+1}]$.

The proof of (4.28) is completed by induction.

Further, by analogous arguments and using H4.7–H4.11, we obtain from (4.24) and (4.30) that

$$
\begin{cases}
\dot{x}_i(t) \geq x_i(t)\left[b_i(t) - f_i(t, x_i(t)) - \sum_{\substack{j=1 \\ j \neq i}}^{n} a_{ij}(t)\mu_i \sup_{-\infty < s \leq t} Q_i(s)\right], \ t \neq t_k \\
x_i(t_k^+) \geq x_i(t_k) + g^L x_i(t_k) + c^L, \ k = 1, 2, \ldots,
\end{cases}
$$

$i = 1, \ldots, n, n \geq 2$, and hence $\varphi_i(s) \geq P_i(t_0^+)$ for $s \leq t_0$ implies that

$$x_i(t) \geq P_i(t) \tag{4.31}$$

for all $t \geq t_0$ and $i = 1, 2, \ldots, n$, where $P_i(t)$ is the minimal solution of the initial value problem

$$\begin{cases} \dot{p}_i(t) = p_i(t)\left[b_i(t) - f_i(t, p_i(t)) - \sum_{\substack{j=1 \\ j \neq i}}^{n} a_{ij}(t)\mu_i \sup_{-\infty < s \leq t} Q_i(s) \right], \ t \neq t_k \\ p_i(t_0^+) = p_{i0} > 0 \\ p_i(t_k^+) = p_i(t_k) + g^L p_i(t_k) + c^L, \ k = 1, 2, \ldots, \end{cases}$$

$$\tag{4.32}$$

$i = 1, \ldots, n$ and $g^L = \min\{g_{ik}\}$ for $1 \leq i \leq n$ and $k = 1, 2, \ldots$. Thus, the proof follows from (4.28) and (4.31). $\qquad\square$

Lemma 4.18. *Let the conditions of Lemma 4.17 hold.*
Then for all $i = 1, 2, \ldots, n$ there exist positive constants α_i and $\beta_i < \infty$ such that

$$\alpha_i \leq x_i(t) \leq \beta_i, \tag{4.33}$$

for all $t \in [t_0, t_1] \cup (t_k, t_{k+1}], k = 1, 2, \ldots$ and if in addition

$$0 < 1 + g_{ik} \leq 1 \quad and \quad -g_{ik}\alpha_i \leq c_i \leq -g_{ik}\beta_i,$$

then the inequalities (4.33) are valid for all $t \geq t_0$ and $1 \leq i \leq n$.

Proof. From Lemma 4.17, we have

$$P_i(t) \leq x_i(t) \leq Q_i(t)$$

for all $t \geq t_0$ and $i = 1, 2, \ldots, n$, where $P_i(t)$ is the minimal solution of the logistic system (4.32) and $Q_i(t)$ is the maximal solution of the logistic system (4.29).

Since [4], under the conditions of Lemma 4.18 for the solutions of (4.32) and (4.29) with initial functions of the form (4.25), it is valid that

$$\alpha_i \leq P_i(t), \quad Q_i(t) \leq \beta_i,$$

$\alpha_i > 0, 0 < \beta_i < \infty$, for all $t \in [t_0, t_1] \cup (t_k, t_{k+1}], k = 1, 2, \ldots$ and $i = 1, 2, \ldots, n$ then

$$\alpha_i \leq x_i(t) \leq \beta_i,$$

$1 \leq i \leq n$, for all $t \in [t_0, t_1] \cup (t_k, t_{k+1}], k = 1, 2, \ldots$.

If in addition $0 < 1 + g_{ik} \leq 1$ and $-g_{ik}\alpha_i < c_i < -g_{ik}\beta_i$, then from the left continuity of $x_i(t)$ at the points t_k we have

$$\alpha_i = (1 + g_{ik})\alpha_i - g_{ik}\alpha_i \leq x_i(t_k^+)$$
$$\leq (1 + g_{ik})\beta_i - g_{ik}\beta_i = \beta_i,$$

hence

$$\alpha_i \leq x_i(t) \leq \beta_i,$$

$1 \leq i \leq n$, for all $t \geq t_0$. $\qquad\qquad\qquad\qquad\qquad\qquad\qquad\qquad\qquad\qquad$ □

Corollary 4.19. *Let the conditions of Lemma 4.18 hold, and the constants g_{ik} and c_i be such that*

$$0 < 1 + g_{ik} \leq 1 \quad and \quad -g_{ik}\alpha_i \leq c_i \leq -g_{ik}\beta_i,$$

$i = 1, 2, \ldots, n, \ k = 1, 2, \ldots$.
  Then:

(1) *The system (4.24) is uniformly ultimately bounded.*

(2) *There exist positive constants m and $M < \infty$ such that the inequalities (4.9) are valid.*

Let $\phi \in CB[(-\infty, 0], \mathbb{R}^n]$, $\phi = \text{col}(\phi_1, \phi_2, \ldots, \phi_n)$ and $x^*(t) = x^*(t; t_0, \phi) = \text{col}(x_1^*(t; t_0, \phi), x_2^*(t; t_0, \phi), \ldots, x_n^*(t; t_0, \phi))$ be a solution of system (4.24), satisfying the initial conditions

$$\begin{cases} x_i^*(t; t_0, \phi) = \phi_i(t - t_0), \ t \in (-\infty, t_0] \\ x_i^*(t_0^+; t_0, \phi) = \phi_i(0), \quad i = 1, 2, \ldots, n. \end{cases}$$

In the following, we shall suppose that

$$x_i(t) = \varphi_i(t - t_0) \geq 0, \quad \sup \varphi_i(s) < \infty, \quad \varphi_i(0) > 0,$$

$$x_i^*(t) = \phi_i(t - t_0) \geq 0, \quad \sup \phi_i(s) < \infty, \quad \phi_i(0) > 0, \quad i = 1, 2, \ldots, n.$$

Theorem 4.20. *Assume that.*

(1) *The conditions of Lemma 4.18 hold.*

(2) *The following inequalities are satisfied:*

$$ma_{ii}(t) > M \sum_{\substack{j=1 \\ j \neq i}}^{n} \mu_i a_{ij}(t)$$

for all $t \in [t_0, \infty)$, $t \neq t_k$, $k = 1, 2, \ldots$ and $i = 1, 2, \ldots, n$.

(3) *For each* $1 \le i \le n$ *and* $k = 1, 2, \ldots$

$$0 < 1 + g_{ik} \le 1 \quad and \quad -g_{ik}m \le c_i \le -g_{ik}M.$$

Then the solution $x^*(t)$ *of (4.24) is uniformly stable.*

Proof. Let $t_0 \in \mathbb{R}$. Given $0 < \varepsilon < M$, choose $\delta = \frac{\varepsilon m}{2nM}$. Let $\alpha = \frac{\varepsilon}{2M}$.

Consider the Lyapunov function defined by (4.10). For all $i = 1, 2, \ldots, n$, we introduce the notations

$$v_i^\alpha = \left\{ (x_i, x_i^*) \in \mathbb{R}_+ \times \mathbb{R}_+ : \left| \ln \frac{x_i}{x_i^*} \right| < \frac{\alpha}{n} \right\},$$

$$\partial v_i^\alpha = \left\{ (x_i, x_i^*) \in \mathbb{R}_+ \times \mathbb{R}_+ : \left| \ln \frac{x_i}{x_i^*} \right| = \frac{\alpha}{n} \right\},$$

$$|\varphi_i|_\infty = \sup_{t \in (-\infty, t_0]} |\varphi_i(t - t_0)|.$$

If $|\varphi_i - \phi_i|_\infty \le \delta$ for all $i = 1, 2, \ldots, n$, we obtain

$$V^i(t_0^+, x(t_0^+)) = |\ln x_i(t_0^+) - \ln x_i^*(t_0^+)| \le \frac{1}{m} |x_i(t_0^+) - x_i^*(t_0^+)|$$

$$= \frac{1}{m} |\varphi_i(0) - \phi_i(0)| \le \frac{1}{m} |\varphi_i - \phi_i|_\infty < \frac{\alpha}{n}. \tag{4.34}$$

Then $(x_i(t_0^+), x_i^*(t_0^+)) \in v_i^\alpha$ for all $i = 1, 2, \ldots, n$.

We shall prove that $(x_i(t), x_i^*(t)) \in v_i^\alpha$ for all $t > t_0$ and all $i = 1, 2, \ldots, n$. Suppose that this is not true.

Note that from $(x_i(t_k), x_i^*(t_k)) \in v_i^\alpha$, $t_k > t_0$, $k = 1, 2, \ldots$, $i = 1, 2, \ldots, n$ it follows, from the condition (3) of Theorem 4.20, that

$$V^i(t_k^+, x(t_k^+)) = \left| \ln \frac{x_i(t_k^+)}{x_i^*(t_k^+)} \right| = \left| \ln \frac{(1 + g_{ik})x_i(t_k) + c_i}{(1 + g_{ik})x_i^*(t_k) + c_i} \right|$$

$$\le \left| \ln \frac{(1 + g_{ik})M - g_{ik}M}{(1 + g_{ik})m - g_{ik}m} \right| = \left| \ln \frac{M}{m} \right|$$

$$= \left| -\ln \frac{M}{m} \right| = \left| \ln \frac{m}{M} \right| \le \left| \ln \frac{x_i(t_k)}{x_i^*(t_k)} \right|, \tag{4.35}$$

i.e. $(x_i(t), x_i^*(t))$ can not leave v_i^α by jump.

Now the assumption that $(x_i(t), x_i^*(t)) \in v_i^\alpha$ for all $t > t_0$ and $i = 1, 2, \ldots, n$ is not true implies the existence of $T > t_0$, $T \neq t_k$, $k = 1, 2, \ldots$ and $l = 1, 2, \ldots, n$ such that $(x_l(t), x_l^*(t)) \in v_l^\alpha$ for all $-\infty < t < T$ and $(x_l(T), x_l^*(T)) \in \partial v_l^\alpha$.

Consider the upper right-hand derivative $D_{(4.24)}^+ V^l(t, x(t))$ of the function $V^l(t, x(t))$. For $t > t_0$ and $t \neq t_k$, $k = 1, 2, \ldots$, we derive the estimate

$$D_{(4.24)}^+ V^l(t, x(t)) = \left(\frac{\dot{x}_l(t)}{x_l(t)} - \frac{\dot{x}_l^*(t)}{x_l^*(t)} \right) \operatorname{sgn}\left(x_l(t) - x_l^*(t)\right)$$

$$\leq \left[-|f_l(t, x_l(t)) - f_l(t, x_l^*(t))| \right.$$

$$\left. + \sum_{\substack{j=1 \\ j \neq l}}^{n} \int_{-\infty}^{t} k_l(t, s)|h_{lj}(t, x_j(s)) - h_{lj}(t, x_j^*(s))| \, ds \right].$$

From hypotheses H4.7, H4.8 and H4.9, we obtain

$$D_{(4.24)}^+ V^l(t, x(t))$$

$$\leq \left[-a_{ll}(t)|x_l(t) - x_l^*(t)| + \sum_{\substack{j=1 \\ j \neq l}}^{n} \int_{-\infty}^{t} k_l(t, s)a_{lj}(t)|x_j(s) - x_j^*(s)| \, ds \right]$$

$$\leq \left[-a_{ll}(t)|x_l(t) - x_l^*(t)| + \sum_{\substack{j=1 \\ j \neq l}}^{n} \mu_l a_{lj}(t) \sup_{\infty < s \leq t} |x_j(s) - x_j^*(s)| \, ds \right].$$

From (4.11) for $t = T$, we deduce the inequality

$$D_{(4.24)}^+ V^l(T, x(T))$$

$$\leq \left[-a_{ll}(T)m|\ln x_l(T) - \ln x_l^*(T)| + M|\ln x_l(T) - \ln x_l^*(T)| \sum_{\substack{j=1 \\ j \neq l}}^{n} \mu_l a_{lj}(T) \right].$$

$$(4.36)$$

Since, by the condition (2) of Theorem 4.20, the inequalities

$$ma_{ll}(t) > M \sum_{\substack{j=1 \\ j \neq l}}^{n} \mu_l a_{lj}(t)$$

are satisfied for all $t > t_0$ and $l = 1, 2, \ldots, n$, we have $D_{(4.24)}^+ V^l(T, x(T)) < 0$. Hence, the assumption that $(x_i(t), x_i^*(t)) \in v_i^\alpha$ for all $-\infty < t < T$ and $(x_i(T), x_i^*(T)) \in \partial v_i^\alpha$ will not be true, i.e. $(x_i(t), x_i^*(t)) \in v_i^\alpha$ for all $t > t_0$ and $i = 1, 2, \ldots, n$.

Then from (4.11) and (4.34) it follows that

$$\|x(t) - x^*(t)\| \leq MV(t, x(t)) < M\alpha < \varepsilon,$$

for all $t \geq t_0$, whenever $\|\varphi - \phi\|_\infty \leq \delta_1 = \frac{\varepsilon m}{2M}$ and $t_0 \in \mathbb{R}$. Since $t_0 \in \mathbb{R}$ is arbitrary, the solution $x^*(t)$ of (4.24) is uniformly stable.   □

Theorem 4.21. *In addition to the assumptions of Theorem 4.20, suppose that:*

(1) *There exist nonnegative continuous functions $\sigma_i(t)$ such that*

$$ma_{ii}(t) - M \sum_{\substack{j=1 \\ j \neq i}}^{n} \mu_i a_{ij}(t) > \sigma_i(t)$$

for all $t \in [t_0, \infty)$, $t \neq t_k$, $k = 1, 2, \ldots$ and $1 \leq i \leq n$.

(2) *The function $\sigma(t) = \min(\sigma_1(t), \ldots, \sigma_n(t))$ is such that*

$$\int_{t_0}^{\infty} \sigma(s)\, ds = \infty.$$

Then the solution $x^*(t)$ of (4.24) is uniformly asymptotically stable.

Proof. Since all conditions of Theorem 4.20 are satisfied, the solution $x^*(t)$ of (4.24) is uniformly stable. We have to prove that

$$\lim_{t \to \infty} \|x(t) - x^*(t)\| = 0.$$

Let $t_0 \in \mathbb{R}$ and $0 < \varepsilon < M$. Choose $\delta = \delta(\varepsilon) = \frac{\varepsilon m}{2nM}$.

We shall prove that there exist $\tau = \tau(\varepsilon) > 0$ and $t^* \in [t_0, t_0 + \tau]$ such that for any solution $x_i(t; t_0, \varphi)$, $i = 1, 2, \ldots, n$, $(t_0, \varphi) \in \mathbb{R} \times CB[(-\infty, 0], \mathbb{R}^n]$ of (4.24) for which $|\varphi_i - \phi_i|_\infty \leq \delta$ the following inequalities are valid:

$$|x_i(t^* + 0) - x_i^*(t^* + 0)| < \delta(\varepsilon), \ i = 1, 2, \ldots, n. \tag{4.37}$$

Suppose that this is not true. Then for any $\tau > 0$ there exists solution $x_i(t; t_0, \varphi)$, $i = 1, 2, \ldots, n$, $(t_0, \varphi) \in \mathbb{R} \times CB[(-\infty, 0], \mathbb{R}^n]$ of (4.24) for which $|\varphi_i - \phi_i|_\infty \leq \delta$ and

$$|x_i(t + 0) - x_i^*(t + 0)| \geq \delta(\varepsilon) \tag{4.38}$$

for any $t \in [t_0, t_0 + \tau]$.

Consider the upper right-hand derivative $D_{(4.24)}^+ V(t, x(t))$ with respect to system (4.24). For $t > t_0$ and $t \neq t_k$, $k = 1, 2, \ldots$ from hypotheses H4.7, H4.8 and H4.9, we have

$$D_{(4.24)}^+ V(t, x(t)) \leq \sum_{i=1}^{n} \left[-a_{ii}(t)|x_i(t) - x_i^*(t)| + \sum_{\substack{j=1 \\ j \neq i}}^{n} \mu_i a_{ij}(t) \sup_{-\infty < s \leq t} |x_j(s) - x_j^*(s)| \right].$$

From the above estimate and from (4.35) we can obtain that for $t > t_0$ and $t \neq t_k$, $k = 1, 2, \dots$

$$V(t, x(t)) - V(t_0^+, x(t_0^+)) \leq \int_{t_0}^{t} D_{(4.24)}^+ V(u, x(u)) du$$

$$\leq \int_{t_0}^{\infty} \sum_{i=1}^{n} \left[-a_{ii}(u)|x_i(u) - x_i^*(u)| \right.$$

$$\left. + \sum_{\substack{j=1 \\ j \neq i}}^{n} \mu_i a_{ij}(u) \sup_{-\infty < s \leq u} |x_j(s) - x_j^*(s)| \right] du. \quad (4.39)$$

From the properties of the function $V(t, x(t))$ in the interval (t_0, ∞) it follows that there exists the finite limit

$$\lim_{t \to \infty} V(t, x(t)) = v_0 \geq 0. \quad (4.40)$$

Then from (4.11), (4.38), (4.39) and (4.40), it follows that

$$\int_{t_0}^{\infty} \sum_{i=1}^{n} \left[a_{ii}(t)|x_i(t) - x_i^*(t)| - \sum_{\substack{j=1 \\ j \neq i}}^{n} \mu_i a_{ij}(t) \sup_{-\infty < s \leq t} |x_j(s) - x_j^*(s)| \right] dt$$

$$\leq V(t_0^+, x(t_0^+)) - v_0 \leq \frac{1}{m} \|\varphi - \phi\|_\infty - v_0 \leq \frac{n\delta}{m} - v_0.$$

From condition (2) of Theorem 4.21, it follows that the number $\tau > 0$ can be chosen so that

$$\int_{t_0}^{t_0 + \tau} \sigma(t) dt > \frac{m(\frac{n\delta}{m} - v_0 + 1)}{\delta}.$$

Then,

$$\frac{n\delta}{m} - v_0$$

$$\geq \int_{t_0}^{\infty} \sum_{i=1}^{n} \left[a_{ii}(t)|x_i(t) - x_i^*(t)| - \sum_{\substack{j=1 \\ j \neq i}}^{n} \mu_i a_{ij}(t) \sup_{-\infty < s \leq t} |x_j(s) - x_j^*(s)| \right] dt$$

$$\geq \int_{t_0}^{t_0 + \tau} \sum_{i=1}^{n} \left[a_{ii}(t)\delta \right.$$

$$\left. - \sum_{\substack{j=1 \\ j \neq i}}^{n} \mu_i a_{ij}(t) \max\{ \sup_{-\infty < s \leq t_0} |x_j(s) - x_j^*(s)|; \sup_{t_0 < s \leq t} |x_j(s) - x_j^*(s)| \} \right] dt$$

$$\geq \int_{t_0}^{t_0+\tau} \sum_{i=1}^{n} \left[a_{ii}(t)\delta \right.$$

$$- \sum_{\substack{j=1 \\ j \neq i}}^{n} \mu_i a_{ij}(t) \max\{\delta; M \sup_{t_0 < s \leq t} |\ln x_j(s) - \ln x_j^*(s)|\} \Big] dt. \qquad (4.41)$$

Since all conditions of Theorem 4.20 are satisfied, we have that (4.34) is true and

$$|\ln x_i(t) - \ln x_i^*(t)| < \frac{\alpha}{n} = \frac{\delta}{m} \qquad (4.42)$$

for all $t \geq t_0$ and $i = 1, 2, \ldots, n$.

From (4.41), (4.42) and from the condition (1) of Theorem 4.21, it follows that

$$\frac{n\delta}{m} - v_0 \geq \frac{\delta(\varepsilon)}{m} \int_{t_0}^{t_0+\tau} \sigma(t) dt > \frac{n\delta}{m} - v_0 + 1.$$

The contradiction obtained shows that there exist $\tau = \tau(\varepsilon) > 0$ and $t^* \in [t_0, t_0 + \tau]$ such that for any solution $x_i(t; t_0, \varphi)$, $i = 1, 2, \ldots, n$, $(t_0, \varphi) \in \mathbb{R} \times CB[(-\infty, 0], \mathbb{R}^n]$ of (4.24) for which $|\varphi_i - \phi_i|_\infty \leq \delta$ the inequalities (4.37) hold.

Then, for $t \geq t^*$ (hence for any $t \geq t_0 + \tau$ as well) the following inequalities are valid:

$$\frac{1}{M} \|x(t) - x^*(t)\| \leq V(t, x(t)) \leq V(t^* + 0, x(t^* + 0))$$

$$\leq \frac{1}{m} \|x(t^* + 0) - x^*(t^* + 0)\| < \frac{n}{m} \delta = \frac{\varepsilon}{2M},$$

from which we have $\lim_{t\to\infty} \|x(t) - x^*(t)\| = 0$.

This shows that the solution $x^*(t)$ of (4.24) is uniformly asymptotically stable. $\square$

Example 4.22. For the nonatonomous competitive Lotka–Volterra system without impulsive perturbations

$$\begin{cases} \dot{x}_1(t) = x_1(t) \left[4 - 7 \ln x_1(t) - \int_{-\infty}^{t} k_1(t, s) \ln x_2(s) \, ds \right] \\[2mm] \dot{x}_2(t) = x_2(t) \left[\frac{1}{2} - \frac{1}{3} \int_{-\infty}^{t} k_2(t, s) \ln x_1(s) \, ds - \frac{4}{3} \ln x_2(t) \right], \end{cases} \qquad (4.43)$$

where $x : \mathbb{R} \to \mathbb{R}_+$ and $\int_{-\infty}^{t} k_1(t, s) \, ds = 2$, $\int_{-\infty}^{t} k_2(t, s) \, ds = 1$, one can show that the point $(x_1^*, x_2^*) = (\sqrt{e}, \sqrt[4]{e})$ is an equilibrium, and it is uniformly asymptotically stable [4].

Now, we consider the impulsive nonatonomous competitive Lotka–Volterra system

$$
\begin{cases}
\dot{x}_1(t) = x_1(t)\left[4 - 7\ln x_1(t) - \displaystyle\int_{-\infty}^{t} k_1(t,s)\ln x_2(s)\,ds \right], \ t \neq t_k \\[2mm]
\dot{x}_2(t) = x_2(t)\left[\dfrac{1}{2} - \dfrac{1}{3}\displaystyle\int_{-\infty}^{t} k_2(t,s)\ln x_1(s)\,ds - \dfrac{4}{3}\ln x_2(t) \right], \ t \neq t_k \\[3mm]
x_1(t_k^+) = \dfrac{2\sqrt{e} + x_1(t_k)}{3}, \ k = 1,2,\dots \\[3mm]
x_2(t_k^+) = \dfrac{3\sqrt[4]{e} + x_2(t_k)}{4}, \ k = 1,2,\dots,
\end{cases}
\tag{4.44}
$$

where $t_k < t_{k+1} < \cdots, k = 1,2,\dots,$ $\lim_{k\to\infty} t_k = \infty$.

For the system (4.44), the point $(x_1^*, x_2^*) = (\sqrt{e}, \sqrt[4]{e})$ is an equilibrium and all conditions of Theorem 4.21 are satisfied. We also have that

$$
0 < 1 + g_{1k} = \frac{1}{3} < 1, \qquad 0 < 1 + g_{2k} = \frac{1}{4} < 1
$$

and

$$
\frac{2}{3}m = \frac{2}{3}.1 < c_1 = \frac{2}{3}\sqrt{e} < \frac{2}{3}e = \frac{2}{3}M,
$$

$$
\frac{3}{4}m = \frac{3}{4}.1 < c_2 = \frac{3}{4}\sqrt[4]{e} < \frac{3}{4}e = \frac{3}{4}M.
$$

Therefore, the equilibrium $(x_1^*, x_1^*) = (\sqrt{e}, \sqrt[4]{e})$ is an uniformly asymptotically stable solution of (4.44).

If, in the system (4.44), we change the impulsive perturbations as follows:

$$
\begin{cases}
x_1(t_k^+) = 3\sqrt{e} - 2x_1(t_k), \ k = 1,2,\dots \\[3mm]
x_2(t_k^+) = \dfrac{3\sqrt[4]{e} + x_2(t_k)}{4}, \ k = 1,2,\dots,
\end{cases}
\tag{4.45}
$$

we obtain the following system:

$$
\begin{cases}
\dot{x}_1(t) = x_1(t)\left[4 - 7\ln x_1(t) - \displaystyle\int_{-\infty}^{t} k_1(t,s)\ln x_2(s)\,ds \right], \ t \neq t_k \\[2mm]
\dot{x}_2(t) = x_2(t)\left[\dfrac{1}{2} - \dfrac{1}{3}\displaystyle\int_{-\infty}^{t} k_2(t,s)\ln x_1(s)\,ds - \dfrac{4}{3}\ln x_2(t) \right], \ t \neq t_k \\[3mm]
x_1(t_k^+) = 3\sqrt{e} - 2x_1(t_k), \ k = 1,2,\dots \\[3mm]
x_2(t_k^+) = \dfrac{3\sqrt[4]{e} + x_2(t_k)}{4}, \ k = 1,2,\dots,
\end{cases}
\tag{4.46}
$$

The point $(x_1^*, x_2^*) = (\sqrt{e}, \sqrt[4]{e})$ is again an equilibrium for the system (4.46), but there is nothing we can say about its uniform asymptotic stability, because $1 + g_{1k} = -2 < 0$.

The example shows that impulses have played an important role in stabilizing a Lotka–Volterra system.

4.2   Neural networks

Neural networks have been successfully employed in various areas such as pattern recognition, associative memory and combinatorial optimization [72–75, 104, 187, 216].

One of the most investigated problems in the study of neural networks is the global asymptotic stability of the equilibrium point. If an equilibrium of a neural network is globally asymptotically stable, it means that the domain of attraction of the equilibrium point is the whole space and the convergence is in real time. This is significant both theoretically and practically. Such neural networks are known to be well-suited for solving some class of optimization problems. In fact, a globally asymptotically stable neural network is guaranteed to compute the global optimal solution independently of the initial condition, which in turn implies that the network is devoid of spurious suboptimal responses.

While an artificial neural network has been known insofar for its transient processing behavior, its circuit design has never been disentangled from destabilizing factors such as delays and impulses. In hardware implementation, time delays occur due to finite switching speed of the amplifiers and communication time. Time delays will affect the stability of designed neural networks and may lead to some complex dynamic behaviors such as periodic oscillation, bifurcation or chaos. Therefore, the study of neural dynamics with consideration of the delayed problem becomes extremely important to manufacture high-quality neural networks. The delayed neural networks have been widely studied and some progress has been made [66–68, 76, 79, 92, 106, 107, 117, 154, 160, 170, 192, 227, 232, 233].

Impulses can make unstable systems stable so they have been widely used in many fields such as physics, chemistry, biology, population dynamics, and industrial robotics. The abrupt changes in the voltages produced by faulty circuit elements are exemplary of impulse phenomena that can affect the transient behavior of the network. Some results for impulsive neural networks have been given, for example, see [9, 10, 15, 101, 199, 209, 221] and references therein.

In this section of Chapter 4, we shall investigate stability of the equilibrium states in neural network with finite and infinite delays, and neural states that are subject to impulsive state displacements at fixed instants of time. By applying the Lyapunov–Razumikhin method, sufficient conditions for global asymptotic stability and global exponential stability of such neural networks will be obtained.

Impulsive neural networks with finite delays

Gopalsamy and Leung [92] considered the following scalar autonomous delay equation with dynamical thresholds

$$\dot{x}(t) = -x(t) + a \cdot \tanh\left(x(t) - bx(t-\tau) - c\right), \quad t \geq 0,$$

where $x : \mathbb{R}_+ \to \mathbb{R}$; a is a positive constant; b, c and τ are nonnegative constants. By using Lyapunov functions, Gopalsamy and Leung established a sufficient condition for global asymptotic stability of the equilibrium $x^* = 0$ for the case $c = 0$.

For the case $c \neq 0$ some stability criteria are investigated in [232] for the equilibrium of the following more general model

$$\dot{x}(t) = -x(t) + af\left(x(t) - bx(t-\tau) - c\right), \quad t \geq 0. \tag{4.47}$$

In this part of Section 4.2, we shall study the global asymptotic stability of the impulsive generalization of the equation (4.47).

We consider the following impulsive delayed neural network with dynamical thresholds

$$\begin{cases} \dot{x}(t) = -x(t) + af\left(x(t) - bx(t - \tau(t)) - c\right), \ t \neq t_k, \ t \geq 0 \\ \Delta x(t_k) = x(t_k + 0) - x(t_k) = I_k(x(t_k)), \ k = 1, 2, \ldots, \end{cases} \tag{4.48}$$

where $a > 0$; b and c are nonnegative constants; $f : \mathbb{R} \to \mathbb{R}$; $\tau(t)$ corresponds to the transmission delay and satisfies $0 \leq \tau(t) \leq \tau$ ($\tau = $ const); $t - \tau(t) \to \infty$ as $t \to \infty$; $I_k : \mathbb{R} \to \mathbb{R}$, $k = 1, 2, \ldots$; $t_k < t_{k+1} < \cdots$, $\lim_{k \to \infty} t_k = \infty$.

Let $J \subset \mathbb{R}$ be an interval. Define the following class of functions:

$$PCB[J, \mathbb{R}] = \{\sigma \in PC[J, \mathbb{R}] : \sigma(t) \text{ is bounded on } J\}.$$

Let $\varphi \in PCB[[-\tau, 0], \mathbb{R}]$. Denote by $x(t) = x(t; 0, \varphi)$, $x \in \mathbb{R}$ the solution of equation (4.48), satisfying the initial conditions

$$\begin{cases} x(t; 0, \varphi) = \varphi(t), & -\tau \leq t \leq 0 \\ x(0^+; 0, \varphi) = \varphi(0). \end{cases} \tag{4.49}$$

Introduce the following notation:

$$|\varphi|_\tau = \sup_{s \in [-\tau, 0]} |\varphi(s)| \text{ is the norm of the function } \varphi \in PCB[[-\tau, 0], \mathbb{R}].$$

We introduce the following conditions:

H4.12. There exists a constant $L > 0$ such that

$$|f(u) - f(v)| \leq L|u - v|$$

for all $u, v \in \mathbb{R}$.

H4.13. There exists a constant $M > 0$ such that for all $u \in \mathbb{R}$

$$|f(u)| \leq M < \infty.$$

H4.14. $a > 0$, $b \geq 0$, $a(1 - b) < 1$.

H4.15. For any $k = 1, 2, \ldots$ the functions I_k are continuous in $\mathbb{R}$.

H4.16. $0 < t_1 < t_2 < \cdots < t_k < t_{k+1} < \cdots$ and $t_k \to \infty$ as $k \to \infty$.

The main results here are obtained by means of piecewise continuous Lyapunov functions $V : [0, \infty) \times \mathbb{R} \to \mathbb{R}_+$ such that $V \in V_0$. The derivatives of the functions $V \in V_0$ are estimated by the elements of the set

$$\Omega_1 = \left\{ x \in PC[[0, \infty), \mathbb{R}] : V(s, x(s)) \leq V(t, x(t)),\ t - \tau \leq s \leq t \right\}.$$

Let $y(t) = x(t) - bx(t - \tau(t)) - c$. We transform (4.48) to the form

$$\begin{cases} \dot{y}(t) = -y(t) - c + af(y(t)) - abf(y(t - \tau(t))),\ t \neq t_k,\ t \geq 0 \\ \Delta y(t_k) = J_k(y(t_k)),\ k = 1, 2, \ldots, \end{cases} \tag{4.50}$$

where $J_k(y(t_k)) = I_k(y(t_k) + bx(t_k - \tau(t_k)) + c) - I_k(bx(t_k - \tau(t_k)) + c)$, $k = 1, 2, \ldots$.

In the proof of the main results we shall use the following lemma.

Lemma 4.23. *Let the conditions H4.12–H4.16 hold. Then:*

(1) *There exists a unique equilibrium x^* of the equation* (4.48) *defined on the interval* $[0, \infty)$.

(2) $\lim_{t \to \infty} x(t) = x^*$ *as* $\lim_{t \to \infty} y(t) = y^*$, *where y^* is the equilibrium of* (4.50).

Proof of Assertion 1. Under the hypotheses H4.12–H4.14, the equation without impulses

$$\dot{x}(t) = -x(t) + af\ (x(t) - bx(t - \tau(t)) - c),\ t \geq 0,$$

has [98, 99, 232] a unique equilibrium x^* on the interval $[0, \infty)$. That means that the solution x^* of problem (4.48), (4.49) is defined on $[0, t_1] \cup (t_{k-1}, t_k]$, $k = 1, 2, \ldots$. From the conditions H4.15 and H4.16 we conclude that it is continuable for $t \geq 0$.

Proof of Assertion 2. The proof of Assertion 2 follows from the corresponding assertion for the continuous case [232] and from the fact that if y^* denotes an equilibrium of the equation (4.50), then $J_k(y^*) = 0$, $k = 1, 2, \ldots$. □

Set $u(t) = y(t) - y^*$ and consider the following equation

$$\begin{cases} \dot{u}(t) = -u(t) + a[f(u(t) + y^*) - f(y^*)] \\ \qquad\quad -ab[f(u(t - \tau(t)) + y^*) - f(y^*)], \; t \neq t_k, \; t \geq 0 \\ \Delta u(t_k) = P_k(u(t_k)), \; k = 1, 2, \ldots, \end{cases} \qquad (4.51)$$

where $P_k(u) = J_k(u + y^*) - J_k(y^*) = J_k(u + y^*), k = 1, 2, \ldots$.

Theorem 4.24. *Assume that:*

(1) *Conditions H4.12–H4.16 hold.*

(2) *There exists a constant $d > 0$ such that*

$$0 < d \leq 1 - La(1 + b).$$

(3) *The functions P_k are such that*

$$P_k(u(t_k)) = -\sigma_k u(t_k), \quad 0 < \sigma_k < 2, \; k = 1, 2, \ldots.$$

Then the equilibrium x^* of (4.48) is globally equi-asymptotically stable.

Proof. We define a Lyapunov function

$$V(t, u) = \frac{1}{2} u^2.$$

Then for $t = t_k$, from the condition (3) of Theorem 4.24, we obtain

$$V(t_k + 0, u(t_k) + P_k(u(t_k))) = \frac{1}{2}(u(t_k) + P_k(u(t_k)))^2$$
$$= \frac{1}{2}(1 - \sigma_k)^2 u^2(t_k) < V(t_k, u(t_k)), \; k = 1, 2, \ldots.$$

Let $t \geq 0$ and $t \neq t_k$. Then for the upper right-hand derivative $D^+_{(4.51)} V(t, u(t))$ of V with respect to equation (4.51) we get

$$D^+_{(4.51)} V(t, u(t)) = u(t)\dot{u}(t)$$
$$= u(t)\big(-u(t) + a\big[f(u(t) + y^*) - f(y^*)\big] - ab\big[f(u(t - \tau(t)) + y^*) - f(y^*)\big]\big)$$
$$= -u^2(t) + au(t)\big[f(u(t) + y^*) - f(y^*)\big] - abu(t)\big[f(u(t - \tau(t)) + y^*) - f(y^*)\big].$$

Since for the function f assumption H4.12 is true, we have

$$f(u(t) + y^*) - f(y^*) = u(t)f'(\xi_1(t)),$$

$$f(u(t - \tau(t)) + y^*) - f(y^*) = u(t - \tau(t))f'(\xi_2(t)),$$

where $\xi_1(t)$ lies between y^* and $u(t) + y^*$, and $\xi_2(t)$ lies between y^* and $u(t - \tau(t))$ and

$$D^+_{(4.51)} V(t, u(t)) \leq -u^2(t) + aLu^2(t) + abLu(t)u(t - \tau(t)), \ t \neq t_k, \ k = 1, 2, \ldots.$$

From the above estimate for any solution $u(t)$ of (4.51) such that $u \in \Omega_1$ for $t \neq t_k$, $k = 1, 2, \ldots$, we have

$$D^+_{(4.51)} V(t, u(t)) \leq (-1 + aL(1 + b))V(t, u(t)) \leq -dV(t, u(t)),$$

where, by condition (2) of Theorem 4.24, $d > 0$.

Thus, since all the conditions of Theorem 2.32 are satisfied, the zero solution of equation (4.51) is globally equi-asymptotically stable, and hence the equilibrium x^* of (4.48) is globally equi-asymptotically stable.   □

Let $t_0 \in \mathbb{R}_+$ and let $\|x\| = \sum_{i=1}^{n} |x_i|$ define the norm of $x \in \mathbb{R}^n$. We consider the following impulsive cellular neural network with time-varying delays:

$$
\begin{cases}
\dot{x}_i(t) = -c_i x_i(t) + \sum_{j=1}^{n} a_{ij} f_j\left(x_j(t)\right) \\
\\
\quad + \sum_{j=1}^{n} b_{ij} f_j\left(x_j(t - \tau_j(t))\right) + I_i, \ t \neq t_k, \ t \geq t_0 \\
\\
\Delta x_i(t_k) = x_i(t_k + 0) - x_i(t_k) = P_{ik}(x_i(t_k)), \ k = 1, 2, \ldots,
\end{cases}
\tag{4.52}
$$

where $i = 1, 2, \ldots, n$; n corresponds to the numbers of units in the neural network; $x_i(t)$ corresponds to the state of the ith unit at time t; $f_j(x_j(t))$ denotes the output of the jth unit at time t; a_{ij}, b_{ij}, I_i, c_i are constants, a_{ij} denotes the strength of the jth unit on the ith unit at time t, b_{ij} denotes the strength of the jth unit on the ith unit at time $t - \tau_j(t)$, I_i denotes the external bias on the ith unit, $\tau_j(t)$ corresponds to the transmission delay along the axon of the jth unit and satisfies $0 \leq \tau_j(t) \leq \tau$ ($\tau = \text{const}$), c_i represents the rate with which the ith unit will reset its potential to the resting state in isolation when disconnected from the network and external inputs; t_k, $k = 1, 2, \ldots$ are the moments of impulsive perturbations and satisfy $t_0 < t_1 < t_2 < \cdots$ and $\lim_{k \to \infty} t_k = \infty$; $x_i(t_k) = x_i(t_k - 0)$ and $x_i(t_k + 0)$ are the states of the ith unit before and after the impulsive perturbation at t_k, respectively; and $P_{ik}(x_i(t_k))$ represents the abrupt change of the state $x_i(t)$ at the impulsive moment t_k.

Let $\varphi \in CB[[-\tau, 0], \mathbb{R}^n]$. Denote by $x(t) = x(t; t_0, \varphi)$, $x \in \mathbb{R}^n$, the solution of system (4.52), satisfying the initial condition

$$
\begin{cases}
x(t; t_0, \varphi) = \varphi(t - t_0), t_0 - \tau \leq t \leq t_0 \\
x(t_0 + 0; t_0, \varphi) = \varphi(0).
\end{cases}
\tag{4.53}
$$

Let $\|\varphi\|_\tau = \max_{t \in [t_0-\tau, t_0]} \|\varphi(t - t_0)\|$ be the norm of the function $\varphi \in CB[[-\tau, 0],$ $\mathbb{R}^n]$.

We introduce the following conditions:

H4.17. There exist constants $L_i > 0$ such that

$$|f_i(u) - f_i(v)| \le L_i |u - v|$$

for all $u, v \in \mathbb{R}$, $i = 1, 2, \ldots, n$.

H4.18. There exist constants $M_i > 0$ such that for all $u \in \mathbb{R}$ and $i = 1, 2, \ldots, n$

$$|f_i(u)| \le M_i < \infty.$$

H4.19. The functions P_{ik} are continuous on $\mathbb{R}$, $i = 1, 2, \ldots, n$, $k = 1, 2, \ldots$.

H4.20. $t_0 < t_1 < t_2 < \cdots < t_k < t_{k+1} < \cdots$ and $t_k \to \infty$ as $k \to \infty$.

H4.21. There exists a unique equilibrium

$$x^* = \mathrm{col}(x_1^*, x_2^*, \ldots, x_n^*)$$

of the system (4.52) such that

$$c_i x_i^* = \sum_{j=1}^n a_{ij} f_j(x_j^*) + \sum_{j=1}^n b_{ij} f_j\left(x_j^*\right) + I_i,$$

$$P_{ik}(x_i^*) = 0, \quad i = 1, 2, \ldots, n, \quad k = 1, 2, \ldots.$$

Remark 4.25. The problems of existence and uniqueness of equilibrium states of neural networks without impulses have been investigated in [106, 107, 170, 232, 233]. Efficient sufficient conditions for the existence and uniqueness of an equilibrium of systems of type (4.52) are given in [9, 221].

Further on we shall use piecewise continuous Lyapunov functions $V : [t_0, \infty) \times \mathbb{R}^n \to \mathbb{R}_+$ such that $V \in V_0$.

One can derive from (4.52) that $y_i(t) = x_i(t) - x_i^*$ satisfies

$$
\left\{
\begin{aligned}
\dot{y}_i(t) &= -c_i y_i(t) + \sum_{j=1}^n a_{ij} [f_j\left(x_j^* + y_j(t)\right) - f_j(x_j^*)] \\
&\quad + \sum_{j=1}^n b_{ij} [f_j\left(x_j^* + y_j(t - \tau_j(t))\right) - f_j(x_j^*)], \ t \ne t_k, \ t \ge t_0 \\
\Delta y_i(t_k) &= Q_{ik}(y_i(t_k)), \ k = 1, 2, \ldots,
\end{aligned}
\right.
\tag{4.54}
$$

where $Q_{ik}(y_i(t_k)) = P_{ik}(y_i(t_k) + x_i^*)$, $i = 1, 2, \ldots, n, k = 1, 2, \ldots$.

Theorem 4.26. *Assume that:*

(1) *Conditions H4.17–H4.21 hold.*

(2) *The system parameters c_i, a_{ij} and b_{ij} $(i, j = 1, 2, \ldots, n)$ satisfy the following inequalities:*

$$\min_{1 \le i \le n} \left(c_i - L_i \sum_{j=1}^{n} |a_{ji}| \right) > \max_{1 \le i \le n} \left(L_i \sum_{j=1}^{n} |b_{ji}| \right) > 0.$$

(3) *The functions P_{ik} are such that*

$$P_{ik}(x_i(t_k)) = -\sigma_{ik}(x_i(t_k) - x_i^*), \quad 0 < \sigma_{ik} < 2,$$

$i = 1, 2, \ldots, n, \ k = 1, 2, \ldots.$

Then the equilibrium x^* of (4.52) is globally exponentially stable.

Proof. We define a Lyapunov function

$$V(t, y) = \sum_{i=1}^{n} |y_i(t)|.$$

Then for $t = t_k$, from the condition (3) of Theorem 4.26, we obtain

$$V(t_k + 0, y(t_k) + \Delta y(t_k)) = \sum_{i=1}^{n} |y_i(t_k) + Q_{ik}(y_i(t_k))|$$

$$= \sum_{i=1}^{n} |x_i(t_k) - x_i^* - \sigma_{ik}(x_i(t_k) - x_i^*)|$$

$$= \sum_{i=1}^{n} |1 - \sigma_{ik}| |x_i(t_k) - x_i^*|$$

$$< \sum_{i=1}^{n} |x_i(t_k) - x_i^*| = V(t_k, y(t_k)), \quad k = 1, 2, \ldots.$$

$$(4.55)$$

Let $t \ge t_0$ and $t \ne t_k$, $k = 1, 2, \ldots$. Then for the upper right-hand derivative

$D^+_{(4.54)} V(t, y(t))$ of V with respect to system (4.54) we get

$$D^+_{(4.54)} V(t, y(t))$$

$$\leq \sum_{i=1}^{n} \left[-c_i |y_i(t)| + \sum_{j=1}^{n} L_j |a_{ij}| \, |y_j(t)| + \sum_{j=1}^{n} L_j |b_{ij}| \, |y_j(t - \tau_j(t))| \right]$$

$$= - \sum_{i=1}^{n} \left[c_i - L_i \sum_{j=1}^{n} |a_{ji}| \right] |y_i(t)| + \sum_{j=1}^{n} \sum_{i=1}^{n} L_j |b_{ij}| \, |y_j(t - \tau_j(t))|$$

$$\leq - \min_{1 \leq i \leq n} \left(c_i - L_i \sum_{j=1}^{n} |a_{ji}| \right) \sum_{i=1}^{n} |y_i(t)|$$

$$+ \max_{1 \leq i \leq n} \left(L_i \sum_{j=1}^{n} |b_{ji}| \right) \sum_{i=1}^{n} |y_i(t - \tau_i(t))|$$

$$\leq -k_1 V(t, y(t)) + k_2 \sup_{t - \tau \leq s \leq t} V(s, y(s)),$$

where

$$k_1 = \min_{1 \leq i \leq n} \left(c_i - L_i \sum_{j=1}^{n} |a_{ji}| \right) > 0,$$

$$k_2 = \max_{1 \leq i \leq n} \left(L_i \sum_{j=1}^{n} |b_{ji}| \right) > 0.$$

From the above estimate for any solution $y(t)$ of (4.54) such that

$$V(s, y(s)) \leq V(t, y(t)), \quad t - \tau \leq s \leq t,$$

we have

$$D^+_{(4.54)} V(t, y(t)) \leq -(k_1 - k_2) V(t, y(t)), \ t \neq t_k, \ k = 1, 2, \dots .$$

By virtue of condition (2) of Theorem 4.26 there exists a real number $\alpha > 0$ such that

$$k_1 - k_2 \geq \alpha,$$

and it follows that

$$D^+_{(4.54)} V(t, y(t)) \leq -\alpha V(t, y(t)), \ t \neq t_k, \ t \geq t_0. \tag{4.56}$$

Then using (4.56), (4.55), we get

$$V(t, y(t)) \leq e^{-\alpha(t - t_0)} V(t_0 + 0, y(t_0 + 0)), \quad t \geq t_0.$$

So,

$$\sum_{i=1}^{n} |x_i(t) - x_i^*| \leq e^{-\alpha(t-t_0)} \sum_{i=1}^{n} |x_i(t_0 + 0) - x_i^*|$$

$$\leq e^{-\alpha(t-t_0)} \max_{s \in [t_0-\tau, t_0]} \left(\sum_{i=1}^{n} |x_i(s) - x_i^*| \right), \quad t \geq t_0,$$

and this completes the proof of the theorem.   □

Theorem 4.27. *Assume that:*

(1) *Conditions (1) and (3) of Theorem 4.26 hold.*

(2) *The system parameters* c_i, a_{ij} *and* b_{ij} $(i, j = 1, 2, \ldots, n)$ *satisfy the following inequalities:*

$$\min_{1 \leq i \leq n} \left(2c_i - \sum_{j=1}^{n} (L_j(|a_{ij}| + |b_{ij}|) + L_i|a_{ji}|) \right) > \max_{1 \leq i \leq n} \left(L_i \sum_{j=1}^{n} |b_{ji}| \right) > 0.$$

Then the equilibrium x^* *of (4.52) is globally exponentially stable.*

Proof. We define a Lyapunov function

$$V(t, y) = \frac{1}{2} \sum_{i=1}^{n} y_i^2(t).$$

Then for $t = t_k$ from the condition (3) of Theorem 4.26, we obtain

$$V(t_k + 0, y(t_k) + \Delta y(t_k)) = \frac{1}{2} \sum_{i=1}^{n} (y_i(t_k) + Q_{ik}(y_i(t_k)))^2$$

$$= \frac{1}{2} \sum_{i=1}^{n} (x_i(t_k) - x_i^* - \sigma_{ik}(x_i(t_k) - x_i^*))^2$$

$$= \frac{1}{2} \sum_{i=1}^{n} (1 - \sigma_{ik})^2 (x_i(t_k) - x_i^*)^2$$

$$< \frac{1}{2} \sum_{i=1}^{n} (x_i(t_k) - x_i^*)^2 = V(t_k, y(t_k)), \quad k = 1, 2, \ldots.$$

$$(4.57)$$

Let $t \geq t_0$ and $t \neq t_k$, $k = 1, 2, \ldots$. Then for the upper right-hand derivative $D^+_{(4.54)} V(t, y(t))$ of $V(t, y(t))$ with respect to system (4.54) we get

$$
D^+_{(4.54)} V(t, y(t))
$$

$$
= \sum_{i=1}^{n} \left[y_i(t)(-c_i y_i(t) + \sum_{j=1}^{n} a_{ij}(f_j(x_j^* + y_j(t)) - f_j(x_j^*)) \right.
$$

$$
\left. + \sum_{j=1}^{n} b_{ij}(f_j(x_j^* + y_j(t - \tau_j(t))) - f_j(x_j^*)) \right]
$$

$$
\leq \sum_{i=1}^{n} \left[-c_i y_i^2(t) + \sum_{j=1}^{n} L_j |a_{ij}| |y_i(t)| |y_j(t)| \right.
$$

$$
\left. + \sum_{j=1}^{n} L_j |b_{ij}| |y_i(t)| |y_j(t - \tau_j(t))| \right]
$$

$$
\leq \sum_{i=1}^{n} \left[-c_i y_i^2(t) + \frac{1}{2} \sum_{j=1}^{n} L_j |a_{ij}| (y_i^2(t) + y_j^2(t)) \right.
$$

$$
\left. + \frac{1}{2} \sum_{j=1}^{n} L_j |b_{ij}| (y_i^2(t) + y_j^2(t - \tau_j(t))) \right]
$$

$$
= -\frac{1}{2} \sum_{i=1}^{n} \left[\left(2c_i - \sum_{j=1}^{n} (L_j(|a_{ij}| + |b_{ij}|) + L_i |a_{ji}|) \right) y_i^2(t) \right]
$$

$$
+ \frac{1}{2} \sum_{i=1}^{n} \sum_{j=1}^{n} L_i |b_{ji}| y_i^2(t - \tau_i(t))
$$

$$
\leq - \min_{1 \leq i \leq n} \left(2c_i - \sum_{j=1}^{n} (L_j(|a_{ij}| + |b_{ij}|) + L_i |a_{ji}|) \right) \frac{1}{2} \sum_{i=1}^{n} y_i^2(t)
$$

$$
+ \max_{1 \leq i \leq n} \left(L_i \sum_{j=1}^{n} |b_{ji}| \right) \frac{1}{2} \sum_{i=1}^{n} y_i^2(t - \tau_i(t))
$$

$$
\leq -k_1 V(t, y(t)) + k_2 \sup_{t-\tau \leq s \leq t} V(s, y(s)),
$$

where

$$
k_1 = \min_{1 \leq i \leq n} \left(2c_i - \sum_{j=1}^{n} (L_j(|a_{ij}| + |b_{ij}|) + L_i |a_{ji}|) \right) > 0,
$$

$$k_2 = \max_{1 \le i \le n} \left(L_i \sum_{j=1}^{n} |b_{ji}| \right) > 0.$$

From the above estimate for any solution $y(t)$ of (4.54) such that

$$V(s, y(s)) \le V(t, y(t)), \quad t - \tau \le s \le t,$$

we have

$$D_{(4.54)}^+ V(t, y(t)) \le -(k_1 - k_2) V(t, y(t)), \quad t \ne t_k, \ k = 1, 2, \ldots.$$

By virtue of condition (2) of Theorem 4.27 there exists a real number $\alpha > 0$ such that

$$k_1 - k_2 \ge \alpha,$$

and it follows that

$$D_{(4.54)}^+ V(t, y(t)) \le -\alpha V(t, y(t)), \quad t \ne t_k, \ t \ge t_0. \tag{4.58}$$

Then using (4.57), (4.58), we get

$$V(t, y(t)) \le e^{-\alpha(t-t_0)} V(t_0 + 0, y(t_0 + 0)), \quad t \ge t_0.$$

So,

$$\sum_{i=1}^{n} |x_i(t) - x_i^*| \le e^{-\frac{\alpha}{2}(t-t_0)} \sum_{i=1}^{n} |x_i(t_0 + 0) - x_i^*|$$

$$\le e^{-\frac{\alpha}{2}(t-t_0)} \max_{s \in [t_0-\tau, t_0]} \left(\sum_{i=1}^{n} |x_i(s) - x_i^*| \right), \quad t \ge t_0,$$

and this completes the proof of the theorem.   □

In the following, we shall give three examples to show our results.

Example 4.28. Consider the impulsive neural network with time-varying delays

$$\begin{cases} \dot{x}_i(t) = -c_i x_i(t) + \sum_{j=1}^{n} a_{ij} f_j \left(x_j(t) \right) \\ \\ \quad + \sum_{j=1}^{n} b_{ij} f_j \left(x_j(t - \tau_j(t)) \right) + I_i, \quad t \ne t_k, \ t \ge 0, \end{cases} \tag{4.59}$$

where $n = 2$; $I_1 = 1$, $I_2 = 1$; $c_1 = c_2 = 3$; $f_i(x_i) = \frac{1}{2}(|x_i + 1| - |x_i - 1|)$; $0 \le \tau_i(t) \le \tau$ $(\tau = 1)$, $i = 1, 2$;

$$(a_{ij})_{2 \times 2} = \begin{pmatrix} a_{11} & a_{12} \\ a_{21} & a_{22} \end{pmatrix} = \begin{pmatrix} 1 & 1 \\ -1 & 1 \end{pmatrix};$$

$$(b_{ij})_{2\times 2} = \begin{pmatrix} b_{11} & b_{12} \\ b_{21} & b_{22} \end{pmatrix} = \begin{pmatrix} 0.9 & -0.8 \\ -0.05 & 0.15 \end{pmatrix};$$

with impulsive perturbations of the form

$$\begin{cases} x_1(t_k + 0) = \dfrac{1.8262806 + x_1(t_k)}{3}, \ k = 1, 2, \ldots \\ x_2(t_k + 0) = \dfrac{0.0668151 + x_2(t_k)}{4}, \ k = 1, 2, \ldots, \end{cases} \tag{4.60}$$

where the impulsive moments are such that $0 < t_1 < t_2 < \cdots$, and $\lim_{k\to\infty} t_k = \infty$.

It is easy to verify that the condition (2) of Theorem 4.26 is satisfied for $L_1 = L_2 = 1$, $k_1 = 1$, $k_2 = 0.95$, and the condition (2) of Theorem 4.27 is not satisfied. We also have that

$$0 < \sigma_{1k} = \frac{2}{3} < 2, \qquad 0 < \sigma_{2k} = \frac{3}{4} < 2.$$

According to Theorem 4.26, the unique equilibrium

$$x^* = (x_1^*, x_2^*)^T = (0.9131403, 0.0222717)^T \tag{4.61}$$

of (4.59), (4.60) is globally exponentially stable.

If we consider again system (4.59) but with impulsive perturbations of the form

$$\begin{cases} x_1(t_k + 0) = 2.7394209 - 2x_1(t_k), \ k = 1, 2, \ldots \\ x_2(t_k + 0) = \dfrac{0.0668151 + x_2(t_k)}{4}, \ k = 1, 2, \ldots, \end{cases} \tag{4.62}$$

the point (4.61) will be again an equilibrium of (4.59), (4.62), but there is nothing we can say about its exponential stability, because $\sigma_{1k} = 3 > 2$.

Example 4.29. Consider the impulsive neural network with time-varying delays (4.59), where $n = 2$; $I_1 = I_2 = 1$; $c_1 = c_2 = 4$; $f_i(x_i) = \frac{1}{2}(|x_i + 1| - |x_i - 1|)$; $0 \le \tau_i(t) \le \tau \ (\tau = 1)$, $i = 1, 2$;

$$(a_{ij})_{2\times 2} = \begin{pmatrix} a_{11} & a_{12} \\ a_{21} & a_{22} \end{pmatrix} = \begin{pmatrix} 2 & 1 \\ 0 & 2 \end{pmatrix};$$

$$(b_{ij})_{2\times 2} = \begin{pmatrix} b_{11} & b_{12} \\ b_{21} & b_{22} \end{pmatrix} = \begin{pmatrix} 0.4 & -0.6 \\ -0.4 & 0.4 \end{pmatrix};$$

with impulsive perturbations of the form

$$\begin{cases} x_1(t_k + 0) = \dfrac{0.7352941 + x_1(t_k)}{2}, \ k = 1, 2, \ldots \\ x_2(t_k + 0) = \dfrac{1.3235295 + 2x_2(t_k)}{5}, \ k = 1, 2, \ldots, \end{cases} \tag{4.63}$$

where the impulsive moments are such that $0 < t_1 < t_2 < \cdots$, and $\lim_{k\to\infty} t_k = \infty$.

It is easy to verify that the condition (2) of Theorem 4.27 is satisfied for $L_1 = L_2 = 1$, $k_1 = 2$, $k_2 = 1$, and the condition (2) of Theorem 4.26 is not satisfied. We also have that

$$0 < \sigma_{1k} = \frac{1}{2} < 2, \qquad 0 < \sigma_{2k} = \frac{3}{5} < 2.$$

According to Theorem 4.27, the unique equilibrium

$$x^* = (x_1^*, x_2^*)^T = (0.7352941, 0.4411765)^T \tag{4.64}$$

of (4.59), (4.63) is globally exponentially stable.

If we consider again system (4.59) but with impulsive perturbations of the form

$$
\begin{cases}
x_1(t_k + 0) = \dfrac{0.7352941 + x_1(t_k)}{2}, \ k = 1, 2, \ldots \\[2mm]
x_2(t_k + 0) = 5x_2(t_k) - 1.764706, \ k = 1, 2, \ldots,
\end{cases}
\tag{4.65}
$$

the point (4.64) will be again an equilibrium of (4.59), (4.65), but there is nothing we can say about its exponential stability, because $\sigma_{2k} = -4 < 0$.

Example 4.30. Consider again the impulsive neural network with time-varying delays (4.59), where $n = 2$; $I_1 = I_2 = 1$; $c_1 = c_2 = 4$; $f_i(x_i) = \frac{1}{2}(|x_i + 1| - |x_i - 1|)$; $0 \le \tau_i(t) \le \tau$ ($\tau = 1$), $i = 1, 2$;

$$(a_{ij})_{2\times 2} = \begin{pmatrix} a_{11} & a_{12} \\ a_{21} & a_{22} \end{pmatrix} = \begin{pmatrix} 2 & 1 \\ -1 & 2 \end{pmatrix};$$

$$(b_{ij})_{2\times 2} = \begin{pmatrix} b_{11} & b_{12} \\ b_{21} & b_{22} \end{pmatrix} = \begin{pmatrix} 0.2 & -0.4 \\ -0.2 & 0.2 \end{pmatrix};$$

with impulsive perturbations of the form

$$
\begin{cases}
x_1(t_k + 0) = \dfrac{1.8181818 - x_1(t_k)}{2}, \ k = 1, 2, \ldots \\[2mm]
x_2(t_k + 0) = \dfrac{1.0606064 - x_2(t_k)}{6}, \ k = 1, 2, \ldots,
\end{cases}
\tag{4.66}
$$

where the impulsive moments are such that $0 < t_1 < t_2 < \cdots$, and $\lim_{k\to\infty} t_k = \infty$.

It is easy to verify that the conditions (2) of Theorems 4.26 and 4.27 are satisfied for $L_1 = L_2 = 1$. We also have that

$$0 < \sigma_{1k} = \frac{3}{2} < 2, \qquad 0 < \sigma_{2k} = \frac{7}{6} < 2.$$

According to Theorem 4.26 ($k_1 = 1$, $k_2 = 0.6$) and according to Theorem 4.27 ($k_1 = 1.4$, $k_2 = 0.6$), the unique equilibrium

$$x^* = (x_1^*, x_2^*)^T = (0.6060606, 0.1515152)^T \tag{4.67}$$

of (4.59), (4.66) is globally exponentially stable.

If we consider again system (4.59) but with impulsive perturbations of the form

$$\begin{cases} x_1(t_k + 0) = 6x_1(t_k) - 3.030303, \ k = 1, 2, \ldots \\ x_2(t_k + 0) = \dfrac{1.0606064 - x_2(t_k)}{6}, \ k = 1, 2, \ldots, \end{cases} \tag{4.68}$$

the point (4.67) will be again an equilibrium of (4.59), (4.68), but there is nothing we can say about its exponential stability, because $\sigma_{1k} = -5 < 0$.

The examples considered show that by means of appropriate impulsive perturbations we can control stability properties of the neural networks.

Impulsive Bidirectional Associative Memory (BAM) neural network models with time delays

Considering the following BAM impulsive system:

$$\begin{cases} \dot{x}_i(t) = -c_i x_i(t) + \displaystyle\sum_{j=1}^{n} w_{ji} f_j(y_j(t - \tau_{ji})) + I_i, \ t \neq t_k \\ \dot{y}_j(t) = -d_j y_j(t) + \displaystyle\sum_{i=1}^{m} h_{ij} g_i(x_i(t - \sigma_{ij})) + J_j, \ t \neq t_k \\ \Delta x_i(t_k) = P_{ik}(x_i(t_k)), \ \Delta y_j(t_k) = Q_{jk}(y_j(t_k)), \ k = 1, 2, \ldots, \end{cases} \tag{4.69}$$

for $t \geq 0$; $i = 1, 2, \ldots, m$; $j = 1, 2, \ldots, n$; $x_i(t)$ and $y_j(t)$ are the activations; c_i, d_j are positive constants; time delays τ_{ji}, σ_{ij} are nonnegative constants; w_{ji}, h_{ij} are the connection weights; f_j, g_i are activation functions; I_i, J_j, denote external inputs; P_{ik}, Q_{jk} are the abrupt changes of the states at the impulsive moments t_k; and $0 < t_1 < t_2 < \cdots$ is a strictly increasing sequence such that $\lim_{k \to \infty} t_k = \infty$.

Let $\varphi \in CB[[-\sigma, 0], \mathbb{R}^m]$, $\varphi = (\varphi_1, \varphi_2, \ldots, \varphi_m)^T$ and $\phi \in CB[[-\tau, 0], \mathbb{R}^n]$, $\phi = (\phi_1, \phi_2, \ldots, \phi_n)^T$. Denote by $\mathrm{col}(x(t), y(t)) = \mathrm{col}(x(t; 0, \varphi), y(t; 0, \phi)) \in \mathbb{R}^{m+n}$, $\mathrm{col}(x(t; 0, \varphi), y(t; 0, \phi)) = (x_1(t; 0, \varphi), \ldots, x_m(t; 0, \varphi), y_1(t; 0, \phi), \ldots, y_n(t; 0, \phi))^T$ the solution of system (4.69), satisfying the initial conditions

$$\begin{cases} x_i(t; 0, \varphi) = \varphi_i(t), -\sigma \leq t \leq 0, \ i = 1, 2, \ldots, m \\ y_j(t; 0, \phi) = \phi_j(t), -\tau \leq t \leq 0, \ j = 1, 2, \ldots, n \\ x_i(0^+, 0, \varphi) = \varphi_i(0), \ y_j(0^+, 0, \phi) = \phi_i(0), \end{cases} \tag{4.70}$$

where $\tau = \max_{1 \leq i \leq m, 1 \leq j \leq n} \tau_{ji}$ and $\sigma = \max_{1 \leq i \leq m, 1 \leq j \leq n} \sigma_{ij}$.

Note that at the moments of impulse effects t_k, $k = 1, 2, \ldots$ the following relations are satisfied:

$$\begin{cases} x_i(t_k + 0) = x_i(t_k) + P_{ik}(x_i(t_k)), \ i = 1, 2, \ldots, m \\ y_j(t_k + 0) = y_j(t_k) + Q_{jk}(y_j(t_k)), \ j = 1, 2, \ldots, n \\ x_i(t_k - 0) = x_i(t_k), \quad y_j(t_k - 0) = y_j(t_k), \ k = 1, 2, \ldots. \end{cases} \quad (4.71)$$

Let $\|z\| = (\sum_{l=1}^{m+n} z_l^2)^{1/2}$ define the norm of $z \in \mathbb{R}^{m+n}$.

We introduce the following conditions:

H4.22. There exist constants $a_j > 0$ such that

$$|f_j(u) - f_j(v)| \le a_j|u - v|$$

for all $u, v \in \mathbb{R}$, $j = 1, 2, \ldots, n$.

H4.23. There exist constants $b_i > 0$ such that

$$|g_i(u) - g_i(v)| \le b_i|u - v|$$

for all $u, v \in \mathbb{R}$, $i = 1, 2, \ldots, m$.

H4.24. The functions P_{ik} are continuous on $\mathbb{R}$, $i = 1, 2, \ldots, m$, $k = 1, 2, \ldots.$

H4.25. The functions Q_{jk} are continuous on $\mathbb{R}$, $j = 1, 2, \ldots, n$, $k = 1, 2, \ldots.$

H4.26. $0 < t_1 < t_2 < \cdots < t_k < t_{k+1} < \cdots$ and $t_k \to \infty$ as $k \to \infty$.

H4.27. There exists a unique equilibrium

$$\mathrm{col}(x^*, y^*) = \mathrm{col}(x_1^*, x_2^*, \ldots, x_m^*, y_1^*, y_2^*, \ldots, y_n^*)$$

of the system (4.69) such that

$$c_i x_i^* = \sum_{j=1}^{n} w_{ji} f_j(y_j^*) + I_i, \quad d_j y_j^* = \sum_{i=1}^{m} h_{ij} g_i(x_i^*) + J_j,$$

$$P_{ik}(x_i^*) = 0, \ Q_{jk}(y_j^*) = 0, \ i = 1, 2, \ldots, m, \ j = 1, 2, \ldots, n, \ k = 1, 2, \ldots.$$

Further on we shall use piecewise continuous Lyapunov functions $V : [0, \infty) \times \mathbb{R}^{m+n} \to \mathbb{R}_+$ such that $V \in V_0$.

We introduce the following notations:

$$x(t) = (x_1(t), x_2(t), \ldots, x_m(t))^T, \quad y(t) = (y_1(t), y_2(t), \ldots, y_n(t))^T,$$

$$f(t - \tau) = (f_j(y_j(t - \tau_{ji})))_{n \times m}, \quad g(t - \sigma) = (g_i(x_i(t - \sigma_{ij})))_{m \times n},$$

$$C = \text{diag}(c_1, c_2, \dots, c_m), \quad D = \text{diag}(d_1, d_2, \dots, d_n), \quad W = (w_{ji})_{n \times m},$$

$$H = (h_{ij})_{m \times n}, \quad I = (I_1, I_2, \dots, I_m)^T, \quad J = (J_1, J_2, \dots, J_n)^T,$$

$$A = (a_1, a_2, \dots, a_n)^T, \quad B = (b_1, b_2, \dots, b_m)^T,$$

$\lambda_{\min}(P)$ is the smallest eigenvalue of matrix P,

$\lambda_{\max}(P)$ is the greatest eigenvalue of matrix P,

$\|P\| = [\lambda_{\max}(P^T P)]^{\frac{1}{2}}$ is the norm of matrix P.

Theorem 4.31. *Assume that:*

(1) *Conditions H4.22–H4.27 hold.*

(2) *There exist symmetric, positive definite matrices $P_{m \times m}$ and $Q_{n \times n}$ such that*

$$-2\lambda_{\min}(CP) + \|W\| \|A\| \|P\| \left(\frac{\lambda_{\max}(P) + \lambda_{\min}(Q)}{\lambda_{\min}(Q)} \right)$$

$$+ \|H\| \|B\| \|Q\| \frac{\lambda_{\max}(P)}{\lambda_{\min}(P)} \leq -p,$$

$$-2\lambda_{\min}(DQ) + \|H\| \|B\| \|Q\| \left(\frac{\lambda_{\min}(P) + \lambda_{\max}(Q)}{\lambda_{\min}(P)} \right)$$

$$+ \|W\| \|A\| \|P\| \frac{\lambda_{\max}(Q)}{\lambda_{\min}(Q)} \leq -q,$$

where $p, q = \text{const} > 0$.

(3) *The functions P_{ik} and Q_{jk} are such that*

$$P_{ik}(x_i(t_k)) = -\gamma_{ik}(x_i(t_k) - x_i^*), \quad 0 < \gamma_{ik} < 2,$$

$$Q_{jk}(y_j(t_k)) = -\delta_{jk}(y_j(t_k) - y_j^*), \quad 0 < \delta_{jk} < 2,$$

$i = 1, 2, \dots, m, \ j = 1, 2, \dots, n, \ k = 1, 2, \dots.$

Then the equilibrium $\text{col}(x^*, y^*)$ of (4.69) is uniformly globally asymptotically stable.

Proof. Set $u(t) = x(t) - x^*, v(t) = y(t) - y^*$ and consider the following system:

$$\begin{cases} \dot{u}_i(t) = -c_i u_i(t) + \sum_{j=1}^{n} w_{ji}[f_j(y_j^* + v_j(t - \tau_{ji})) - f_j(y_j^*)], \ t \neq t_k \\[2mm] \dot{v}_j(t) = -d_j v_j(t) + \sum_{i=1}^{m} h_{ij} g_i(x_i^* + u_i(t - \sigma_{ij})) - g_i(x_i^*)], \ t \neq t_k \\[2mm] \Delta u_i(t_k) = I_{ik}(u_i(t_k)), \ \Delta v_j(t_k) = J_{jk}(v_j(t_k)), \ k = 1, 2, \dots, \end{cases} \quad (4.72)$$

where $I_{ik}(u_i(t_k)) = P_{ik}(u_i(t_k) + x_i^*)$ and $J_{jk}(v_j(t_k)) = Q_{jk}(v_j(t_k) + y_j^*), i = 1, 2, \dots, m, \ j = 1, 2, \dots, n, \ k = 1, 2, \dots.$

We define a Lyapunov function

$$V(t, u(t), v(t)) = u^T(t)Pu(t) + v^T(t)Qv(t). \tag{4.73}$$

Then for $t = t_k$, from the condition (3) of Theorem 4.31, we obtain

$V(t_k + 0, u(t_k + 0), v(t_k + 0))$

$= u^T(t_k + 0)Pu(t_k + 0) + v^T(t_k + 0)Qv(t_k + 0)$

$= ((1 - \gamma_{1k})u_1(t_k), \ldots, (1 - \gamma_{mk})u_m(t_k))^T P((1 - \gamma_{1k})u_1(t_k), \ldots, (1 - \gamma_{mk})u_m(t_k))$

$\quad + ((1 - \delta_{1k})v_1(t_k), \ldots, (1 - \delta_{nk})v_n(t_k))^T Q((1 - \delta_{1k})v_1(t_k), \ldots, (1 - \delta_{nk})v_n(t_k))$

$< u^T(t_k)Pu(t_k) + v^T(t_k)Qv(t_k)$

$= V(t_k, u(t_k), v(t_k)), \quad k = 1, 2, \ldots .$

Let $t \geq 0$ and $t \neq t_k$, $k = 1, 2, \ldots$. Then from H4.22 and H4.23, for the upper right-hand derivative $D^+_{(4.72)}V(t, u(t), v(t))$ of the function $V(t, u(t), v(t))$ with respect to system (4.72) we get

$D^+_{(4.72)}V(t, u(t), v(t)) = \dot{u}^T(t)Pu(t) + u^T(t)P\dot{u}(t) + \dot{v}^T(t)Qv(t) + v^T(t)Q\dot{v}(t)$

$\quad \leq (-Cu(t) + WAv(t - \tau))^T Pu(t) + u^T(t)P(-Cu(t) + WAv(t - \tau))$

$\quad + (-Dv(t) + HBu(t - \sigma))^T Qv(t) + v^T(t)Q(-Dv(t) + HBu(t - \sigma)).$

From the last estimate and from the inequalities

$$\lambda_{\min}(CP)\|u(t)\|^2 \leq u^T(t)CPu(t) \leq \lambda_{\max}(CP)\|u(t)\|^2,$$

$$\lambda_{\min}(DQ)\|v(t)\|^2 \leq v^T(t)DQv(t) \leq \lambda_{\max}(DQ)\|v(t)\|^2,$$

we obtain

$D^+_{(4.72)}V(t, u(t), v(t)) \leq -2\lambda_{\min}(CP)\|u(t)\|^2 - 2\lambda_{\min}(DQ)\|v(t)\|^2$

$\quad\quad\quad + 2\|W\|\,\|A\|\,\|P\|\,\|v(t - \tau)\|\,\|u(t)\|$

$\quad\quad\quad + 2\|H\|\,\|B\|\,\|Q\|\,\|u(t - \sigma)\|\,\|v(t)\|$

$\quad\quad \leq -2\lambda_{\min}(CP)\|u(t)\|^2 - 2\lambda_{\min}(DQ)\|v(t)\|^2$

$\quad\quad\quad + \|W\|\,\|A\|\,\|P\|(\|v(t - \tau)\|^2 + \|u(t)\|^2)$

$\quad\quad\quad + \|H\|\,\|B\|\,\|Q\|(\|u(t - \sigma)\|^2 + \|v(t)\|^2), \tag{4.74}$

for $t \neq t_k$, $k = 1, 2, \ldots$.

Since for the function $V(t, u(t), v(t))$ we have

$\lambda_{\min}(P)\|u(t)\|^2 + \lambda_{\min}(Q)\|v(t)\|^2 \leq u^T(t)Pu(t) + v^T(t)Qv(t)$

$\quad\quad\quad\quad \leq \lambda_{\max}(P)\|u(t)\|^2 + \lambda_{\max}(Q)\|v(t)\|^2, \; t \geq 0,$

for $u(t)$ and $v(t)$ such that

$$V(s, u(s), v(s)) \leq V(t, u(t), v(t)), \quad t - \max\{\tau, \sigma\} \leq s \leq t,$$

we obtain

$$\lambda_{\min}(P)\|u(s)\|^2 + \lambda_{\min}(Q)\|v(s)\|^2 \leq u^T(s)Pu(s) + v^T(s)Qv(s)$$
$$\leq u^T(t)Pu(t) + v^T(t)Qv(t)$$
$$\leq \lambda_{\max}(P)\|u(t)\|^2 + \lambda_{\max}(Q)\|v(t)\|^2,$$

and hence

$$\begin{cases} \|u(s)\|^2 \leq \dfrac{\lambda_{\max}(P)\|u(t)\|^2 + \lambda_{\max}(Q)\|v(t)\|^2}{\lambda_{\min}(P)}, \\[3mm] \|v(s)\|^2 \leq \dfrac{\lambda_{\max}(P)\|u(t)\|^2 + \lambda_{\max}(Q)\|v(t)\|^2}{\lambda_{\min}(Q)}, \end{cases} \tag{4.75}$$

for $t - \max\{\tau, \sigma\} \leq s \leq t, t \geq 0$.

From (4.74) and (4.75), we obtain

$$D_{(4.72)}^+ V(t, u(t), v(t))$$
$$\leq -2\lambda_{\min}(CP)\|u(t)\|^2 - 2\lambda_{\min}(DQ)\|v(t)\|^2$$
$$+ \|W\|\,\|A\|\,\|P\|\left(\frac{\lambda_{\max}(P)\|u(t)\|^2 + \lambda_{\max}(Q)\|v(t)\|^2}{\lambda_{\min}(Q)} + \|u(t)\|^2\right)$$
$$+ \|H\|\,\|B\|\,\|Q\|\left(\frac{\lambda_{\max}(P)\|u(t)\|^2 + \lambda_{\max}(Q)\|v(t)\|^2}{\lambda_{\min}(P)} + \|v(t)\|^2\right)$$
$$= \left[-2\lambda_{\min}(CP)\right.$$
$$\left. + \|W\|\,\|A\|\,\|P\|\left(\frac{\lambda_{\max}(P) + \lambda_{\min}(Q)}{\lambda_{\min}(Q)}\right) + \|H\|\,\|B\|\,\|Q\|\frac{\lambda_{\max}(P)}{\lambda_{\min}(P)}\right]\|u(t)\|^2$$
$$+ \left[-2\lambda_{\min}(DQ)\right.$$
$$\left. + \|H\|\,\|B\|\,\|Q\|\left(\frac{\lambda_{\min}(P) + \lambda_{\max}(Q)}{\lambda_{\min}(P)}\right) + \|W\|\,\|A\|\,\|P\|\frac{\lambda_{\max}(Q)}{\lambda_{\min}(Q)}\right]\|v(t)\|^2,$$

for $t \neq t_k, k = 1, 2, \ldots$.

From the condition (2) of Theorem 4.31, we derive

$$D_{(4.72)}^+ V(t, u(t), v(t)) \leq -p\|u(t)\|^2 - q\|v(t)\|^2, \ t \neq t_k, k = 1, 2, \ldots.$$

Thus, since all the conditions of Theorem 2.34 are satisfied, the zero solution of system (4.72) is uniformly globally asymptotically stable, and hence the equilibrium $\mathrm{col}(x^*, y^*)$ of (4.69) is uniformly globally asymptotically stable.   □

Example 4.32. Consider the impulsive BAM neural network

$$\begin{cases} \dot{x}(t) = -Cx(t) + Wf(y(t-0.2)) + I, \; t \neq t_k, \; t \geq 0 \\ \dot{y}(t) = -Dy(t) + Hg(x(t-0.1)) + J, \; t \neq t_k, \; t \geq 0 \end{cases} \tag{4.76}$$

with impulsive perturbations of the form

$$\begin{cases} \Delta x_1(t_k) = -\frac{1}{2}(x_1(t_k) - 0.12542), \; k = 1, 2, \ldots \\[2mm] \Delta x_2(t_k) = -\frac{2}{3}(x_2(t_k) - 0.12542), \; k = 1, 2, \ldots \\[2mm] \Delta y_1(t_k) = -\frac{2}{5}(y_1(t_k) - 0.25006), \; k = 1, 2, \ldots \\[2mm] \Delta y_2(t_k) = -\frac{1}{3}(y_2(t_k) - 0.25006), \; k = 1, 2, \ldots, \end{cases} \tag{4.77}$$

where the impulsive moments are such that $0 < t_1 < t_2 < \cdots$, and $\lim_{k \to \infty} t_k = \infty$,

$$x(t) = \begin{pmatrix} x_1(t) \\ x_2(t) \end{pmatrix}, \; y(t) = \begin{pmatrix} y_1(t) \\ y_2(t) \end{pmatrix}, \; C = \begin{pmatrix} 9 & 0 \\ 0 & 9 \end{pmatrix}, \; W = \begin{pmatrix} 0.5 & -0.5 \\ 0.5 & 0.5 \end{pmatrix},$$

$$f(y(t-0.2)) = \begin{pmatrix} \sin\left(\frac{1}{3}y_1(t-0.2)\right) + \frac{2}{3}y_1(t-0.2) \\[3mm] \sin\left(\frac{1}{3}y_2(t-0.2)\right) + \frac{2}{3}y_2(t-0.2) \end{pmatrix},$$

$$g(x(t-0.1)) = \begin{pmatrix} \sin\left(\frac{1}{3}x_1(t-0.1)\right) + \frac{2}{3}x_1(t-0.1) \\[3mm] \sin\left(\frac{1}{3}x_2(t-0.1)\right) + \frac{2}{3}x_2(t-0.1) \end{pmatrix},$$

$$D = \begin{pmatrix} 5 & 0 \\ 0 & 5 \end{pmatrix}, \; H = \begin{pmatrix} -1/3 & 1/3 \\ 1/3 & 1/3 \end{pmatrix}, \; I = \begin{pmatrix} I_1 \\ I_2 \end{pmatrix}, \; J = \begin{pmatrix} J_1 \\ J_2 \end{pmatrix},$$

and initial conditions

$$\begin{cases} x_i(t; 0, \varphi) = \varphi_i(t), \; t \in [-0.1, 0], \; i = 1, 2 \\ y_j(t; 0, \phi) = \phi_j(t), \; t \in [-0.2, 0], \; j = 1, 2. \end{cases}$$

For $I_1 = 1.12878$, $I_2 = 0.96062$, $J_1 = 1.2503$ and $J_2 = 1.194071$ system (4.76), (4.77) has an equilibrium $x_1^* = x_2^* = 0.12542$, $y_1^* = y_2^* = 0.25006$.

Let $P = \begin{pmatrix} 2 & 0 \\ 0 & 2 \end{pmatrix}$ and $Q = \begin{pmatrix} 1 & 0 \\ 0 & 1 \end{pmatrix}$. Since $A = B = \begin{pmatrix} 1 & 0 \\ 0 & 1 \end{pmatrix}$, we have

$$- 2\lambda_{\min}(CP) + \|W\| \|A\| \|P\| \left(\frac{\lambda_{\max}(P) + \lambda_{\min}(Q)}{\lambda_{\min}(Q)} \right)$$

$$+ \|H\| \|B\| \|Q\| \frac{\lambda_{\max}(P)}{\lambda_{\min}(P)} = -36 + \frac{10\sqrt{2}}{3} < 0,$$

$$- 2\lambda_{\min}(DQ) + \|H\| \|B\| \|Q\| \left(\frac{\lambda_{\min}(P) + \lambda_{\max}(Q)}{\lambda_{\min}(P)} \right)$$

$$+ \|W\| \|A\| \|P\| \frac{\lambda_{\max}(Q)}{\lambda_{\min}(Q)} = -10 + \frac{3\sqrt{2}}{2} < 0.$$

Also, $\gamma_{1k} = \frac{1}{2}$, $\gamma_{2k} = \frac{2}{3}$, $\delta_{1k} = \frac{2}{5}$, $\delta_{2k} = \frac{1}{3}$.

Since all the conditions of Theorem 4.31 are satisfied, the equilibrium $x_1^* = x_2^* = 0.12542$, $y_1^* = y_2^* = 0.25006$ of (4.76), (4.77) is uniformly globally asymptotically stable.

Impulsive neural networks with infinite delays

We consider the following impulsive delayed neural network with dynamical thresholds

$$\begin{cases} \dot{x}(t) = -x(t) + af\left(x(t) - b \int_0^\infty m(s)x(t-s)\,ds - c \right), \ t \neq t_k, \ t \geq 0 \\ \Delta x(t_k) = x(t_k + 0) - x(t_k) = I_k(x(t_k)), \ k = 1, 2, \ldots, \end{cases}$$

(4.78)

where $x : \mathbb{R}_+ \to \mathbb{R}$; $m : \mathbb{R}_+ \to \mathbb{R}_+$ is the delayed ker-function; $a > 0$; b and c are nonnegative constants; $f : \mathbb{R} \to \mathbb{R}$; $I_k : \mathbb{R} \to \mathbb{R}$, $k = 1, 2, \ldots$; $0 < t_1 < t_2 < \cdots < t_k < t_{k+1} < \cdots$; $\lim_{k \to \infty} t_k = \infty$.

Let $\varphi \in PCB[(-\infty, 0], \mathbb{R}]$. Denote by $x(t) = x(t; 0, \varphi)$, $x \in \mathbb{R}$, the solution of equation (4.78), satisfying the initial conditions

$$\begin{cases} x(t; 0, \varphi) = \varphi(t), \ -\infty < t \leq 0 \\ x(0^+; 0, \varphi) = \varphi(0). \end{cases}$$

(4.79)

Let $|\varphi|_\infty = \sup_{s \in (-\infty, 0]} |\varphi(s)|$ be the norm of the function $\varphi \in PCB[(-\infty, 0], \mathbb{R}]$.

We introduce the following conditions:

H4.28. $\displaystyle\int_0^\infty m(s)\,ds = 1.$

H4.29. $\displaystyle\int_0^\infty sm(s)\,ds < \infty.$

Further on we shall use piecewise continuous Lyapunov functions $V : \mathbb{R}_+ \times \mathbb{R} \to \mathbb{R}_+$ such that $V \in V_0$. The derivatives of the functions $V \in V_0$ are estimated by the elements of the set

$$\Omega_1 = \left\{ x \in PC[[0, \infty), \mathbb{R}] : V(s, x(s)) \le V(t, x(t)), -\infty < s \le t \right\}.$$

Lemma 4.33. *Let the conditions H4.12–H4.16, H4.28 and H4.29 hold. Then there exists a unique equilibrium x^* of the equation* (4.78).

Proof. The proof of Lemma 4.33 is similar to the proof of Lemma 4.23.   □

Set $y(t) = x(t) - x^*$ and consider the following equation:

$$\begin{cases} \dot{y}(t) = -y(t) + af\left(y(t) - b\displaystyle\int_0^\infty m(s)(y(t-s) + x^*)\,ds\right. \\[2mm] \qquad\qquad \left.+x^* - c\right) - x^*,\ t \ne t_k,\ t \ge 0 \\[2mm] \Delta y(t_k) = J_k(y(t_k)),\ k = 1, 2, \ldots, \end{cases} \tag{4.80}$$

where $J_k(y) = I_k(y + x^*), k = 1, 2, \ldots$.

Theorem 4.34. *Assume that:*

(1) *Conditions H4.12–H4.16, H4.28 and H4.29 hold.*

(2) $La(1 + b) < 1.$

(3) *The functions J_k are such that*

$$|y(t_k) + J_k(y(t_k))| \le |y(t_k)|,\ y \in \mathbb{R},\ t_k > 0.$$

Then the equilibrium x^* of (4.78) *is globally equi-asymptotically stable.*

Proof. We define a Lyapunov function

$$V(t, y) = |y|.$$

Then for $t = t_k$, from the condition (3) of Theorem 4.34, we obtain

$$V(t_k + 0, y(t_k) + J_k(y(t_k))) = |y(t_k) + J_k(y(t_k))| \le |y(t_k)| = V(t_k, y(t_k)),$$

for $k = 1, 2, \ldots$.

Let $t \geq 0$ and $t \neq t_k$. Then for the upper right-hand derivative $D^+_{(4.80)}V(t, y(t))$ of $V(t, y(t))$ with respect to (4.80) we get

$$D^+_{(4.80)}V(t, y(t)) = \operatorname{sgn}(y(t))\, \dot{y}(t)$$

$$= \operatorname{sgn}(y(t))\left(-y(t) + af\left(y(t) - b\int_0^\infty m(s)(y(t-s) + x^*)\, ds + x^* - c\right) - x^*\right).$$

Since x^* is the equilibrium of (4.78), for $t \geq 0$ and $t \neq t_k$ it satisfies the equation

$$-x^* + af\left(x^* - b\int_0^\infty m(s)x^*\, ds - c\right) = 0. \qquad (4.81)$$

From (4.81) and condition H4.12, it follows

$$D^+_{(4.80)}V(t, y(t)) \leq -|y(t)| + aL\left|y(t) - b\int_0^\infty m(s)y(t-s)\, ds\right|$$

$$\leq -|y(t)| + aL|y(t)| + abL\int_0^\infty m(s)|y(t-s)|\, ds.$$

From the above estimate for any solution $y(t)$ of (4.80) such that $y \in \Omega_1$ for $t \geq 0$ and $t \neq t_k$, we have

$$D^+_{(4.80)}V(t, y(t)) \leq (-1 + aL(1+b))V(t, y(t)) = -\beta V(t, y(t)),$$

where, by condition (2) of Theorem 4.34, $\beta > 0$.

Thus, since all the conditions of Theorem 2.32 are satisfied, the zero solution of equation (4.80) is globally equi-asymptotically stable, and hence the equilibrium x^* of (4.78) is globally equi-asymptotically stable.   □

Impulsive delay neural networks of a general type

Let $t_0 \in \mathbb{R}_+$ and let $\|x\| = \sum_{i=1}^n |x_i|$ define the norm of $x \in \mathbb{R}^n$. We consider the following impulsive nonautonomous cellular neural network with bounded and unbounded delays:

$$\begin{cases} \dot{x}_i(t) = -d_i(t)x_i(t) + \sum_{j=1}^n a_{ij} f_j\left(x_j(t)\right) + \sum_{j=1}^n b_{ij} f_j\left(x_j(t - \tau_j(t))\right) \\ \qquad\qquad + \sum_{j=1}^n c_{ij}\int_{-\infty}^t m_j(t,s) f_j\left(x_j(s)\right) ds + I_i,\ t \neq t_k,\ t \geq t_0 \\ \Delta x_i(t_k) = x_i(t_k + 0) - x_i(t_k) = P_{ik}(x_i(t_k)),\ k = 1, 2, \ldots, \end{cases} \qquad (4.82)$$

where $i = 1, 2, \ldots, n$; $x_i(t)$ corresponds to the state of the ith unit at time t; $A_{n \times n} = (a_{ij})_{n \times n}$, $B_{n \times n} = (b_{ij})_{n \times n}$, $C_{n \times n} = (c_{ij})_{n \times n}$ denote the connection weight matrices;

$0 \leq \tau_j(t) \leq \tau$; I_i is the external bias on the ith neuron; $d_i(t)$ represents the rate with which the ith unit will reset its potential to the resting state in isolation when disconnected from the network and external inputs; the delay kernel $m_j(t,s) = m_j(t-s)$ $(j = 1, 2, \ldots, n)$ is of convolution type; t_k, $k = 1, 2, \ldots$ are the moments of impulsive perturbations and satisfy $t_0 < t_1 < t_2 < \cdots$ and $\lim_{k \to \infty} t_k = \infty$; and $P_{ik}(x_i(t_k))$ represents the abrupt change of the state $x_i(t)$ at the impulsive moment t_k.

Let $\varphi \in CB[(-\infty, 0], \mathbb{R}^n]$. Denote by $x(t) = x(t; t_0, \varphi), x \in \mathbb{R}^n$, the solution of system (4.82), satisfying the initial conditions

$$\begin{cases} x(t; t_0, \varphi) = \varphi(t - t_0), -\infty < t \leq t_0 \\ x(t_0 + 0; t_0, \varphi) = \varphi(0). \end{cases} \tag{4.83}$$

Let $\|\varphi\|_\infty = \max_{t \in (-\infty, t_0]} \|\varphi(t - t_0)\|$ be the norm of the function $\varphi \in CB[(-\infty, 0]$, $\mathbb{R}^n]$.

We introduce the following conditions:

H4.30. The delay kernel $m_i : \mathbb{R}^2 \to \mathbb{R}_+$ is continuous, and there exist positive numbers μ_i such that

$$\int_{-\infty}^t m_i(t, s)\, ds \leq \mu_i < \infty$$

for all $t \geq t_0, t \neq t_k, k = 1, 2, \ldots$ and $i = 1, 2, \ldots, n$.

H4.31. There exists a unique equilibrium

$$x^* = \mathrm{col}(x_1^*, x_2^*, \ldots, x_n^*)$$

of the system (4.82) such that

$$d_i(t)x_i^* = \sum_{j=1}^n a_{ij} f_j(x_j^*) + \sum_{j=1}^n b_{ij} f_j(x_j^*) + \sum_{j=1}^n c_{ij} \int_{-\infty}^t m_j(t, s) f_j(x_j^*) + I_i,$$

$$P_{ik}(x_i^*) = 0, \quad i = 1, 2, \ldots, n, \quad k = 1, 2, \ldots.$$

Further on we shall use piecewise continuous Lyapunov functions $V : [t_0, \infty) \times \mathbb{R}^n \to \mathbb{R}_+$ such that $V \in V_0$.

Set $y_i(t) = x_i(t) - x_i^*$, $i = 1, 2, \ldots, n$. Then $y_i(t)$ satisfies

$$
\begin{cases}
\dot{y}_i(t) = -d_i(t)(y_i(t) + x_i^*) + \displaystyle\sum_{j=1}^{n} a_{ij} f_j \left(x_j^* + y_j(t) \right) \\[2ex]
\qquad + \displaystyle\sum_{j=1}^{n} b_{ij} f_j \left(x_j^* + y_j(t - \tau_j(t)) \right) \\[2ex]
\qquad + \displaystyle\sum_{j=1}^{n} c_{ij} \int_{-\infty}^{t} m_j(t, s) f_j \left(x_j^* + y_j(s) \right) ds + I_i, \ t \neq t_k, \ t \geq t_0 \\[2ex]
\Delta y_i(t_k) = Q_{ik}(y_i(t_k)), \ k = 1, 2, \ldots,
\end{cases}
$$

$$(4.84)$$

where $Q_{ik}(y_i(t_k)) = P_{ik}(y_i(t_k) + x_i^*)$, $i = 1, 2, \ldots, n$, $k = 1, 2, \ldots$.

Theorem 4.35. *Assume that:*

(1) *Conditions H4.17–H4.20, H4.30 and H4.31 hold.*

(2) *For $t \geq t_0$, $t \neq t_k$, $k = 1, 2, \ldots$ the inequalities*

$$
\min_{1 \leq i \leq n} \left(d_i(t) - L_i \sum_{j=1}^{n} |a_{ji}| \right) > \max_{1 \leq i \leq n} \left(L_i \left(\sum_{j=1}^{n} |b_{ji}| + \mu_i \sum_{j=1}^{n} |c_{ji}| \right) \right) > 0
$$

   are valid.

(3) *The functions P_{ik} are such that*

$$
P_{ik}(x_i(t_k)) = -\sigma_{ik}(x_i(t_k) - x_i^*), \ 0 < \sigma_{ik} < 2, \ i = 1, 2, \ldots, n, \ k = 1, 2, \ldots.
$$

Then the equilibrium x^* of system (4.82) is globally exponentially stable.

Proof. We define a Lyapunov function

$$
V(t, y) = \sum_{i=1}^{n} |y_i(t)|.
$$

Then for $t = t_k$, $k = 1, 2, \ldots$, from the condition (3) of Theorem 4.35, we obtain

$$
V(t_k + 0, y(t_k) + \Delta y(t_k)) = \sum_{i=1}^{n} |y_i(t_k) + Q_{ik}(y(t_k))|
$$

$$
= \sum_{i=1}^{n} |x_i(t_k) - x_i^* - \sigma_{ik}(x_i(t_k) - x_i^*)|
$$

$$= \sum_{i=1}^{n} |1 - \sigma_{ik}||x_i(t_k) - x_i^*|$$

$$< \sum_{i=1}^{n} |x_i(t_k) - x_i^*| = V(t_k, y(t_k)), \quad k = 1, 2, \dots. \tag{4.85}$$

Let $t \geq t_0$ and $t \neq t_k$, $k = 1, 2, \dots$. Then for the upper right-hand derivative $D_{(4.84)}^{+} V(t, y(t))$ with respect to system (4.84) we get

$$D_{(4.84)}^{+} V(t, y(t)) = \sum_{i=1}^{n} \operatorname{sgn}(y_i(t)) \dot{y}_i(t)$$

$$= \sum_{i=1}^{n} \operatorname{sgn}(y_i(t)) \left[-d_i(t)(y_i(t) + x_i^*) + \sum_{j=1}^{n} a_{ij} f_j \left(x_j^* + y_j(t) \right) \right.$$

$$+ \sum_{j=1}^{n} b_{ij} f_j \left(x_j^* + y_j(t - \tau_j(t)) \right)$$

$$\left. + \sum_{j=1}^{n} c_{ij} \int_{-\infty}^{t} m_j(t, s) f_j \left(x_j^* + y_j(s) \right) ds + I_i \right].$$

From the conditions H4.30 and H4.31, we obtain

$$D_{(4.84)}^{+} V(t, y(t))$$

$$\leq \sum_{i=1}^{n} \left[-d_i(t)|y_i(t)| + \sum_{j=1}^{n} L_j |a_{ij}||y_j(t)| + \sum_{j=1}^{n} L_j |b_{ij}||y_j(t - \tau_j(t))| \right.$$

$$\left. + \sum_{j=1}^{n} L_j |c_{ij}| \int_{-\infty}^{t} m_j(t, s)|y_j(s)|ds \right]$$

$$= -\sum_{i=1}^{n} \left[d_i(t) - L_i \sum_{j=1}^{n} |a_{ji}| \right] |y_i(t)| + \sum_{j=1}^{n} \sum_{i=1}^{n} L_j |b_{ij}||y_j(t - \tau_j(t))|$$

$$+ \sum_{j=1}^{n} \sum_{i=1}^{n} L_j \mu_j |c_{ij}| \sup_{-\infty < s \leq t} |y_j(s)|$$

$$\leq -k_1 V(t, y(t)) + k_2 \sup_{-\infty < s \leq t} V(s, y(s)),$$

where

$$k_1 = \min_{1 \leq i \leq n} \left(d_i(t) - L_i \sum_{j=1}^{n} |a_{ji}| \right) > 0,$$

$$k_2 = \max_{1 \le i \le n} \left(L_i \left(\sum_{j=1}^{n} |b_{ji}| + \mu_i \sum_{j=1}^{n} |c_{ji}| \right) \right) > 0.$$

From the above estimate for any solution $y(t)$ of (4.84) such that

$$V(s, y(s)) \le V(t, y(t)), \quad -\infty < s \le t,$$

we have

$$D^+_{(4.84)} V(t, y(t)) \le -(k_1 - k_2) V(t, y(t)), \ t \ne t_k, \ k = 1, 2, \dots .$$

By virtue of condition (2) of Theorem 4.35 there exists a real number $\alpha > 0$ such that

$$k_1 - k_2 \ge \alpha,$$

and it follows that

$$D^+_{(4.84)} V(t, y(t)) \le -\alpha V(t, y(t)), \ t \ne t_k, \ t \ge t_0. \tag{4.86}$$

Then using (4.86) and (4.85), we get

$$V(t, y(t)) \le e^{-\alpha(t-t_0)} V(t_0 + 0, y(t_0 + 0)), \quad t \ge t_0.$$

So,

$$\sum_{i=1}^{n} |x_i(t) - x_i^*| \le e^{-\alpha(t-t_0)} \sum_{i=1}^{n} |x_i(t_0 + 0) - x_i^*|$$

$$\le e^{-\alpha(t-t_0)} \max_{s \in (-\infty, t_0]} \left(\sum_{i=1}^{n} |x_i(s) - x_i^*| \right), \quad t \ge t_0,$$

and this completes the proof of the theorem.   □

Example 4.36. Consider the impulsive neural network of type (4.82), where $n = 2$; $t_0 = 0$; $I_1 = 0.35$, $I_2 = 5.825$; $d_1(t) = d_2(t) = 3$; $f_i(x_i) = \frac{1}{2}(|x_i + 1| - |x_i - 1|)$; $0 \le \tau_i(t) \le \tau \ (\tau = 1)$, $m_i(s) = e^{-s}$, $i = 1, 2$;

$$(a_{ij})_{2 \times 2} = \begin{pmatrix} a_{11} & a_{12} \\ a_{21} & a_{22} \end{pmatrix} = \begin{pmatrix} 0.5 & 0.5 \\ 0.5 & 0.5 \end{pmatrix};$$

$$(b_{ij})_{2 \times 2} = \begin{pmatrix} b_{11} & b_{12} \\ b_{21} & b_{22} \end{pmatrix} = \begin{pmatrix} 0.9 & -0.8 \\ 0.05 & 0.15 \end{pmatrix};$$

$$(c_{ij})_{2 \times 2} = \begin{pmatrix} c_{11} & c_{12} \\ c_{21} & c_{22} \end{pmatrix} = \begin{pmatrix} 0.5 & 0.5 \\ 0.5 & 0.5 \end{pmatrix};$$

with impulsive perturbations of the form

$$\begin{cases} x_1(t_k + 0) = \dfrac{1.5 + x_1(t_k)}{4}, \ k = 1, 2, \dots \\[3mm] x_2(t_k + 0) = \dfrac{2.5 + x_2(t_k)}{2}, \ k = 1, 2, \dots, \end{cases} \tag{4.87}$$

where the impulsive moments are such that $0 < t_1 < t_2 < \cdots$, and $\lim_{k \to \infty} t_k = \infty$.

It is easy to verify that the condition (2) of Theorem 4.35 is satisfied for $L_1 = L_2 = 1$, $k_1 = 2$, $k_2 = 1.95$ and the condition H4.30 is satisfied, since $\int_0^\infty e^{-s} ds = 1$. We also have that

$$0 < \sigma_{1k} = \frac{3}{4} < 2, \quad 0 < \sigma_{2k} = \frac{1}{2} < 2.$$

According to Theorem 4.35, the unique equilibrium

$$x^* = (x_1^*, x_2^*)^T = (0.5, 2.5)^T \tag{4.88}$$

of (4.82), (4.87) is globally exponentially stable.

If we consider again system (4.82) but with impulsive perturbations of the form

$$\begin{cases} x_1(t_k + 0) = \dfrac{1.5 + x_1(t_k)}{4}, \ k = 1, 2, \dots \\[3mm] x_2(t_k + 0) = 12.5 - 5x_2(t_k), \ k = 1, 2, \dots, \end{cases} \tag{4.89}$$

the point (4.88) will be again an equilibrium of (4.82), (4.89), but there is nothing we can say about its exponential stability, because $\sigma_{2k} = 3 > 2$.

The example again shows that, by means of appropriate impulsive perturbations, we can control stability properties of the neural networks.

4.3   Economic models

The Solow impulsive model with endogenous delay

The centrality of the Solow neo-classical growth model (1956) [86, 96, 188] for economic theory is witnessed by the current persistency of new contributions simulated by his work.

The original Solow growth model is defined by the ordinary differential equation

$$\dot{k} = sf(k) - n_s k, \tag{4.90}$$

where $k = K/L$ denotes the capital-labour ratio; $f(k)$ is the production per unit of labour; s is the saving rate $(0 < s < 1)$; $n_s > 0$ is the rate of change of the labour supply $\dot{L}/L$, which was initially assumed exogenous by Solow. Contrary to most subsequent developments, where the supply of labour was treated as exogenously determined, Solow also tried to endogenise it. He wrote the rate of change of the labour supply as a function of the current level of per-capita income: $n_s = n_s(f(k))$.

It is known that the current rate of change of the labour supply is related to past fertility, and thus to past levels of wage, following a prescribed pattern of delay. There are two main alternatives: fixed delays and distributed delays. The former is better suited when there is no variability in the process of transmission of the past into the future; for instance: when we assume that all individuals are recruited in the labour force at the same fixed age. Conversely, when recruitment may occur at different ages, i.e. with different delays (for instance because the time needed to complete formal education is heterogeneous within the population), distributed delays appear more suitable. The introduction of a distributed delay in the population term in (4.90) leads to the following integro-differential equation

$$\dot{k}(t) = sf(k(t)) - \left[\int_{-\infty}^{t} n_s\big(f(k(\tau))\big)g(t-\tau)\,d\tau\right]k(t), \qquad (4.91)$$

where the term $n_s\big(f(k(\tau))\big)$, $\tau < t$ captures past (rather than present), income-related fertility, and $g(t-\tau)$ is the corresponding delaying kernel.

Integral models of type (4.91) are one of the major economic applications, known as Vintage Capital Models (VCMs). The VCMs bring a new type of stability and optimization problems that involve the optimal control of an endogenous delay [1, 62, 78, 86, 96, 105, 113, 176, 177].

On the other hand, the state of economic processes is often subject to instantaneous perturbations at certain instants, which may be caused by population changes, technological and financial-structural changes, that is, do exhibit impulsive effects. For instance, considering the present empirical results in the German time series Emmenegger and Stamova [83] show that during the process of growth, the capital can be subject to short-term perturbations at certain moments of time.

Therefore, VCMs with delay and impulsive effects should be more accurate in describing the evolutionary process of the systems. Since delays and impulses can affect the dynamical behaviors of the system, it is necessary to investigate both delay and impulsive effects on the stability of economic models.

Let $t_0 \in \mathbb{R}_+$. We consider the following impulsive Solow growth model with endogenous delay:

$$\begin{cases} \dot{k}(t) = sf(k(t)) \left[\int_{-\infty}^{t} n_s\big(f(k(\tau))\big)g(t-\tau)\,d\tau\right]k(t), \ t \geq t_0, \ t \neq t_i \\ \Delta k(t_i) = k(t_i + 0) - k(t_i) = P_i k(t_i), \ i = 1, 2, \ldots, \end{cases} \qquad (4.92)$$

where $k : [t_0, \infty) \to \mathbb{R}$; $f : \mathbb{R} \to \mathbb{R}$; $g : \mathbb{R} \to \mathbb{R}_+$ is the delay kernel function; $0 < s < 1$; $n_s : \mathbb{R} \to \mathbb{R}$; $t_i < t_{i+1} < \cdots$ ($i = 1, 2, \ldots$) are the moments of impulsive perturbations, due to which the capital-labour ratio k changes from position $k(t_i)$ to position $k(t_i + 0)$; P_i are constants, which represent the magnitude of the impulse effect at the moments t_i and $\lim_{i \to \infty} t_i = \infty$.

The equation (4.92) is a generalization of the Solow growth equation with endogenous delay (4.91). It can be used in the economic studies of business cycles in situation when the capital-labour ratio $k(t)$ is subject to shock effects [81, 177]. Particularly, the case $\Delta k(t_i) < 0$ corresponds to instantaneous reduction of the capital-labour ratio at times t_i, while the case $\Delta k(t_i) > 0$ describes heavy intensification of the capital-labour ratio.

Let $k_0 \in CB[(-\infty, 0], \mathbb{R}]$. Denote by $k(t) = k(t; t_0, k_0), k \in \mathbb{R}$, the solution of system (4.92), satisfying the initial conditions

$$
\begin{cases}
k(t; t_0, k_0) = k_0(t - t_0), \ t \in (-\infty, t_0] \\
k(t_0 + 0; t_0, k_0) = k_0(0),
\end{cases}
\tag{4.93}
$$

and by $J^+ = J^+(t_0, k_0)$ the maximal interval of type $[t_0, \beta)$ in which the solution $k(t; t_0, k_0)$ is defined.

Let $|k_0|_\infty = \max_{t \in (-\infty, t_0]} |k_0(t - t_0)|$ be the norm of the function $k_0 \in CB[(-\infty, 0], \mathbb{R}]$.

Introduce the following conditions:

H4.32. The delay kernel $g : \mathbb{R} \to \mathbb{R}_+$ is continuous, and there exists a positive number μ such that

$$
\int_{-\infty}^t g(t - \tau) \, d\tau \le \mu < \infty
$$

for all $t \in [t_0, \infty), t \ne t_i, i = 1, 2, \ldots$.

H4.33. The function f is continuous on $\mathbb{R}$, $f(k) > 0$ for $k > 0$, $f(0) = 0$, and there exists a positive continuous function $a(t)$ such that

$$
\left(\frac{f(k_1(t))}{k_1(t)} - \frac{f(k_2(t))}{k_2(t)} \right) \frac{1}{k_1(t) - k_2(t)} \le -a(t)
$$

for all $k_1, k_2 \in \mathbb{R}$, $k_1, k_2 \ne 0$, $k_1 \ne k_2$ and for all $t \in [t_0, \infty), t \ne t_i$, $i = 1, 2, \ldots$.

H4.34. The function n_s is continuous on $\mathbb{R}$, $n_s(f(k)) > 0$ for $k > 0$, $n_s(f(0)) = 0$, and there exists a positive continuous function $b(t)$ such that

$$
|n_s(f(k_1(t))) - n_s(f(k_2(t)))| \le b(t)|k_1(t) - k_2(t)|
$$

for all $k_1, k_2 \in \mathbb{R}$ and $b(t)$ is non-increasing for $t \in [t_0, \infty), t \ne t_i, i = 1, 2, \ldots$.

H4.35. $t_0 < t_1 < t_2 < \cdots$ and $\lim_{i \to \infty} t_i = \infty$.

In the proofs of our main theorems we shall use the following lemmas.

Lemma 4.37. *Let the conditions* H4.32–H4.35 *hold, and*

$$\int_{-\infty}^{t} n_s(f(k(\tau)))g(t - \tau)d\tau$$

be continuous for all $t \geq t_0$.
    Then $J^+(t_0, k_0) = [t_0, \infty)$.

Proof. If $\int_{-\infty}^{t} n_s(f(k(\tau)))g(t - \tau)d\tau$ is continuous for all $t \geq t_0$, then under the hypotheses H4.32–H4.34, the equation (4.91) has a unique solution $k(t) = k(t; t_0, k_0)$ with $k_0 \in CB[(-\infty, t_0], \mathbb{R}]$ on the interval $[t_0, \infty)$ [98, 99, 134]. This means that the solution $k(t) = k(t; t_0, k_0)$ of problem (4.92), (4.93) is defined on $[t_0, t_1] \cup (t_i, t_{i+1}]$, $i = 1, 2, \ldots$. From the hypothesis H4.35, we conclude that it is continuable for $t \geq t_0$. ☐

Lemma 4.38. *Assume that:*

(1) *The conditions of Lemma* 4.37 *hold.*

(2) $k(t) = k(t; t_0, k_0)$ *is a solution of* (4.92), (4.93) *such that*

$$k(t) = k_0(t - t_0) \geq 0, \quad \sup k_0(s) < \infty, \quad k_0(0) > 0.$$

(3) *For each* $i = 1, 2, \ldots$
$$1 + P_i > 0.$$

Then
$$k(t) > 0, \quad t \geq t_0.$$

Proof. The proof of Lemma 4.38 is analogous to the proof of Lemma 4.16. ☐

Theorem 4.39. *Assume that:*

(1) *Conditions* (1) *and* (2) *of Lemma* 4.38 *hold.*

(2) $-1 < P_i \leq 0$ *for each* $i = 1, 2, \ldots$.

Then the equation (4.92) *is uniformly ultimately bounded.*

Proof. From the conditions (1) and (2) of Lemma 4.38, it follows [4, 134] that for $t \in [t_0, t_1] \cup (t_i, t_{i+1}]$, $i = 1, 2, \ldots$ there exist positive constants m_i^* and $M_i^* < \infty$ such that

$$m_i^* \leq k(t) \leq M_i^*.$$

If we set $M^* = \max_i M_i^*$, $i = 1, 2, \ldots$, then by Lemma 4.38 and condition (2) of Theorem 4.39, we have

$$0 < k(t_i + 0) = (1 + P_i)k(t_i) \leq k(t_i) \leq M^*.$$

☐

Corollary 4.40. *Let the conditions of Theorem 4.39 hold.*
   Then there exist positive constants m and M $< \infty$ *such that*

$$m \leq k(t) \leq M, \quad t \in [t_0, \infty). \tag{4.94}$$

Let $\tilde{k}_0 \in CB[(-\infty, 0], \mathbb{R}]$, and let $\tilde{k}(t) = \tilde{k}(t; t_0, \tilde{k}_0)$ be a solution of (4.92) for all $t \geq t_0$ with initial conditions

$$\tilde{k}(t; t_0, \tilde{k}_0) = \tilde{k}_0(t - t_0), \ t \in (-\infty, t_0]; \quad \tilde{k}(t_0 + 0) = \tilde{k}_0(0).$$

In the following, we shall suppose that

$$k(t) = k_0(t - t_0) \geq 0, \quad \sup k_0(s) < \infty, \quad k_0(0) > 0;$$

$$\tilde{k}(t) = \tilde{k}_0(t - t_0) \geq 0, \quad \sup \tilde{k}_0(s) < \infty, \quad \tilde{k}_0(0) > 0.$$

Theorem 4.41. *Assume that:*

(1) *The conditions of Theorem 4.39 hold.*

(2) *There exists a nonnegative constant L such that*

$$Lm + M\mu \max_{\tau \in (-\infty, t]} b(\tau) \leq msa(t), \quad t \geq t_0, \quad t \neq t_i, \quad i = 1, 2, \ldots.$$

Then the solution $k(t)$ of (4.92) is uniformly asymptotically stable.

Proof. Define the Lyapunov function

$$V(t, k, \tilde{k}) = \left| \ln \frac{k}{\tilde{k}} \right|. \tag{4.95}$$

By the Mean Value Theorem, it follows that for any closed interval contained in $[t_0, t_1] \cup (t_i, t_{i+1}), i = 1, 2, \ldots$, we have

$$\frac{1}{M}|k(t) - \tilde{k}(t)| \leq |\ln k(t) - \ln \tilde{k}(t)| \leq \frac{1}{m}|k(t) - \tilde{k}(t)|. \tag{4.96}$$

If $|k_0 - \tilde{k}_0|_\infty < \delta < \infty$, then we obtain from the inequalities (4.96)

$$V(t_0 + 0, k(t_0 + 0), \tilde{k}(t_0 + 0)) = |\ln k(t_0 + 0) - \ln \tilde{k}(t_0 + 0)|$$

$$\leq \frac{1}{m}|k_0(t_0 + 0) - \tilde{k}_0(t_0 + 0)|$$

$$\leq \frac{1}{m}|k_0 - \tilde{k}_0|_\infty < \delta < \infty. \tag{4.97}$$

Consider the upper right-hand derivative $D^+_{(4.92)}V(t, k(t), \tilde{k}(t))$ of the function $V(t, k(t), \tilde{k}(t))$ with respect to (4.92). For $t \geq t_0$ and $t \neq t_i$, $i = 1, 2, \ldots$, we derive the estimate

$$D^+_{(4.92)}V(t, k(t), \tilde{k}(t)) = \left(\frac{\dot{k}(t)}{k(t)} - \frac{\dot{\tilde{k}}(t)}{\tilde{k}(t)} \right) \operatorname{sgn}\left(k(t) - \tilde{k}(t) \right)$$

$$\leq s \left(\frac{f(k(t))}{k(t)} - \frac{f(\tilde{k}(t))}{\tilde{k}(t)} \right) \frac{|k(t) - \tilde{k}(t)|}{k(t) - \tilde{k}(t)}$$

$$+ \int_{-\infty}^{t} |n_s(f(k(\tau))) - n_s(f(\tilde{k}(\tau)))| g(t - \tau) \, d\tau$$

$$\leq -sa(t)|k(t) - \tilde{k}(t)| + \int_{-\infty}^{t} b(\tau)|k(\tau) - \tilde{k}(\tau)| g(t - \tau) \, d\tau$$

$$\leq -sa(t)|k(t) - \tilde{k}(t)|$$

$$+ \max_{\tau \in (-\infty, t]} b(\tau) \int_{-\infty}^{t} |k(\tau) - \tilde{k}(\tau)| g(t - \tau) \, d\tau.$$

From (4.96), using the Razumikhin condition $V(\tau, k(\tau), \tilde{k}(\tau)) \leq V(t, k(t), \tilde{k}(t))$, $\tau \in (-\infty, t]$, $t > t_0$, we have

$$\frac{1}{M}|k(\tau) - \tilde{k}(\tau)| \leq V(\tau, k(\tau), \tilde{k}(\tau))$$

$$\leq V(t, k(t), \tilde{k}(t)) \leq \frac{1}{m}|k(t) - \tilde{k}(t)|, \quad \tau \in (-\infty, t],$$

and hence

$$|k(\tau) - \tilde{k}(\tau)| \leq \frac{M}{m}|k(t) - \tilde{k}(t)|, \quad \tau \in (-\infty, t], \, t \neq t_i, \, i = 1, 2, \ldots. \quad (4.98)$$

Then, from (4.98) and from condition (2) of Theorem 4.41, we obtain

$$D^+_{(4.92)}V(t, k(t), \tilde{k}(t)) \leq -L|k(t) - \tilde{k}(t)| \leq -LmV(t, k(t), \tilde{k}(t)), \quad (4.99)$$

for $t \geq t_0$ and $t \neq t_i$, $i = 1, 2, \ldots$.

Also, for $t = t_i$, $i = 1, 2, \ldots$, we have

$$V(t_i + 0, k(t_i + 0), \tilde{k}(t_i + 0)) = \left| \ln \frac{k(t_i + 0)}{\tilde{k}(t_i + 0)} \right|$$

$$= \left| \ln \frac{(1 + P_i)k(t_i)}{(1 + P_i)\tilde{k}(t_i)} \right| = V(t_i, k(t_i), \tilde{k}(t_i)). \quad (4.100)$$

From (4.99) and (4.100), we obtain

$$V(t, k(t), \tilde{k}(t)) \leq V(t_0 + 0, k(t_0 + 0), \tilde{k}(t_0 + 0)) \exp\{-Lm(t - t_0)\} \quad (4.101)$$

for all $t \geq t_0$.

Then, from (4.101), (4.96) and (4.97), we deduce the inequality

$$V(t, k(t), \tilde{k}(t)) \le \frac{1}{m}|k_0 - \tilde{k}_0|_\infty \exp\{-Lm(t - t_0)\}, \ t \ge t_0.$$

This shows that the solution $k(t)$ of equation (4.92) is uniformly asymptotically stable. $\qquad\square$

When a Cobb–Douglas production function $f(k) = k^\alpha$, where $0 < \alpha < 1$, is chosen [77], the model (4.91) becomes

$$\dot{k}(t) = sk^\alpha(t) - \left[\int_{-\infty}^t n_s\left(k^\alpha(\tau)\right)g(t - \tau)\,d\tau\right]k(t).$$

As we are essentially interested in the effects of forces of "fundamental" nature, in what follows we assume, for simplicity, that the function n_s is linear and increasing, i.e. we consider $n_s(k^\alpha) = n_s k^\alpha$, where $n_s \ge 0$ is a constant parameter, tuning the reaction of the rate of change of the labour supply to changes in per-capita income. We therefore have the model

$$\dot{k}(t) = sk^\alpha(t) - n_s\left[\int_{-\infty}^t k^\alpha(\tau)g(t - \tau)d\tau\right]k(t). \tag{4.102}$$

We consider the impulsive generalization of model (4.102)

$$\begin{cases} \dot{k}(t) = sk^\alpha(t) - n_s\left[\int_{-\infty}^t k^\alpha(\tau)g(t - \tau)d\tau\right]k(t), \ t \ge t_0, \ t \ne t_i \\ \Delta k(t_i) = k(t_i + 0) - k(t_i) = Q_i(k(t_i)), \ i = 1, 2, \ldots, \end{cases} \tag{4.103}$$

where $t_0 \in \mathbb{R}_+$; $t_i, i = 1, 2, \ldots$ are the moments of impulsive perturbations and satisfy $t_0 < t_1 < t_2 < \cdots$ and $\lim_{i \to \infty} t_i = \infty$, $Q_i(k(t_i))$ represents the abrupt change of the state $k(t)$ at the impulsive moment t_i.

Introduce the following condition:

H4.36. $\displaystyle\int_{-\infty}^t g(t - \tau)d\tau = 1, t \in \mathbb{R}_+.$

It is well known [86, 176] that under the hypothesis H4.36, the equation (4.102) has the zero equilibrium E_0 and a positive equilibrium $E_1 = s/n$ on the interval $[t_0, \infty)$. Now, we shall investigate the discontinuous case.

Lemma 4.42. *Assume that:*

(1) *Conditions H4.35 and H4.36 hold.*

(2) *The functions $Q_i : \mathbb{R} \to \mathbb{R}$ are such that*

$$Q_i(k(t_i)) = -\sigma_i(k(t_i) - E_1), \ \sigma_i = \mathrm{const} > 0, \ i = 1, 2, \ldots.$$

Then there exists a positive equilibrium of the equation (4.103).

Proof. From H4.36, it follows that $E_1 = s/n$ is a positive equilibrium of (4.102) on the interval $[t_0, \infty)$. This means that the solution E_1 of (4.103) is defined on $[t_0, t_1] \cup (t_i, t_{i+1}]$, $i = 1, 2, \ldots$. Also, we have that the functions Q_i are such that

$$Q_i(k(t_i)) = -\sigma_i(k(t_i) - E_1).$$

So,

$$\Delta E_1(t_i) = Q_i(E_1(t_i)) = 0, \quad i = 1, 2, \ldots,$$

i.e., E_1 is an equilibrium of (4.103). From condition H4.35, we conclude that it is continuable for $t \geq t_0$. $\qquad\square$

Theorem 4.43. *Assume that:*

(1) *Conditions H4.35 and H4.36 hold.*

(2) *There exists a positive continuous function $a(t)$ such that*

$$\frac{k^{\alpha-1}(t) - E_1^{\alpha-1}}{k(t) - E_1} \leq -a(t)$$

for all $k \in \mathbb{R}$, $k \neq E_1$ and for all $t \in [t_0, \infty)$, $t \neq t_i$, $i = 1, 2, \ldots$.

(3) *The functions $Q_i : \mathbb{R} \to \mathbb{R}$ are such that*

$$Q_i(k(t_i)) = -\sigma_i(k(t_i) - E_1), \quad 0 < \sigma_i < 2,\ i = 1, 2, \ldots.$$

(4) *There exists a nonnegative constant L such that*

$$L + \alpha n_s m^{\alpha-1} \leq sa(t), \quad t \geq t_0,\ t \neq t_i,\ i = 1, 2, \ldots.$$

Then the equilibrium E_1 of (4.103) is uniformly asymptotically stable.

Proof. Consider the Lyapunov function

$$V(t, k) = |k - E_1|.$$

If $|k_0 - E_1|_\infty < \delta < \infty$, we obtain

$$V(t_0 + 0, k(t_0 + 0)) = |k(t_0 + 0) - E_1| \leq |k_0 - E_1|_\infty < \delta < \infty. \qquad (4.104)$$

Consider the upper right-hand derivative $D^+_{(4.103)}V(t, k(t))$ of the function $V(t, k(t))$ with respect to (4.103). For $t \geq t_0$ and $t \neq t_i$, $i = 1, 2, \ldots$, we derive the estimate

$$D^+_{(4.103)}V(t, k(t)) = \dot{k}(t)\operatorname{sgn}(k(t) - E_1)$$

$$= k(t)\left[sk^{\alpha-1}(t) - n_s \int_{-\infty}^{t} g(t - \tau)k^\alpha(\tau)d\tau\right]\operatorname{sgn}(k(t) - E_1).$$

Since E_1 is an equilibrium of (4.103), we obtain

$$D^+_{(4.103)}V(t, k(t))$$

$$\leq k(t)\left[|k(t) - E_1|s\frac{k^{\alpha-1}(t) - E_1^{\alpha-1}}{k(t) - E_1} + n_s\int_{-\infty}^{t} g(t - \tau)|k^{\alpha}(\tau) - E_1^{\alpha}|d\tau\right],$$

for $t \neq t_i$, $i = 1, 2, \ldots$.

The function $k^{\alpha}(t)$ is differentiable on any closed interval contained in $[t_0, t_1] \cup (t_i, t_{i+1}]$, $i = 1, 2, \ldots$, and the inequalities $m \leq k(t) \leq M$ are valid for all $t \geq t_0$, $t \neq t_i, i = 1, 2, \ldots$.

Therefore,

$$|k_1^{\alpha}(t) - k_2^{\alpha}(t)| = |\alpha||k^{\alpha-1}(t)||k_1(t) - k_2(t)| \leq \alpha m^{\alpha-1}|k_1(t) - k_2(t)|$$

for $k_1(t) \leq k(t) \leq k_2(t)$, $k_1, k_2 \in \mathbb{R}$ and for all $t \geq t_0, t \neq t_i, i = 1, 2, \ldots$.

From the last estimate, we obtain

$$D^+_{(4.103)}V(t, k(t)) \leq k(t)\left[-sa(t)|k(t) - E_1| + n_s\int_{-\infty}^{t} g(t-\tau)\alpha m^{\alpha-1}|k(\tau) - E_1|d\tau\right],$$

for $t \neq t_i$, $i = 1, 2, \ldots$.

From condition (4) of Theorem 4.43, for any solution k of (4.103) such that $V(\tau) \leq V(t), \tau \in (-\infty, t]$, $t \geq t_0$, we have

$$D^+_{(4.103)}V(t, k(t)) \leq -Lk(t)|k(t) - E_1| \leq -LmV(t, k(t)), \qquad (4.105)$$

$t \geq t_0$ and $t \neq t_i, i = 1, 2, \ldots$.

Also, for $t > t_0$ and $t = t_i$ from the condition (3) of Theorem 4.43, we have

$$V(t_i + 0, k(t_i + 0)) = |k(t_i + 0) - E_1| = |k(t_i) - \sigma_i(k(t_i) - E_1) - E_1|$$
$$= |1 - \sigma_i||k(t_i) - E_1| < |k(t_i) - E_1| = V(t_i, k(t_i)). \quad (4.106)$$

Then for $t \geq t_0$ from (4.106), (4.104) and (4.105), we deduce the inequality

$$|k(t) - E_1| \leq |k_0 - E_1|_{\infty}e^{-Lm(t-t_0)},$$

which shows that the equilibrium E_1 of equation (4.103) is uniformly asymptotically stable.   □

Impulsive functional differential equations modeling price fluctuations in single commodity markets

First, in a single good market, there are three variables: the quantity demanded q_d, the quantity supplied q_s and its price p. The equilibrium is attained when the excess demand is zero, $q_d - q_s = 0$, that is, the market is cleared. But generally, the market is not in equilibrium and at an initial time t_0 the price p_0 is not at the equilibrium value $\overline{p}$, that is, $p_0 \neq \overline{p}$. In such a situation the variables q_d, q_s and p must change over time and are considered as functions of time. The dynamic question is: given sufficient time, how has the adjustment process $p(t) \to \overline{p}$ as $t \to \infty$ to be described?

The dynamic process of attaining an equilibrium in a single good market model is tentatively described by differential equations, on the basis of considerations on price changes, governing the relative strength of the demand and supply forces. In a first approach and for the sake of simplicity, the rate of price change with respect to time is assumed to be proportional to the the excess demand $q_d - q_s$. Moreover, definitive relationships between the market price p of a commodity, the quantity demanded and the quantity supplied are assumed to exist. These relationships are called the demand curve and the supply curve, occasionally modeled by a demand function $q_d = q_d(p)$ or a supply function $q_s = q_s(p)$, both dependant of the price variable p. In the case where the rate of price change with respect to time is assumed to be proportional to the excess demand, the differential equation belongs to the class

$$\frac{1}{p}\frac{dp}{dt} = f(q_s(p), q_d(p)), \tag{4.107}$$

of differential equations. The question that arises is about the nature of the time path $p(t)$, resulting from equation (4.107).

Many authors precisely considered the model (4.107) and its generalizations in order to study the dynamics of the prices, production and consumption for a particular commodity (see [206] and the references cited therein).

In [162] Muresan studied a special case of a fluctuation model for the price with delay of the form

$$\dot{p}(t) = \left(\frac{a}{b + p^q(t)} - \frac{cp^r(g(t))}{d + p^r(g(t))}\right)p(t) \tag{4.108}$$

where $a, b, c, d, r > 0$, $q \in [1, \infty)$, $g \in C[\mathbb{R}_+, \mathbb{R}_+]$ and proved that there exists a positive, bounded, unique solution.

Rus and Iancu [175] generalized the model (4.108) and studied a model of the form

$$\begin{cases} \dot{p}(t) = F(p(t), p(t - \tau))p(t), \ t \geq 0 \\ p(t) = \psi(t), \ t \in [-\tau, 0]. \end{cases} \tag{4.109}$$

They proved the existence and uniqueness of the equilibrium solution of the model considered and established some relations between this solution and coincidence points.

An empirical time series analysis [83] of German macroeconomic data emphasized to model capital intensity, subject to short-term perturbations at certain moments of time. Then it is not reasonable to expect a regular solution of the equation (4.109). Instead, the solution must have some jumps and the jumps follow a specific pattern.

In the long-term planning an adequate mathematical model of this case will be the following impulsive functional differential equation:

$$\begin{cases} \dot{p}(t) = F(p(t), p_t)p(t), \ t \geq t_0, \ t \neq t_i \\ \Delta p(t_i) = p(t_i + 0) - p(t_i) = P_i(p(t_i)), \ i = 1, 2, \ldots, \end{cases} \qquad (4.110)$$

where $t_0 \in \mathbb{R}_+$; $t_0 < t_1 < t_2 < \cdots$, $\lim_{i \to \infty} t_i = \infty$; Ω be a domain in $\mathbb{R}_+$ containing the origin; $F : \Omega \times PC[[-\tau, 0], \Omega] \to \mathbb{R}$; $P_i : \Omega \to \mathbb{R}, i = 1, 2, \ldots$ are functions which characterize the magnitude of the impulse effect at the times t_i; $p(t_i)$ and $p(t_i + 0)$ are respectively the price levels before and after the impulse effects and for $t \geq t_0$; $p_t \in PC[[-\tau, 0], \Omega]$ is degined by $p_t(s) = p(t + s), -\tau \leq s \leq 0$.

Let $p_0 \in CB[[-\tau, 0], \Omega]$. Denote by $p(t) = p(t; t_0, p_0), \ p \in \Omega$, the solution of equation (4.110), satisfying the initial conditions

$$\begin{cases} p(t; t_0, p_0) = p_0(t - t_0), \ t_0 - \tau \leq t \leq t_0 \\ p(t_0 + 0; t_0, p_0) = p_0(0), \end{cases} \qquad (4.111)$$

$J^+(t_0, p_0)$ the maximal interval of type $[t_0, \beta)$ in which the solution $p(t; t_0, p_0)$ is defined, and by $|p_0|_\tau = \max_{t \in [t_0 - \tau, t_0]} |p_0(t - t_0)|$ the norm of the function $p_0 \in CB[[-\tau, 0], \Omega]$.

Introduce the following conditions:

H4.37. The function F is continuous on $\Omega \times PC[[-\tau, 0], \Omega]$.

H4.38. The function F locally Lipschitz continuous with respect to its second argument on $\Omega \times PC[[-\tau, 0], \Omega]$.

H4.39. There exists a constant $M > 0$ such that

$$|F(p, p_t)| \leq M < \infty \text{ for } (p, p_t) \in \Omega \times PC[[-r, 0], \Omega].$$

H4.40. $P_i \in C[\Omega, \mathbb{R}], \ i = 1, 2, \ldots$.

H4.41. The functions $(I + P_i) : \Omega \to \Omega, \ i = 1, 2, \ldots$ where I is the identity in Ω.

H4.42. $t_0 < t_1 < t_2 < \cdots$ and $\lim_{i \to \infty} t_i = \infty$.

Let $p_1 \in CB[[-\tau, 0], \Omega]$. Denote by $p^*(t) = p^*(t; t_0, p_1), \ p^* \in \Omega$ the solution of equation (4.110), satisfying the initial conditions

$$\begin{cases} p^*(t; t_0, p_1) = p_1(t - t_0), \ t_0 - \tau \leq t \leq t_0 \\ p^*(t_0 + 0; t_0, p_1) = p_1(0). \end{cases}$$

In our subsequent analysis, we shall use piecewise continuous functions $V : [t_0, \infty) \times \Omega \to \mathbb{R}_+$ which belong to the class V_0 and for which the following condition is satisfied:

H4.43. $V(t, p^*(t)) = 0, \ t \geq t_0$.

⋅

For a function $V \in V_0$ and for some $t \geq t_0$, we shall use also the class

$$\Omega_1 = \{p \in PC[[t_0, \infty), \Omega] : V(s, p(s)) \leq V(t, p(t)), \ t - \tau \leq s \leq t\}.$$

The next theorems follow directly from Theorems 2.7, 2.8 and 2.9.

Theorem 4.44. *Assume that:*

(1) *Conditions H4.37–H4.42 hold.*

(2) *There exists a function $V \in V_0$ such that H4.43 holds,*

$$a(|p - p^*(t)|) \leq V(t, p), \ (t, p) \in [t_0, \infty) \times \Omega, \ a \in K,$$

$$V(t + 0, p + I_k(p)) \leq V(t, p), \quad p \in \Omega, \ t = t_i, \ i = 1, 2, \ldots,$$

and the inequality

$$D^+_{(4.110)} V(t, p(t)) \leq 0, \ t \neq t_i, \ i = 1, 2, \ldots$$

is valid for $t \in [t_0, \infty), \ p \in \Omega_1.$

Then the solution $p^*(t)$ of equation (4.110) is stable.

Theorem 4.45. *Let the conditions of Theorem 4.44 hold, and let a function $b \in K$ exist such that*

$$V(t, p) \leq b(|p - p^*(t)|), \ (t, p) \in [t_0, \infty) \times \Omega.$$

Then the solution $p^*(t)$ of equation (4.110) is uniformly stable.

Theorem 4.46. *Assume that:*

(1) *Conditions H4.37–H4.42 hold.*

(2) *There exists a function $V \in V_0$ such that H4.43 holds,*

$$a(|p - p^*(t)|) \leq V(t, x) \leq b(|p - p^*(t)|), \ (t, p) \in [t_0, \infty) \times \Omega, \ a, b \in K,$$

$$V(t + 0, p + I_k(p)) \leq V(t, p), \quad p \in \Omega, \ t = t_i, \ i = 1, 2, \ldots,$$

and the inequality

$$D^+_{(4.110)} V(t, p(t)) \leq -c(|p(t) - p^*(t)|), \ t \neq t_i, \ i = 1, 2, \ldots$$

is valid for $c \in K, t \in [t_0, \infty)$ and $p \in \Omega_1.$

Then the solution $p^*(t)$ of equation (4.110) is uniformly asymptotically stable.

Example 4.47. Let for a, b, c, $d > 0$, a linear demand function $q_d = a - bp$ and a linear supply function $q_s = -c + dp$ be given and the function $f = \alpha(q_d - q_s)$, $\alpha > 0$. They can be put into (4.107), giving the linear non homogenous differential equation $\frac{dp}{dt} = \alpha(a + c - p(b + d)) = p\alpha(\frac{a+c}{p} - (b+d))$, corresponding to a special type of the differential equation (4.107). Its complementary and particular solutions are immediate.

A special case of the model studied by Mackey and Belair [151] is the following equation:

$$\dot{p}(t) = \alpha\left[\frac{a+c}{p(t)} - b - d\frac{p(t - \sigma(t))}{p(t)}\right]p(t),$$

where $0 \leq \sigma(t) \leq \tau$ and τ is a constant.

If at the moments $t_1, t_2, \ldots$ ($t_0 < t_1 < t_2 < \cdots < t_i < t_{i+1} < \cdots$ and $\lim_{i \to \infty} t_i = \infty$) the above equation is subject to impulsive perturbations then the adequate mathematical model is the following impulsive equation:

$$\begin{cases} \dot{p}(t) = \alpha\left[\dfrac{a+c}{p(t)} - b - d\dfrac{p(t - \sigma(t))}{p(t)}\right]p(t), \ t \neq t_i, \ t \geq t_0 \\ \Delta p(t_i) = -\delta_i\left(p(t_i) - \dfrac{a+c}{b+d}\right), \ i = 1, 2, \ldots, \end{cases} \tag{4.112}$$

where $t_0 \in \mathbb{R}_+$, $p(t)$ represents the price at the moment t, $\delta_i \in \mathbb{R}$ are constants, $i = 1, 2, \ldots$.

It is easy to verify that the point $p^* = \frac{a+c}{b+d}$ is an equilibrium of (4.112).

We shall show that, if there exists a constant $\beta > 0$ such that $d \leq b - \beta$ and the inequalities $0 < \delta_i < 2$ are valid for $i = 1, 2, \ldots$, then the equilibrium p^* of (4.112) is uniformly asymptotically stable.

Let $V(t, p) = \frac{1}{2}(p - p^*)^2$. Then the set

$$\Omega_1 = \{p \in PC[[t_0, \infty), (0, \infty)] : (p(s) - p^*)^2 \leq (p(t) - p^*)^2, \ t - \tau \leq s \leq t\}.$$

For $t \geq t_0, t \neq t_i$ we have

$$D^+_{(4.112)}V(t, p(t)) = \alpha(p(t) - p^*)[a - bp(t) + c - dp(t - \sigma(t))].$$

Since p^* is an equilibrium of (4.112), we have

$$D^+_{(4.112)}V(t, p(t)) = \alpha(p(t) - p^*)[-b(p(t) - p^*) - d(p(t - \sigma(t)) - p^*)].$$

From the last relation for $t \geq t_0, t \neq t_i$ and $p \in \Omega_1$ we obtain the estimate

$$D^+_{(4.112)}V(t, p(t)) \leq \alpha[-b + d](p(t) - p^*)^2 \leq -\alpha\beta(p(t) - p^*)^2.$$

Also, if $0 < \delta_i < 2$ for all $i = 1, 2, \ldots$, then

$$V(t_i + 0, p(t_i + 0)) = \frac{1}{2}\Big[(1 - \delta_i)p(t_i) + \delta_i\, p^* - p^*\Big]^2$$

$$= \frac{1}{2}(1 - \delta_i)^2\Big[p(t_i) - p^*\Big]^2 < V(t_i, p(t_i)).$$

Since all conditions of Theorem 4.46 are satisfied, the equilibrium p^* of (4.112) is uniformly asymptotically stable.

If the constants δ_i are such that $\delta_i < 0$ or $\delta_i > 2$, then condition (4) of Theorem 4.46 is not satisfied and we can not make any conclusion about the asymptotic stability of the equilibrium p^*.

The example again demonstrates the utility of the second method of Lyapunov. The main characteristic of the method is the introduction of a function, namely, Lyapunov function which defines a generalized distance between $p(t)$ and the equilibrium value p^*.

By means of piecewise continuous functions we give the conditions for uniform asymptotic stability of the price p^*. A technique is applied, based on certain minimal subsets of a suitable space of piecewise continuous functions, by the elements of which the derivatives of the piecewise continuous auxiliary functions of Lyapunov are estimated.

It is shown, also, that the role of impulses in changing the behavior of solutions of impulsive differential equations is very important.

Notes and Comments

Theorems 4.6–4.13 are new. For related results for impulsive Lotka–Volterra models without delay see Ahmad and Stamova [7] and for Lotka–Volterra models without impulses see Fan, Wang and Jiang [85] and Wei and Wang [214]. Lemmas 4.15–4.17 and Theorems 4.20 and 4.21 are taken from Ahmad and Stamova [8].

Theorem 4.24 is new. Theorems 4.26 and 4.27 are adapted from Ahmad and Stamova [9]. Theorem 4.31 is new. Theorem 4.34 is taken from Stamova [199]. Theorem 4.35 is new.

Economic models discussed in Section 4.3 are due to Stamova and Emmenegger [83] and [206]. Theorems 4.39–4.43 are new. Theorems 4.44–4.46 are adapted from Stamova and Emmenegger [206].

Bibliography

[1] P. Aghion and P. Howitt, *Endogenous Growth Theory*, MIT Press, Cambridge, MA, 1998.

[2] S. Ahmad and A.C. Lazer, *On Persistence and Extinction of Species*, Centro International de Matematica, Lisbon, Portugal, 2002.

[3] S. Ahmad and A.C. Lazer, *Average conditions for global asymptotic stability in a nonautonomous Lotka–Volterra system*, Nonlinear Analysis 40 (2000), pp. 37–49.

[4] S. Ahmad and M.Rama M.Rao, *Asymptotic periodic solutions of n-competing species problem with time delay*, J. Math. Anal. Appl. 186 (1994), pp. 559–571.

[5] S. Ahmad and I.M. Stamova, *Almost necessary and sufficient condition for survival of species*, Nonlinear Analysis: Real World Applications 5/1 (2004), pp. 219–229.

[6] S. Ahmad and I.M. Stamova, *Partial persistence and extinction in n-dimensional competitive systems*, Nonlinear Analysis 60 (2005), pp. 821–836.

[7] S. Ahmad and I.M. Stamova, *Asymptotic stability of an n-dimensional impulsive competitive system*, Nonlinear Analysis: Real World Applications 8 (2007), pp. 654–663.

[8] S. Ahmad and I.M. Stamova, *Asymptotic stability of competitive systems with delays and impulsive perturbations*, J. Math. Anal. Appl. 334 (2007), pp. 686–700.

[9] S. Ahmad and I.M. Stamova, *Global exponential stability for impulsive cellular neural networks with time-varying delays*, Nonlinear Analysis 69 (2008), pp. 786–795.

[10] H. Akca, R. Alassar, V. Covachev, Z. Covacheva and E. Al-Zahrani, *Continuous-time additive Hopfield-type neural networks with impulses*, J. Math. Anal. Appl. 290 (2004), pp. 436–451.

[11] A.A. Andronov, A.A. Witt and S.E. Haykin, *Oscillation Theory*, Nauka, Moscow, 1981(in Russian).

[12] A.V. Anokhin, *Linear impulsive systems for functional differential equations*, Reports Acad. Sci. SSSR 286 (1986), pp. 1037–1040 (in Russian).

[13] A.V. Anokhin, L. Berezansky and E. Braverman, *Exponential stability of linear delay impulsive differential equations*, J. Math. Anal. Appl. 193 (1995), pp. 923–941.

[14] A.V. Anokhin, L. Berezansky and E. Braverman, *Stability of linear delay impulsive differential equations*, Dynamic Systems and Applications 4 (1995), pp. 173–178.

[15] M.A. Arbib, *Branins, Machines and Mathematics*, New York, Springer-Verlag, 1987.

[16] S. Arik and V. Tavsanoglu, *On the global asymptotic stability of delayed cellular neural networks*, IEEE Trans. Circuits Syst. I 47 (2000), pp. 571–574.

[17] M. Bachar and O. Arino, *Stability of a general linear delay-differential equation with impulses*, Dynamics of Continuous, Discrete and Impulsive Systems Series A: Mathematical Analysis 10 (2003), pp. 973–990.

[18] D.D. Bainov and V. Covachev, *Impulsive Differential Equations with a Small Parameter*, World Scientific, Singapore, 1994.

[19] D.D. Bainov, V. Covachev and I.M. Stamova, *Estimates of the solutions of impulsive quasilinear functional differential equations*, Annales de la Faculté des Sciences de Toulouse, Vol. XII 2 (1991), pp. 149–161.

[20] D.D. Bainov, V. Covachev and I.M. Stamova, *Stability under persistent disturbances of impulsive differential-difference equations of neutral type*, J. Math. Anal. Appl. 187 (1994), pp. 790–808.

[21] D.D. Bainov and A.B. Dishliev, *Quasiuniqueness, uniqueness and continuability of the solutions of impulsive functional differential equations*, Rendiconti di Matematica 15 (1995), pp. 391–404.

[22] D.D. Bainov and A.B. Dishliev, *The phenomenon "beating" of the solutions of impulsive functional differential equations*, Communications in Applied Analysis 1 (1997), pp. 435–441.

[23] D.D. Bainov, A.B. Dishliev and I.M. Stamova, *Asymptotic equivalence of a linear system of impulsive differential equations and a system of differential-difference equations*, Annali Dell' Universita' di Ferrara, Vol. XLI (1995), pp. 45–54.

[24] D.D. Bainov, A.B. Dishliev and I.M. Stamova, *Lipschitz quasistability of impulsive differential-difference equations with variable impulsive perturbations*, Journal of Computational and Applied Mathematics 70 (1996), pp. 267–277.

[25] D.D. Bainov, A.B. Dishliev and I.M. Stamova, *Continuous dependence of solutions of impulsive systems of differential-difference equations on initial data and parameter*, Bol. Soc. Paran. Mat. 18 (1/2), (1998), pp. 21–34.

[26] D.D. Bainov, A.B. Dishliev and I.M. Stamova, *Practical stability of the solutions of impulsive systems of differential-difference equations via the method of comparison and some applications to population dynamics*, ANZIAM J. 43 (2002), pp. 525–539.

[27] D.D. Bainov, G.K. Kulev and I.M. Stamova, *Global stability of the solutions of impulsive differential-difference equations*, SUT Journal of Mathematics 1 (1995), pp. 55–71.

[28] D.D. Bainov, G.K. Kulev and I.M. Stamova, *Instability of solutions of impulsive systems of differential equations*, International Journal of Theoretical Physics 8 (1996), pp. 1799–1804.

[29] D.D. Bainov, E. Minchev and I.M. Stamova, *Present state of the stability theory for impulsive differential equations*, Communications in Applied Mathematics 2 (1998), pp. 197–226.

[30] D.D. Bainov and P.S. Simeonov, *Systems with Impulsive Effect: Stability Theory and Applications*, Ellis Horwood, Chichester, John Wiley and Sons, Inc., New York, 1989.

[31] D.D. Bainov and P.S. Simeonov, *Integral Inequalities and Applications*, Kluwer, 1992.

[32] D.D. Bainov and P.S. Simeonov, *Impulsive Differential Equations: Periodic Solutions and Applications*, Longman Scientific and Technical, Harlow, 1993.

[33] D.D. Bainov and P.S. Simeonov, *Impulsive Differential Equations: Asymptotic Properties of the Solutions*, World Scientific, Singapore, 1995.

[34] D.D. Bainov and I.M. Stamova, *Lipschitz stability of impulsive systems of differential-difference equations*, Note di Matematica 2 (1994), pp. 167–178.

[35] D.D. Bainov and I.M. Stamova, *Uniform asymptotic stability of impulsive differential-difference equations of neutral type by Lyapunov's direct method*, Journal of Computational and Applied Mathematics 62 (1995), pp. 359–369.

[36] D.D. Bainov and I.M. Stamova, *Lipschitz stability of linear impulsive differential-difference equations*, Note di Matematica 2 (1995), pp. 137–142.

[37] D.D. Bainov and I.M. Stamova, *On the practical stability of the solutions of impulsive systems differential-difference equations with variable impulsive perturbations*, J. Math. Anal. Appl. 200 (1996), pp. 272–288.

[38] D.D. Bainov and I.M. Stamova, *Applications of Lyapunov's direct method to the investigation of the stability of the solutions of impulsive differential-difference equations with variable impulsive perturbations*, Appl. Anal. 63 (1996), pp. 253–269

[39] D.D. Bainov and I.M. Stamova, *Asymptotic equivalence of ordinary and impulsive functional differential equations*, Compt. Rend. Acad. Bulg. Sci. 49 (1996), pp. 21–23.

[40] D.D. Bainov and I.M. Stamova, *An application of Lyapunov's direct method investigating stability of solutions to impulsive functional-differential equations*, Compt. Rend. Acad. Bulg. Sci. 50 (1997), pp. 9–10.

[41] D.D. Bainov and I.M. Stamova, *Global stability of the solutions of impulsive differential-difference equations with variable impulsive perturbations*, COMPEL 16 (1997), pp. 3–16.

[42] D.D. Bainov and I.M. Stamova, *Second method of Lyapunov and comparison principle for impulsive differential-difference equations*, J. Austral. Math Soc. Ser. B 38 (1997), pp. 489–505.

[43] D.D. Bainov and I.M. Stamova, *Second method of Lyapunov and existence of periodic solutions of linear impulsive differential-difference equations*, PanAmerican Mathematical Journal 7 (1997), pp. 27–35.

[44] D.D. Bainov and I.M. Stamova, *Stability of sets for impulsive differential-difference equations with variable impulsive perturbations*, Communications on Applied Nonlinear Analysis 5 (1998), pp. 69–81.

[45] D.D. Bainov and I.M. Stamova, *Lyapunov functions and asymptotic stability of impulsive differential-difference equations*, PanAmerican Mathematical Journal 9 (1999), pp. 23–34.

[46] D.D. Bainov and I.M. Stamova, *Strong stability of impulsive differential-difference equations*, PanAmerican Mathematical Journal 9 (1999), pp. 87–95.

[47] D.D. Bainov and I.M. Stamova, *Existence, uniqueness and continuability of solutions of impulsive differential-difference equations*, Journal of Applied Mathematics and Stochastic Analysis 12:3 (1999), pp. 293–300.

[48] D.D. Bainov and I.M. Stamova, *Global stability of the solutions of impulsive functional differential equations*, Kyungpook Mathematical Journal 39 (1999), pp. 239–249.

[49] D.D. Bainov and I.M. Stamova, *Global stability of sets for impulsive differential-difference equations by Lyapunov's direct method*, Acta Matematica et Informatica Universitatis Ostraviensis 7 (1999), pp. 7–22.

[50] D.D. Bainov and I.M. Stamova, *Lipschitz stability of impulsive functional differential equations*, ANZIAM J. 42 (2001), pp. 504–515.

[51] D.D. Bainov and I.M. Stamova, *Vector Lyapunov functions and conditional stability for systems of impulsive differential-difference equations*, ANZIAM J. 42 (2001), pp. 341–353.

[52] D.D. Bainov and I.M. Stamova, *Stability of the solutions of impulsive functional differential equations by Lyapunov's direct method*, ANZIAM J. 43 (2001), pp. 269–278.

[53] D.D. Bainov, I.M. Stamova and A. Vatsala, *Global stability of sets for linear impulsive differential-difference equations with variable impulsive perturbations*, Appl. Anal. 62 (1996), pp. 149–160.

[54] D.D. Bainov, I.M. Stamova and A. Vatsala, *Second method of Lyapunov for stability of linear impulsive differential-difference equations with variable impulsive perturbations*, Journal of Applied Mathematics and Stochastic Analysis 11 (1998), pp. 209–216.

[55] E.A. Barbashin, *Introduction to the Stability Theory*, Nauka, Moscow, 1967 (in Russian).

[56] E.A. Barbashin, *Lyapunov Functions*, Nauka, Moscow, 1970 (in Russian).

[57] N. Bautin, *The theory of point transformations and dynamical theory of clockworks*, Qualitative Methods in the Theory of Non-linear Vibrations, (Proceedings of the International Symposium on Non-linear Vibrations II), Acad. Sci. of Ukr. SSR, Kiev, (1963), pp. 29–54.

[58] R. Bellman and K. Cooke, *Differential-Difference Equations*, Academic Press, New York, 1963.

[59] L. Berezansky and E. Braverman, *Explicit conditions of exponential stability for a linear impulsive delay differential equations*, J. Math. Anal. Appl. 214 (1997), pp. 439–458.

[60] L. Berezansky and L. Idels, *Exponential stability of some scalar delay differential equations*, Communications in Applied Analysis 2 (2000), pp. 301–308.

[61] N.N. Bogolyubov and Y.A. Mitropolskii, *Asymptotic Methods in the Theory of Nonlinear Variations*, Nauka, Moscow, 1974 (in Russian).

[62] R. Boucekkine, O. Licandro and P. Christopher, *Differential-difference equations in economics: On the numerical solutions of vintage capital growth model*, Journal of Economic Dynamics and Control 21 (1997), pp. 347–362.

[63] T.A. Burton, *Stability and Periodic Solutions of Ordinary and Functional Differential Equations*, Academic Press, New York, 1985.

[64] T.A. Burton and B. Zhang, *Uniform ultimate boundedness and periodicity in functional differential equations*, Tahoku Math. Journal 42 (1990), pp. 93–100.

[65] G. Butler, H.I. Freedman and P. Waltman, *Uniformly persistent systems*, Proc. AMS 96 (1986), pp. 425–430.

[66] J. Cao, *On stability of delayed cellular neural networks*, Phys. Lett. A 261 (1999), pp. 303–308.

[67] J. Cao and J. Wang, *Global exponential stability and periodicity of recurrent neural networks with times delays*, IEEE Trans. Circuits Syst. I 52 (2005), pp. 920–931.

[68] T. Chen, *Global exponential stability of delayed Hopfield neural networks*, Neutral Netw. 14 (2001), pp. 977–980.

[69] G. Chen and J. Shen, *Boundedness and periodicity for impulsive functional differential equations with applications to impulsive delayed Hopfield neuron networks*, Dynamics of Continuous Discrete and Impulsive Systems, Ser. A. 14 (2007), pp. 177–188.

[70] M.-P. Chen, J. S. Yu and J. H. Shen, *The persistence of nonoscillatory solutions of delay differential equations under impulsive perturbations*, Computers with Mathematics and Applications 27 (1994), pp. 1–6.

[71] N.G. Chetayev, *Stability of Motion*, Nauka, Moscow, 1965 (in Russian).

[72] L. Chua, *CNN: A Paradigm for Complexity*, World Scientific, Singapore, 1988.

[73] L. Chua and T. Roska, *Stability of a class of nonreciprocal cellular neural networks*, IEEE Trans. Circuits Syst. I 37 (1990), pp. 1520–1527.

[74] L. Chua and L. Yang, *Cellular neural networks: Theory*, IEEE Trans. Circuits Syst. CAS 35 (1988), pp. 1257–1272.

[75] L. Chua and L. Yang, *Cellular neural networks: Applications*, IEEE Trans. Circuits Syst. CAS 35, (1988), pp. 1273–1290.

[76] P.P. Civalleri and M. Gilli, *A set of stability criteria for delayed cellular neural networks*, IEEE Trans. Circuits Syst. I 48 (2001), pp. 494–498.

[77] C.W. Cobb and P.H. Douglas, *A theory of production*, American Economic Review 18 (1928), pp. 139–165.

[78] J.E. Cohen, *Population growth and earth's human carrying capacity*, Science 269 (1995), pp. 341–346.

[79] W. Cui, *Global stability of a class of neural networks model under dynamical thresholds with delay*, Journal of Biomathematics 15 (2000), pp. 420–424.

[80] F. Dannan and S. Elaydi, *Lipschitz stability of nonlinear systems of differential equations*, J. Math. Anal. Appl. 113 (1986), pp. 562–577.

[81] D. Dejong, B. Ingram and C. Whiteman, *Keynesian impulses versus Solow residuals: identifying sources of business cycle fluctuation*, Journal of Applied Econometrics 15 (2000), pp. 311–329.

[82] R. Driver, *Ordinary and Delay Differential Equations*, Springer-Verlag, New York, 1977.

[83] G.-F. Emmenegger and I.M. Stamova, *Shock to capital intensity make the Solow equation an impulsive differential equation*, International Journal of Differential Equations and Applications 6 (2002), pp. 93–110.

[84] M. Fan, J. Dishen, Q. Wan and K. Wang, *Stability and boundedness of solutions of neutral functional differential equations with finite delay*, J. Math. Anal. Appl. 276 (2002), pp. 545–560.

[85] M. Fan, K. Wang and D. Jiang, *Existence and global attractivity of positive periodic solutions of periodic species Lotka–Volterra competition systems with several deviating arguments*, Mathematical Biosciences 160 (1999), pp. 47–61.

[86] L. Fanti and P. Manfredi, *The Solow's model with endogenous population: a neoclassical growth cycle model*, Journal of Economic Development 28 (2003), pp. 103–115.

[87] X. Fu and L. Zhang, *On boundedness of solutions of impulsive integro-differential systems with fixed moments of impulse effects*, Acta Mathematica Scientia 17 (1997), pp. 219–229.

[88] R. Gaines and J. Mawhin, *Coincidence Degree and Nonlinear Differential Equations*, Springer-Verlag, Berlin, 1977.

[89] G. Garashchenko and V. Pichkur, *Properties of optimal sets of practical stability of differential inclusions. Part I. Part II*, Journal of Automation and Information Sciences 38 (2006), pp. 1–11.

[90] T.R. Gichev and N.H. Rosov, *A nonlinear controllable impulse system. Dependence of the solution on the initial state and parameters*, Diff. Equations 16 (1980), pp. 208–213 (in Russian).

[91] K. Gopalsamy, *Stability and Oscillations in Delay Differential Equations of Population Dynamics*, Kluwer Academic Publishers, Dordrecht, Netherlands, 1992.

[92] K. Gopalsamy and Issic K. C. Leung, *Convergence under dynamical thresholds with delays*, IEEE Transactions on Neural Netw. 8 (1997), pp. 341–348.

[93] K. Gopalsamy and B. Zhang, *On delay differential equations with impulses*, J. Math. Anal. Appl. 139 (1989), pp. 110–122.

[94] R. Grimmer and G. Seifert, *Stability properties of Volterra integro-differential equations*, J. Diff. Equations 19 (1975), pp. 142–166.

[95] K. Gu, V.L. Kharitonov and J. Chen, *Stability of Time-Delay Systems*, Birkhäuser, 2003.

[96] L. Guerrini, *The Solow–Swan model with a bounded population growth rate*, Journal of Mathematical Economics 42 (2006), pp. 14–21.

[97] S.I. Gurgulla and N.A. Perestyuk, *On Lyapunov's second method in systems with impulse effect*, Reports of Acad. Sci. Ukr. SSR, Ser. A. 10 (1982), pp. 11–14 (in Russian).

[98] J.K. Hale, *Theory of Functional Differential Equations*, Springer-Verlag, New York, Heidelberg, Berlin, 1977.

[99] J.K. Hale and V. Lunel, *Introduction to Functional Differential Equations*, Springer-Verlag, 1993.

[100] P. Hartman, *Ordinary Differential Equations*, John Wiley, New York-London-Sydney, 1964.

[101] S. Haykin, *Neural Networks: A Comprehensive Foundation*, Ehglewood Cliffs, NJ: Prentice-Hall, 1998.

[102] A.V.M. Herz, S. Bonhoeffer, R.M. Anderson, R.M. May and M.A. Nowak, *Viral dynamics in vivo: Limitations on estimates of intacellular delay and virus decay*, Proceedings National Academy of Science USA 93 (1996), pp. 7247–7251.

[103] D.W.C. Ho and J. Sun, *Stability of Takagi–Sugeno fuzzy delay systems with impulse*, IEEE Transactions on Fuzzy Systems 15 (2007), pp. 784–790.

[104] J.J. Hopfield, *Neurons with graded response have collective computational properties like those of two-state neurons*, Proceedings National Academy of Science USA 81 (1984), pp. 3088–3092.

[105] N. Hritonenko and Yu. Yatsenko, *Mathematical Modelling in Economics, Ecology, and the Environment*, Kluwer Academic Publishers, Dordrecht, Netherlands, 1999.

[106] D. Hu, H. Zhao and H. Zhu, *Global dynamics of Hopfield neural networks involving variable delays*, Computers and Mathematics with Applications 42 (2001), pp. 39–45.

[107] H. Huang and J. Cao, *On global asymptotic stability of recurrent neural networks with time-varying delays*, Applied Mathematics and Computation 142 (2003), pp. 143–154.

[108] X. Huang and J. Reyn, *On bounded solutions and their stability for some ordinary and functional differential equations of Volterra type*, Reports of the Faculty of Technical Mathematics and Informatics, no. 95-31, Delft, (1995), pp. 1–23.

[109] H.-F. Huo, *Existence of positive periodic solutions of a neutral delay Lotka–Volterra system with impulses*, Computers and Mathematics with Applications 48 (2004), pp. 1833–1846.

[110] G. Jiang and Q. Lu, *Impulsive state feedback control of a predator-prey model*, Journal of Computational and Applied Mathematics 200 (2007), pp. 193–207.

[111] Z. Jin, H. Maoan and L. Guihua, *The persistence in a Lotka–Volterra competition systems with impulsive perturbations*, Chaos, Solutions and Fractals 24 (2005), pp. 1105–1117.

[112] C. Jost, O. Ariono and R. Arditi, *About deterministic extinction in ratio-dependent predator-prey models*, Bull. Math. Biol. 61 (1999), pp. 19–32.

[113] J. Kapur, *Mathematical Modelling*, Wiley, New York, NY. 1988.

[114] J. Kato, *Stability problems in functional differential equation with infinite delays*, Funkcial. Ekvac. 21 (1978), pp. 63–80.

[115] A. Khadra, X. Liu and X. Shen, *Application of impulsive synchronization to communication security*, IEEE Trans. Circuits Syst. 50 (2003), pp. 341–351.

[116] A. Khadra, X. Liu and X. Shen, *Robuts impulsive synchronization and application to communication security*, Dynamics of Continuous and Discrete Impulsive Systems 10 (2003), pp. 403–416.

[117] S. Kim, S. Campbell and X. Liu, *Stability of a class of linear switching systems with time delay*, IEEE Trans. Circuits Syst. I 53 (2006), pp. 384–393.

[118] V.B. Kolmanovskii and A.D. Myshkis, *Applied Theory of Functional Differential Equations*, Kluwer Academic Publishers, Dordrecht, 1992.

[119] V.B. Kolmanovskii and V.R. Nosov, *Stability of Functional Differential Equations*, Academic Press Inc., London, 1986.

[120] M. Konstantinov and D.D. Bainov, *Absolute exponential stability of nonlinear systems of differential equations with several delays*, Soviet Math. (Iz. VUZ), 23 (1979), pp. 30–35.

[121] M. Konstantinov and D.D. Bainov, *Stability and approximate solution of regularly perturbed boundary value problems for differential delayed equations*, Qualitative Theory of Differential Equations, Colloquia Math. Soc. J. Bolyai 30 (1979), pp. 591–613.

[122] M. Konstantinov and D.D. Bainov, *Absolute exponential stability of nonlinear systems of differential equations with lag*, Period. Math. Hungarica 12 (1981), pp. 87–92.

[123] M. Konstantinov, P. Petkov, S. Patarinski and N. Christov, *Absolute stability and stabilization of differential delayed systems*, Arch. Autom. Telemech. 24 (1979), pp. 339–350.

[124] N.N. Krasovskii, *Stability of Motion*, Stanford University Press, 1963.

[125] S. Krishna and A. Anokhin, *Delay differential systems with discontinuous initial data and existence and uniqueness theorems for systems with impulse and delay*, Journal of Applied Mathematics and Stochastic Analysis 7 (1994), pp. 49–67.

[126] S. Krishna, J. Vasundhara and K. Satyavani, *Boundedness and dichotomies for impulsive equations*, J. Math. Anal. Appl. 158 (1991), pp. 352–375.

[127] Y. Kuang, *Delay Differential Equations with Applications in Population Dynamics*, Academic Press, Boston, 1993.

[128] J.L. Lagrange, *Mecanique Celeste*, Dunod, Paris, 1788.

[129] V. Lakshmikantham, D.D. Bainov and P.S. Simeonov, *Theory of Impulsive Differential Equations*, World Scientific, Singapore, New Jersey, London, 1989.

[130] V. Lakshmikantham and S. Leela, *Differential and Integral Inequalities*, Academic Press, New York, 1969.

[131] V. Lakshmikantham, S. Leela and A.A. Martynyuk, *Stability Analysis of Nonlinear Systems*, Marcel Dekker, New York, 1989.

[132] V. Lakshmikantham, S. Leela and A.A. Martynyuk, *Practical Stability Analysis of Nonlinear Systems*, World Scientific Publishers, Singapore, New Jersey, London, Hong Kong, 1990.

[133] V. Lakshmikantham and X. Liu, *Stability Analysis in Terms of Two Measures*, World Scientific Publishing Co. Pte. Ltd., Singapore, 1993.

[134] V. Lakshmikantham and M.Rama M. Rao, *Theory of Integro-Differential Equations*, Gordon and Beach Science Publishers, 1995.

[135] M. Li, Y. Duan, W. Zhang and M. Wang, *The existence of positive periodic solutions of a class of Lotka–Volterra type impulsive systems with infinitely distributed delay*, Computers and Mathematics with Applications 49 (2005), pp. 1037–1044.

[136] C. Li, X. Liao, X. Yang and T. Huang, *Impulsive stabilization and synchronization of a class of chaotic delay systems*, Chaos 15 (2005), 043103.

[137] D. Li and W. Ma, *Asymptotic properties of a HIV-1 infection model with time-delay*, J. Math. Anal. Appl. 335 (2007), pp. 683–691.

[138] J. Li and J. Yan, *Partial permanence and extinction in an n-species nonautonomous Lotka–Volterra competitive system*, Computers and Mathematics with Applications 55 (2008), pp. 76–88.

[139] B. Lisena, *Extinction in three species competitive systems with periodic coefficients*, Dynamics Systems and Applications 14 (2005), pp. 393–406.

[140] J. Liu, *Bounded and periodic solutions of finite delay evolution equations*, Nonlinear Analysis 34 (1998), pp. 101–111.

[141] X. Liu, *Stability results for impulsive differential systems with applications to population growth models*, Dynamics and Stability of Systems 9 (1994), pp. 163–174.

[142] X. Liu, *Stability of impulsive control systems with time delay*, Math. Computer Modelling 39 (2004), pp. 511–519.

[143] X. Liu and G. Ballinger, *Existence and continuability of solutions for differential equations with delays and state-dependent impulses*, Nonlinear Analysis 51 (2002), pp. 633–647.

[144] X. Liu and W. Ge, *Global attractivity in delay "food-limited" models with exponential impulses*, J. Math. Anal. Appl. 287 (2003), pp. 200–216.

[145] X. Liu and K. Rohlf, *Impulsive control of a Lotka–Volterra system*, IMA J. Math. Control and Information 15 (1998), pp. 269–284.

[146] X. Liu and Q. Wang, *The method of Lyapunov functionals and exponential stability of impulsive systems with time delay*, Nonlinear Analysis 66 (2007), pp. 1465–1484.

[147] A. Lotka, *Elements of Physical Biology (1925)*, reprinted by Dover as Elements of Mathematical Biology, 1956.

[148] Z. Luo and J. Shen, *Stability and boundedness for impulsive functional differential equations with infinite delays*, Nonlinear Analysis 46 (2001), pp. 475–493.

[149] Z. Luo and J. Shen, *Stability results for impulsive functional differential equations with infinite delays*, Journal of Computational and Applied Mathematics 131 (2001), pp. 55–64.

[150] A.M. Lyapunov, *General Problem on Stability of Motion*, Grostechizdat, Moscow-Leningrad, 1950 (in Russian).

[151] M.C. Mackey and J. Belair, *Consumer memory and price fluctuation in commodity markets: An integrodifferential model*, Journal of Dynamics and Differential Equations 1 (1989), pp. 299–325.

[152] I.G. Malkin, *Theory of Stability of Motion*, Nauka, Moscow, 1966 (in Russian).

[153] V.P. Marachkov, *On a theorem on stability*, Bull.Soc.Phys.-Math. 12, Kazan, (1940), pp. 171–174 (in Russian).

[154] C. M. Marcus and R.M. Westervelt, *Stability of analog neural networks with delay*, Phys. Rev. A 39 (1989), pp. 347–359.

[155] A. Martynyuk, *Advances in Stability Theory at the End of the 20-th Century (Stability and Control: Theory, Methods and Applications)*, Vol. 13, Taylor and Francis, London, 2003.

[156] R.M. May and R.M. Anderson, *The transmission dynamics of Human Immunodeficiency Virus (HIV)*, Philosophical Trans. Roy. Soc. London, Ser. B., Biological Sciences, (1998), pp. 565–607

[157] J. Maynard-Smith, *Models in Ecology*, Cambridge University Press, Cambridge, 1974.

[158] F. McRae, *Practical stability of impulsive control systems*, J. Math. Anal. Appl. 181 (1994), pp. 656–672.

[159] V.D. Mil'man and A.D. Myshkis, *On the stability of motion in the presence of impulses*, Siberian Mathematical Journal 1 (1960), pp. 233–237 (in Russian).

[160] S. Mohamad, *Global exponential stability of continuous-time and discrete-time delayed bidirectional neural networks*, Phys. D. 159 (2001), pp. 233–251.

[161] Z. Mukandavire, W. Garira and C. Chiyaka, *Asymptotic properties of an HIV/AIDS model with a time delay*, J. Math. Anal. Appl. 330 (2007), pp. 916–933.

[162] A.S. Muresan, *On some models of price fluctuations in a market economy*, Studia Univ. Babes-Bolyai, Math. XXXVIII 2 (1993), pp. 15–19.

[163] A.D. Myshkis, *Linear Differential Equations with a Retarding Argument*, Nauka, Moscow, 1972 (in Russian).

[164] J. Nieto, *Periodic boundary value problems for first-order impulsive ordinary differential equations*, Nonlinear Analysis 51 (2002), pp. 1223–1232.

[165] A.F. Nindjin, M.A. Aziz-Alaoui and M. Cadivel, *Analysis of predator-prey model with modified Leslie-Gower and Holling-type II schemes with time delay*, Nonlinear Analysis 7 (2006), pp. 1104–1118.

[166] Z. Ouyang, X. Liao and S. Zhou, *Permanence of species in nonautonomous discrete Lotka–Volterra competitive system with delays and feedback controls*, Journal of Computational and Applied Mathematics 211 (2008), pp. 1–10.

[167] S.G. Pandit and S.G. Deo, *Differential Systems Involving Impulses*, Springer-Verlag, Berlin-Heidelberg-New York, 1982.

[168] M. Rama M. Rao and V. Raghavendra, *Asymptotic stability properties of Volterra integro-differential equations*, Nonlinear Analysis 11 (1987), pp. 475–480.

[169] B.S. Razumikhin, *Stability of Systems with Retardation*, Nauka, Moscow, 1988 (in Russian).

[170] T. Roska, C.W. Wu, M. Balsi and L.O. Chua, *Stability and dynamics of delay-type general cellular neural networks*, IEEE Trans. Circuits Syst. I 39 (1992), pp. 487–490.

[171] H. Rouche, P. Habets and M. Laloy, *Stability Theory by Lyapunov's Direct Method*, Springer-Verlag, New York, 1977.

[172] N.H. Rozov and T.R. Gichev, *A nonlinear controllable impulse system I*, Diff. Equations 15 (1979), pp. 1933–1939 (in Russian).

[173] N.H. Rozov and T.R. Gichev, *A nonlinear controllable impulse system II*, Diff. Equations 16 (1980), pp. 208–213 (in Russian).

[174] N.H. Rozov and T.R. Gichev, *Singularly perturbed problems with a minimal impulse*, Diff. Equations 19 (1983), pp. 259–269 (in Russian).

[175] A.I. Rus and C. Iancu, *A functional-differential model for price fluctuations in a single commodity market*, Studia Univ. Babes-Bolyai, Math. XXXVIII, 2 (1993), pp. 9–14.

[176] T.L. Saaty and M. Joyce, *Thinking with Models: Mathematical Models in the Physical, Biological, and Social Sciences*, Pergamon, Oxford, UK, 1981.

[177] P. Sakellaris, *Irreversible capital and the stock market response to shocks in profitability*, International Economic Review 38 (1997), pp. 351–379.

[178] A.M. Samoilenko and N.A. Perestyuk, *On the averaging method in systems with impulse effect*, Ukr. Math. J. 26 (1974), pp. 411–419 (in Russian).

[179] A.M. Samoilenko and N.A. Perestyuk, *Stability of the solutions of differential equations with impulse effect*, Diff. Equations 13 (1977), pp. 1981–1992 (in Russian).

[180] A.M. Samoilenko and N.A. Perestyuk, *On the stability of systems with impulse effect*, Diff. Equations 17 (1981), pp. 1995–2001 (in Russian).

[181] A.M. Samoilenko and N.A. Perestyuk, *Differential Equations with Impulse Effect*, World Scientific, Singapore, 1995.

[182] S. Savov and I. Popchev, *Stability analysis of uncertain matrices via parametrised interval system of equations*, Compt. Rend. Acad. Bulg. Sci. 57 (2004), pp. 67–70.

[183] S. Savov and I. Popchev, *Discrete algebraic Lyapunov's equation. Lower and upper matrix bounds*, Compt. Rend. Acad. Bulg. Sci. 58 (2005), pp. 1159–1162.

[184] J. Shen, *Razumikhin techniques in impulsive functional differential equations*, Nonlinear Analysis 36 (1999), pp. 119–130.

[185] J. Shen and J. Li, *Impulsive control for stability of Volterra functional differential equations*, Zeitschrift fur Analysis und ihre Anwendungen 24 (2005), pp. 721–734.

[186] D.D. Siljak, M. Ikeda and Y. Ohta, *Parametric stability*, Proceedings of the Universita di Genova-Ohio State University Joint Conference, Boston, MA, Birkhäuser, (1991), pp. 1–20.

[187] A. Slavova, *Cellular Neural Networks: Dynamics and Modelling*, Kluwer Academic Publishers, 2003.

[188] R. Solow, *A contribution to the theory of economic growth*, Quarterly Journal of Economics 70 (1956), pp. 65–94.

[189] G.T. Stamov and I.M. Stamova, *Second method of Lyapunov and existence of integral manifolds for impulsive differential equations*, SUT Journal of Mathematics 31 (1996), pp. 101–107.

[190] G.T. Stamov and I.M. Stamova, *Second method of Lyapunov and existence of integral manifolds for impulsive differential-difference equations*, J. Math. Anal. Appl. 258 (2001), pp. 371–379.

[191] G.T. Stamov and I.M. Stamova, *Integral manifolds for impulsive differential-difference equations*, Electronic Modelling 4 (2005), pp. 115–120.

[192] G.T. Stamov and I.M. Stamova, *Almost periodic solutions for impulsive neural networks with delay*, Applied Mathematical Modelling 31 (2007), pp. 1263–1270.

[193] I.M. Stamova, *Boundedness of the solutions of impulsive differential-difference equations with respect to the sets*, Journal of the Technical University at Plovdiv, Fundamental Science and Applications, Ser. A 6 (1998), pp. 41–48.

[194] I.M. Stamova, *Global attractivity in a non-autonomous impulsive delay differential equations*, Journal of the Technical University at Plovdiv, Fundamental Science and Applications, Ser. A 6 (1998), pp. 49–54.

[195] I.M. Stamova, *Stability theorems of perturbed linear impulsive equations*, Mathematical Science Research Journal 6 (2002), pp. 96–103.

[196] I.M. Stamova, *On the stability of sets for impulsive functional differential equations*, Nonlinear Studies 12 (2005), pp. 49–58.

[197] I.M. Stamova, *Lyapunov method for boundedness of solutions of nonlinear impulsive functional differential equations*, Dynamic Systems and Applications 14 (2005), pp. 561–568.

[198] I.M. Stamova, *On the stability of an impulsive differential-difference population model*, Communications in Applied Analysis 10 (2006), pp. 285–291.

[199] I.M. Stamova, *Global asymptotic stability of impulse delayed cellular neural networks with dynamical threshold*, Nonlinear Studies 13 (2006), pp. 113–122.

[200] I.M. Stamova, *Razumikhin-type theorems on stability in terms of two measures for impulsive functional differential systems*, Note di Matematica 26 (2006), pp. 69–80.

[201] I.M. Stamova, *Vector Lyapunov functions for practical stability of nonlinear impulsive functional differential equations*, J. Math. Anal. Appl. 325 (2007), pp. 612–623.

[202] I.M. Stamova, *Second method of Lyapunov for boundedness in terms of two measures for impulsive functional differential systems*, Electronic Modelling 30 (2008), pp. 1–50.

[203] I.M. Stamova, *Parametric stability of impulsive functional differential equations*, Journal of Dynamical and Control Systems 14 (2008), pp. 235–250.

[204] I.M. Stamova, *Boundedness of impulsive functional differential equations with variable impulsive perturbations*, Bulletin of the Australian Mathematical Society 77 (2008), pp. 331–345.

[205] I.M. Stamova and J. Eftekhar, *Razumikhin technique and stability of impulsive differential-difference equations in terms of two measures*, Journal of Concrete and Applicable Mathematics 2 (NOVA), (2004), pp. 233–248.

[206] I.M. Stamova and G.-F. Emmenegger, *Stability of the solutions of impulsive functional differential equations modeling price fluctuations in single commodity markets*, International Journal of Applied Mathematics 15 (2004), pp. 271–290.

[207] I.M. Stamova and G.T. Stamov, *Lyapunov–Razumikhin method for impulsive functional differential equations and applications to the population dynamics*, Journal of Computational and Applied Mathematics 130 (2001), pp. 163–171.

[208] I.M. Stamova and G.T. Stamov, *On the conditional stability of impulsive functional differential equations*, Applied Mathematics Research eXpress 2006 (2006), pp. 1–13.

[209] I.M. Stamova and G.T. Stamov, *Asymptotic stability of impulsive neural networks with time-varying delay*, International Journal: Mathematical Manuscripts 1 (2007), pp. 158–168.

[210] I.M. Stamova and G.T. Stamov, *Lyapunov–Razumikhin method for asymptotic stability of sets for impulsive functional differential equations*, Electronic Journal of Differential Equations 48 (2008), pp. 1–10.

[211] Y. Takeuchi, *Global Dynamical Properties of Lotka–Volterra Systems*, World Scientific, Singapore, 1996.

[212] V. Volterra, *Fluctuations in the abundance of a species considered mathematically*, Nature 118 (1926), pp. 558–560.

[213] Q. Wang and X. Liu, *Exponential stability for impulsive delay differential equations by Razumikhin method*, J. Math. Anal. Appl. 309 (2005), pp. 462–473.

[214] F. Wei and K. Wang, *Asymptotically periodic solution of n-species cooperation system with time delay*, Nonlinear Analysis: Real World Applications 7 (2006), pp. 591–596.

[215] J. Widjaja and M.J. Bottema, *Existence of solutions of diffusive logistic equations with impulses and time delay and stability of the steady-states*, Dynamics of Continuous, Discrete and Impulsive Systems Series A: Mathematical Analysis 12 (2005), pp. 563–578.

[216] S. Wolfram, *Theory and Applications of Cellular Automata*, World Scientific, New York, 1986.

[217] S.J. Wu and D. Han, *Exponential stability of functional differential systems with impulsive effect on random moments*, Computers and Mathematics with Applications 50 (2005), pp. 321–328.

[218] Y. Xia, *Positive periodic solutions for a neutral impulsive delayed Lotka–Volterra competition system with the effect of toxic substance*, Nonlinear Analysis: Real World Applications 8 (2007), pp. 204–221.

[219] D. Xiao and S. Ruan, *Global dynamics of a ratio-dependent predator-prey system*, Mathematical Biology 43 (2001), pp. 268–290.

[220] Y. Xue, J. Wang and Z. Jin, *The persistent threshold of single population under pulse input of environmental toxin*, WSEAS Transactions on Mathematics 6 (2007), pp. 22–29.

[221] Z. Yang and D. Xu, *Stability analysis of delay neural networks with impulsive effects*, IEEE Trans. Circuits Syst. 52 (2005), pp. 517–521

[222] D. Ye and M. Fan, *Periodicity in impulsive predator-prey system with Holling III functional response*, Kodai Math. Journal 27 (2004), pp. 189–200.

[223] T. Yoshizawa, *Stability Theory by Lyapunov's Second Method*, The Mathematical Society of Japan, 1966.

[224] F. Zanolin, *Permanence and positive periodic solutions for Kolmogorov competing species systems*, Results Math. 21 (1992), pp. 224–250.

[225] B. Zhang, *Boundedness in functional differential equations*, Nonlinear Analysis 22 (1994), pp. 1511–1527.

[226] S. Zhang, L. Dong and L. Chen, *The study of predator-prey system with defensive ability of prey and impulsive perturbations on the predator*, Chaos, Solutions and Fractals 23 (2005), pp. 631–643.

[227] F. Zhang, W. Li and H. Huo, *Global stability of a class of delayed cellular neural networks with dynamical thresholds*, International Journal of Applied Mathematics 13 (2003), pp. 359–368.

[228] Y. Zhang and J. Sun, *Boundedness of the solutions of impulsive functional differential equations with time-varying delay*, Applied Mathematics and Computations 154 (2004), pp. 279–288.

[229] Y. Zhang and J. Sun, *Stability of impulsive delay differential equations with impulses at variable times*, Dynamical Systems 20 (2005), pp. 323–331.

[230] C.J. Zhao, *On a periodic predator-prey system with time-delays*, J. Math. Anal. Appl. 331 (2007), pp. 978–985.

[231] H. Zhao and E. Feng, *Stability of impulsive system by perturbing Lyapunov functions*, Applied Mathematics Letters 20 (2007), pp. 194–198.

[232] W. Zhong, W. Lin and R. Jiong, *The stability in neural networks with delay and dynamical threshold effects*, Annales of Differential Equations 17:1 (2001), pp. 93–101.

[233] D. Zhou and J. Cao, *Globally exponential stability conditions for cellular neural networks with time-varying delays*, Applied Mathematics and Computation 131 (2002), pp. 487–496.

Index

Moments of impulse effect, 10
 fixed, 11, 13, 19, 24, 36, 50, 60,
  66, 70, 78
 variable, 12, 13, 15, 31, 66, 67,
  105, 111
Monotone increasing, 26

N

Neighborhood, 95
ε-neighbourhood, 67, 68
Neural network(s), 8, 144, 172, 173,
 177, 191, 193
 Bidirectional Associative Memory
  neural network, 185, 190
Non-decreasing, 26–28, 96
Non-uniform stability, 118, 133
Noncontinuable, 15, 17, 18
Norm
 function, 32, 146, 160, 173, 177,
  200, 208
 matrix, 51, 187

P

Parametric stability, 66, 94, 142
Parametrically unstable, 95
Parasitoid-host, 2
Periodic solution(s), 14, 152
Population models, 144
Positive invariant, 148
Positively definite, 128
Practical stability, 66, 112, 142
Practically stable, 113
Practically unstable, 113, 116
Predator-prey, 2–4, 145
Price fluctuations, 206

Q

Quadratic form, 52
Quasi-monotone increasing, 26
Quasi-uniformly ultimately bounded, 41
Quasilinear system, 50
Quasiunique, 16

R

Razumikhin condition, 203
Razumikhin technique, 14, 23, 144
Response function, 2
 prey-dependent, 3
 ratio-dependent, 4

S

Saving rate, 198
Second method of Lyapunov, 14, 22,
 23, 211
Single commodity markets, 206
Solow growth model, 198, 199
Solution, 10–12, 15, 19, 21, 27
 maximal, 26, 27, 96, 147, 162
 minimal, 26, 27, 148, 164
 unique, 15, 21, 201
 zero, 32, 55, 87, 89, 120, 121, 129
Stability, 23, 64, 66, 172
 by part of the variables, 129
 in terms of two measures, 66, 128,
  143
 of sets, 66, 67, 142
Stability theory, 14, 66
Stable
 set, 68, 79
 solution, 36, 209
 system, 129
 zero solution, 32, 55
Strictly increasing, 32

T

t- (or α-)uniformly M-bounded, 76
Thresholds, 173, 191

U

Uniform asymptotic stability, 144
Uniform stability, 144
Uniformly asymptotically stable
 equilibrium, 205
 solution, 37, 152, 157, 168, 202,
  209
 zero solution, 33